Eyton's Herd Book of Hereford Cattle
Volume 6 - For 1865

by T. Duckham

with an introduction by Jackson Chambers

This work contains material that was originally published in 1865.

This publication is within the Public Domain.

This edition is reprinted for educational purposes
and in accordance with all applicable Federal Laws.

Introduction Copyright 2018 by Jackson Chambers

Self Reliance Books

Get more historic titles on animal and stock breeding, gardening and old fashioned skills by visiting us at:

http://selfreliancebooks.blogspot.com/

Introduction

I am pleased to present another title in the "Cattle" series.

The work is in the Public Domain and is re-printed here in accordance with Federal Laws.

As with all reprinted books of this age that are intended to perfectly reproduce the original edition, considerable pains and effort had to be undertaken to correct fading and sometimes outright damage to existing proofs of this title. At times, this task is quite monumental, requiring an almost total "rebuilding" of some pages from digital proofs of multiple copies. Despite this, imperfections still sometimes exist in the final proof and may detract from the visual appearance of the text.

I hope you enjoy reading this book as much as I enjoyed making it available to readers again.

Jackson Chambers

PREFACE.

SINCE I last addressed a few lines in these pages to my kind supporters, I have secured the services of a talented artist, and embellished the work with faithful representations of distinguished animals. With one exception, viz., that of Sir Benjamin (1387) in the fifth volume, the animals represented have all been winners of first prizes at the meetings of the Royal Agricultural Society of England, and for the future I purpose strictly adhering to that practice.

I think it right to mention that in subsequent volumes I intend to publish as an appendix the prizes won by animals entered. It will be remembered that this was my original intention, but the numbers returned to me for the 3rd vol. appeared so insignificant that I resolved upon placing the notices under their respective pedigrees—an arrangement which causes a great amount of repetition in consequence of the prizes at local exhibitions being given for animals in lots.

Believing that it would be to the advantage of the breeders, and that the day had arrived when the work should be made a biennial instead of a triennial publication, I last year resolved to make it so; but, after incurring considerable expense and labour, I found the information supplied to me was of too limited a nature to justify its publication, and I therefore reluctantly felt compelled to hold it over until this year.

PREFACE.

As an appendix to this volume, I beg respectfully to present to its readers a copy of a lecture delivered by myself on the "Rise, Progress, and Comparative Merits of the Hereford Breed of Cattle," and I trust the information it conveys may not be found to be void of interest. Through the kindness of a friend, I have been enabled to embellish the lecture with a lithograph of the Smithfield Ox of 1799, from a coloured print in his possession; and, although I have no doubt great allowance should be made for the work of the artist of the past century when compared with that of the present, as displayed in another lithograph of the gold medal Ox of 1863, yet the massive, evenly-covered frame of the latter at only four years old, bears a most satisfactory comparison to that of the former at seven years old.

With these brief remarks I submit the Sixth Volume of the "Herd Book of Hereford Cattle" to the notice of its numerous kind patrons, and I sincerely hope that my earnest endeavours therewith will continue to secure their approbation.

T. DUCKHAM,

Baysham Court, Ross, Herefordshire,
Oct. 5th, 1865.

PATRON.

HER MOST EXCELLENT MAJESTY THE QUEEN.

SUBSCRIBERS.

A.

Ackers, J., Esq., Prinknash Park, Painswick.
Adams, Mr. W., Brinsop, Hereford.
Alington, H., Esq., Little Barford, St. Neotts.
Allen, J. D., Esq., Pyt House, Tisbury, Wilts.
Allen, H., Esq., Oakfield, Hay.
Andrews, Mr. C., Leysters, Tenbury.
Anthony, C., Esq., Hereford.
Apperley, Mr. J. P., Fownhope, Hereford.
Apperley, W. H., Esq., Withington, Hereford.
Arkwright, J. H., Esq., Hampton Court, Leominster.
Asher & Co., Berlin (through Mr. Nutt, 270, Strand, London).
Ashwood, Mr. J., Longdon Hall, Salop.

B.

Bailey, Sir J. Russell, Bart., M.P., Glanusk Park, Brecon.
Bailliére Brothers, New York, America.
Baker, Rev. F., Allensmore, Hereford.
Bache, Mr., Onibury, Salop.
Baldwin, J., Esq., Luddington, Stratford-on-Avon.

SUBSCRIBERS.

Ballard, Mr. P., Leighton Court, Bromyard.
Bannister, J. S., Esq., Weston, Pembridge. Leominster.
Barton, Mr. J., Coln, St. Aldwin's, Fairford.
Bateman, Right Hon. Lord, Shobdon, Leominster.
Bazzand, Mr. J., Kingley, Alcester.
Beamand, Mr. W., Vron End, Newcastle, Clun, Salop.
Bedford, His Grace The Duke of, Woburn, Beds.
Bedford, Mr. G., Hatfield, Leominster.
Bennett, Mr. J., Ingestone, Ross.
Bennett, Mr. G., North Cerney, Cirencester.
Bennett, E. Esq., Bedstone, Salop.
Bennett, Mr. J., The Park, Ross.
Bennett, Mr. T. M., Wellington, Hereford.
Biddulph, M., Esq., M.P., Ledbury.
Bishop, Mr. J. W., Ledbury.
Blythe J., Esq., Woolhampton, Reading.
Blakeway, Mr. C., Shelderton, Clungunford, Salop.
Bosley, Mr. J., Lower Lyde, Hereford.
Bostock, C. R., Esq., 16, Loraine Place, Holloway, London, N.W.
Bourn, Mr. J., Mawley Town Farm, Cleobury Mortimer.
Bowen, Mr. E., Corfton, Ludlow.
Bowen, Mr. P. W., Shrawardine Castle, Salop.
Bowen, W., Esq., Tref Ivon, Talgarth.
Bradstock, Mr. T. S., Cobrey Park, Ross.
Bray, Mr. G., The Haven, Dilwyn, Leominster.
Bridge, Mr. W., Wynford Eagle, Dorset.
Bright, Mr., Kempton, Ludlow.
Britten, Mr. C., The Woodhouse, Shobdon, Leominster.
Broad, Mr. S., Auctioneeer, Hereford.
Broad, Mr. T., The Castle, Madley, Hereford.
Browne, Mr. T., The Weir End, Ross.
Browne, Mr. W., Lewstone, Monmouth.
Brydges, Sir H. J., Bart., Boultibrook, Presteign
Bull, Mr. U., Mells Frome, Somerset.

SUBSCRIBERS.

Bulmer, C., Esq., Holmer, Hereford.
Burlton, Mr. J., Luntley, Leominster.
Burrows, Mr., The Grange, Bosbury, Ledbury.
Burrow, Mr., Bilbrook House, Wolverhampton.

C.

Cadle, Mr. C., Gloucester.
Cadle, Mr. T., Longcroft, Westbury-on-Severn, Gloucester.
Capper, R. H., Esq., The Northgate, Ross.
Carwardine, Mr. J., Cockcroft, Leominster.
Cave, Mr. J., Monkland, Leominster.
Chattock, H. H., Esq., Solihull, Warwick.
Child, Mr. W., Westonbury Leominster.
Clarke, G., Esq., Hyde Hall, East Springfield, Otsega Co., N.Y., America.
Clifford, Lieut-Col., Llantillio Cresseny, Monmouth.
Clive, The Rev. A., Whitfield, Hereford.
Coate, H., Esq., Sherborne, Dorset.
Cocks, The Rev. C., Neen Savage, Bewdley.
Colles, A., Esq., Ballyfellow, Altbay, Meath, Ireland.
Coleman, J., Esq., Woburn Park Farm, Beds.
College, The Royal Agricultural, Cirencester.
Compton, Mr. T., Stockton, Wilts.
Connop, Mr. W., Street, Court Leominster.
Cooke, Mr., H., Widemarsh Street, Hereford.
Cooke, Mr. J. Y., Moreton House, Hereford.
Cooke, Mr. J., Brampton Bryan.
Cornewall, Sir Velters. Bart., Moccas, Hereford.
Cotterell, Sir H. G., Bart., Garnons, Hereford.
Crawshay, Mrs., Dany Park, Crickhowel.

D.

Davenport, The Rev., G. H., Foxley, Hereford.
Davey, R., Esq., M.P., Redruth, Cornwall.
Davey, J. Sydney, Esq., Redruth, Cornwall.

SUBSCRIBERS.

Davies, Mr. T., Burlton Court, Hereford.
Davies, Mr., Dean Park, Tenbury.
Davies, Mr. E., Patton, Much Wenlock.
Davies, Mrs., Chipp's House, Ivington, Leominster.
Davies, Mr. T., Lady Meadow, Leominster.
Davies, Mr., Aberswin, Brecon.
Davies, M. B., Esq., 5, John's Street, Adelphi, London, W.C.
Dawes, Mr. W., New House, Craven Arms, Salop.
Dew, Mr. T., Craddock, Ross.
Dew, Tomkyns, Esq., Whitney Court, Hereford.
Dowle, Mr. A., Bernithon, Ross.
Dowle, Mr. H. Ross.
Downes, Mr. T., Brynich, Brecon.
Downing, Mr. J. B., Holm Lacey, Hereford.
Drinkwater, Mr. E., Treribble, Ross.
Dyott, Lieut-Col., Freeford, Lichfield.

E.

Edwards, Mr. T., Church Farm, Raglan, Monmouth.
Edwards, Mr. P. N., Brinsop Court, Hereford.
Edwards, Mr. T., Wintercott, Leominster.
Edwards, Mr., Skybory, Knighton.
Edwards, Mr. J., Knockalva, Ramble, Jamaica.
Elliott, Mr. S., Bogmarsh, Holm Lacey, Hereford.
Elsmere, Mr. T., Berrington, Salop.
Essery, Mr. H., Werrington, Launceston.
Essery, Mr. W., Tregallar, Launceston.
Evans, Mr. J., Treberfa, Knighton.
Evans, Mr. W., Llandowlas, Usk, Monmouth.
Evans, Mr. H. R., Swanstone Court, Leominster.
Evans, T. Esq., Moreton, Hereford.
Evans, Mr. F., Old Court, Bredwardine, Hereford.
Evans, Mr. T., Sydney, New South Wales.

SUBSCRIBERS.

F.

Farr, Mr. R., The Grange, Weobley.
Farr, Mr. T., Much Dewchurch, Hereford.
Farr, Mr. J., Pontrilas, Hereford.
Featherstonhaugh, R. S., Esq., Rockview, Killucan, Ireland.
Feaver, Mr. J., Willow, Bath.
Feilden, Lieut.-Col., Dulas Court, Hereford.
Fenn, Mr. T., Downton, Ludlow.
Findlay, Mr. J., Garnstone, Hereford.
Fletcher, Mr. G., Shipton, Cheltenham.
Foley, The Right Hon. Lady Emily, Stoke Edith, Hereford.
Forester, G. T., Esq., High Ercal, Wellington, Salop.

G.

Galliers, Mr. T., Wistaston, Weobley.
Gardiner, Mr. G., Pyon, Hereford.
Garnett, W. S., Esq., Williamstown, Kell's Co., Meath, Ireland.
Garrold, Mr. R. H., Kilforge, Ross.
Gibbons, Mr. H., Hampton Bishop, Hereford.
Gilliland, S., Esq., Brook Hall, Londonderry, Ireland.
Goldingham, E. T., Esq., Grimley, Worcester.
Goode, Mr. M. Felton, Bromyard.
Gosford, Vincent, Esq., Tanylan, Holywell, North Wales.
Green, J. B., Esq., Marlow, Leintwardine, Salop.
Greenhouse, Mr., The Harbour, Leominster.
Gregg, J., Esq., Fencote Abbey, Leominster.
Griffiths, J., Esq., Broadway, Worcester.
Grose, Mr. W. R., Penpont, Wadebridge, Cornwall.

H.

Hall, Mr. B., Malvern Wells.
Hall, Mr. H., Ashton, Leominster.
Hall, Mr. W., Ashton, Leominster.
Hands, Mr., Auctioneer, Abergavenny.

SUBSCRIBERS.

Hartland, Mr. J., Little Marcle, Ledbury.
Hatherton, The Right Hon. Lord, Teddesley Park, Penkeridge.
Hawkins, Mr. B., Orelton, Ludlow.
Haynes, Mr. J., Llanrothall, Monmouth.
Hearne, Mr. S. Street, Abbots, Salford, Evesham.
Herbert, J. M., Esq., Rocklands, Monmouth.
Herefordshire Agricultural Society.
Hewer, Mr. J. E., Vern House, Marden, Hereford.
Hewer, Mr. W., Hill House, Northleach.
Heywood, Mr. H. Moccas, Hereford.
Hickman, Mr. R., Bosbury, Ledbury.
Higgins, Mr. H., Woollaston Grange, Chepstow.
Hill, Mr. R., Orleton, Ludlow.
Hill, The Hon. and Rev. Noel, Berrington, Salop.
Hill, Mr. R. C., Marsh House, Newcastle, Stafford.
Hillier, Mr., Sherstone, Malmesbury.
Hinckesman, Mr. C. H., The Poles, Ludlow.
Hollings, Mr. J. A., The Hillend, Hereford.
Hood, Major-Gen., The Hon. A. N., Cumberland Lodge, Windsor.
Hopton, The Rev. J., Canon Frome, Ledbury.
Horton, Mr. T., Harnage Grange, Salop.
Hoskyns, C. Wren, Esq., Harewood, Ross.

J.

Jackson, P. R., Esq., Blackbrook, Monmouth.
James, Mr. R., Monnington, Hereford.
James, Mr. J. W., Mappowder, Blandford, Dorset.
James, Mr. J. C. Burghill, Hereford.
Jancey, J., Esq., Nunnington, Hereford.
Jeffreys, Mrs., The Downs, Much Wenlock, Salop.
Jones, Mr. W., Hill of Eaton, Ross.
Jones, Mr. J., Llywn-y-gaer, Raglan, Monmouthshire.
Jones, Mr. E., The Moat, Knighton.
Jones, Mr. H. A., Belmont, Shrewsbury.

SUBSCRIBERS.

K.

Kearney, P. J , Esq., Miltown House, Clonmellon, Ireland.
Keene, Mr. R., Pencraig, Caerleon, Monmouth.
Knight, A. R. Boughton, Esq., Downton Castle, Ludlow.

L.

Landon, Mr. T., Haywood Lodge, Hereford.
Lane, Mr. H., Cirencester.
Lane, Mr. W., Compton Casey, Cheltenham.
Leonard, Mr. E., Water End, Dursley, Gloucester.
Lewis, Mr. J., Milton, Pembridge, Leominster.
Lewis, Mr. J., Burlton, Hereford.
Lewis, Messrs. R. and T., Stapleton Castle, Presteign.
Lewis, Mr. T., Newchurch, Kinnersley, Kington.
Lisburne, The Right Hon. Earl of, Crosswood Park, Aberystwith.
Lort, W., King's Norton, Birmingham.
Lobb, Mr. G., Lawhitton, Launceston.
Leyshon, Mr. R., Island Bridge Farm, Bridgend, Glamorganshire.
Longmore, Mr. J., Buckton, Leintwardine, Salop.
Loyd, L., Esq., Monk's Orchard, Addington, Surrey.
Lumsden, J., Esq., Auchry House, Turriff, Aberdeenshire.

M.

Marston, Mr. H., Monkhall, Bridgenorth.
Martyn, C., Esq., Sydney, New South Wales (through Matthew Drew and Co., 38, High Holborn, London).
Mason, C. A., Esq., Tarrington, Ledbury.
Mason, Mr. H., Comberton, Ludlow.
Matthews, Mr., Auctioneer, Newent.
Meire, Mr. T. L., Cound Arbour, Salop.
Meire, Mr. J., Berrington, Salop.
Meire, Mr. J., Brockton House, Shiffnall, Salop.
Meredith, Mr. D., Cwmyoy, Hay.
Merryman, J., Esq., Hayfields, Cockeysville, Maryland, America.

SUBSCRIBERS.

Middleton, Mr., Shobdon, Leominster.
Mildmay, H., Esq., Wessington Court, Woolhope, Ledbury
Milner, Mr. J. M., Kington.
Monkhouse, Mr. J., The Stow, Whitney, Hereford.
Moore, Mr. F., Kings Pyon, Hereford.
Morris, Mr. F., Shuckenhill, Hereford.
Morris, Mr. T., Therrow, Llyswen, Hay.
Morris, Mr. P., Whitwick, Bromyard.
Morris, J. Esq., Madley, Hereford.
Morris, Mr. W., Ross.
Morris, E. J. Esq., Pontlarge, Winchcombe, Gloucester.

N.

Naylor, J., Esq., Leighton Hall, Montgomeryshire.
Newbery, Mr. W., Fern Hill, Kenilworth.
Nott, Mr. C., Bury, Wigmore, Leominster.
Nott, Mr., Glasbury, Hay.

O.

Oatley, W. H., Esq., Wroxeter, Salop.
Ockey, Mr. M., Thruxton, Hereford.
Olver, Mr. T., Penhallow, Grampound Cornwall.

P.

Palmer. Mr. J., Hampton-on-the-Hill, Warwick.
Paramore, Mr. J. R., Dinedor Court, Hereford.
Partridge, J., Esq., Bishop's Wood, Ross.
Peploe, Capt., Garnstone, Weobley.
Peren, W. B., Esq., Compton, South Petherton, Somerset.
Perrott, W., Esq., Llangorse, Brecon.
Perry, Mr. W., Cholstrey, Leominster.
Philpotts, Mr. S., Brook House, Bromyard.
Pinches, Mr., J. T., Hardwick, Leominster, Herefordshire.
Pitt, Mr. T., Freetown, Ledbury.
Pitt, Mr. W., Pigeon House, Weston Beggard, Hereford.

SUBSCRIBERS.

Pitt, Mr. F., White House, Canon Frome, Ledbury.
Pollock, J. O. G., Esq., Mountainstown, Navan, Ireland.
Powell, W. H., Esq., Fawley Court, Ross.
Powell, Mr. T., The Bage, Madley, Hereford.
Powell, Mr. T., Castle Frome, Ledbury.
Powell, Mr. W., Eglwysnunydd, Margam, Glamorganshire.
Powell, Mr. J., The Park, Llantillio Cresseny, Monmouthshire.
Powell, Mr. W., White House, Llantillio Cresseny, Monmouthshire.
Powell, E., Esq., Trewythan, Llandinum, Shrewsbury.
Power, Capt., Hill Court, Ross.
Preece, Mr. W. G., Shrewsbury.
Price, Mr. T., The Helm, Ewyas Harold, Hereford.
Price, Mr. N., Llancillo, Abergavenny.
Price, Mr. W., The Vern, Bodenham, Leominster.
Price, R. G., Esq., M.P., Norton, Presteign.
Price, Mr. J., Brindge Wood, Ludlow.
Pritchard, Mr. G., White House, Raglan, Monmouth.
Proctor, Mr. G., Wellington, Hereford.
Proctor, Messrs., Cathay, Bristol.
Prosser, Mr., Honeybourne Grounds, Broadway, Worcester.
Pye, Mr. G., Cublington, Madley, Hereford.

R.

Racster, Mr. W., Withington, Hereford.
Read, Mr. J. M., Elkstone, Cheltenham.
Reynell, R. W. Esq., Killynan, Killucan, West Meath, Ireland.
Richards, Mr. J., Cound, Salop.
Ricketts, Mr. J., Trebarried, Bronllys, Hay.
Ridler, Mr. R. H., Gattertop, Leominster.
Roberts, Mr. T., Ivington Bury, Leominster.
Roberts, Mr., Trippleton, Ludlow.
Roberts, Mr., Burrington, Ludlow.
Roberts, Lloyde, Esq., Corfton Hall, Ludlow.
Robinson, J., Esq., The Moor, Kington.
Rogers, Mr. B., The Grove, Pembridge.

SUBSCRIBERS.

Rogers, Mr. J., Altyr-y-nys Abergavenny.
Rogers, Mr., Pilleth, Knighton.
Rogers, Mr. T., Coxall, Brampton Bryan.
Rogers, Mr. J., Letchmoor, Presteign.
Royal Agricultural Society of England.
Russell, Mr. E., Kingsland, Leominster.

S.

Sampson & Co., Ludgate Hill, London.
Sexty, Mr. W. G., Sufton, Hereford.
Simcoe, J. G., Esq., Newcourt, Ross.
Simpkin & Co., Stationers' Hall Court, London.
Shirley, Mr., R., Baucott, Munslow, Church Stretton.
Smithies, G., Esq., Marlow, Leintwardine, Salop.
Smith, Mr. J., Sevenhampton, Cheltenham.
Smith, Mr. T., Bodenham, Dymock.
Smith, Mr. J., Shelsley Walsh, Worcester.
Smith, Mr. Sword's Farm,, Marcle, Ledbury.
Southern, F., Esq., Kempton, Ludlow.
Sparkman, Mr., J., Little Marcle, Ledbury.
Squarey, E. P., Esq., Salisbury.
Stallard, W., Esq., Brockhampton, Ross.
Stanhope, Sir. E. F. Scudamore, Bart., Holm House, Hereford.
Stedman, Mr. Lucton, Kingsland, Leominster.
Stedman, Mr. W. Bucknal House, Bedstone, Salop.
Stephens, Mr. P., Harpton, Kington.
Stokes, Mrs., Llyd-y-dway, Hay.
Stone, F. W., Esq., Moreton Lodge, Guelph, Canada West.
Stone, J. J., Esq., Scyborwen, Usk, Monmouth.
Street, Mr. W., Chilston, Madley, Hereford.
Sturgeon and Sons, South Ockendon Hall, Romford, Essex.
Sunderland, Mr. J. H., Hereford.
Swinburne, T. W., Esq., Corndean Hall, Winchcomb.
Symonds, Major, Pengethley, Ross.

SUBSCRIBERS.

T.

Tanner, Mr. E., Hopton Castle, Aston-on-Clun, Salop.
Taylor, Mr. W., Credenhill, Hereford.
Taylor, Mr. T., Burleigh Villa, Wellington, Salop.
Taylor, Mr. W., Showle Court, Ledbury.
Taylor, Mr. W., Thingehill Court, Hereford.
Taylor, Mr. J., Stretford Court, Leominster.
Taylor, Mr., Whitton, Leintwardine, Salop.
Taylor, C., Esq., The Green, Bromyard.
Thomas, Mr., St. Hilary, Cowbridge, Glamorganshire.
Thomas, Mr. J., Cholstrey, Leominster.
Thomas, Mr. E., The Farm, Bleddfa, Knighton.
Trew, Mr. T. Ship-street, Brecon.
Tudge, Mr. W., Adforton, Leintwardine, Herefordshire.
Turner, Mr. P., The Leen, Pembridge, Leominster.
Turner, Mr. W. D., Lynch Court, Leominster.
Turner, Mr. J., Eardisley Park, Hereford.

U.

Urwick, Mr. S. W., Leinthall, Ludlow.
Urwick, Mr., Felhampton, Ludlow.

V.

Vevers, Mr. C., Ivington, Leominster.
Villar, Mr. J., Portland-street, Cheltenham.
Vaughan, Messrs., Lawton, Leominster.

W.

Wadlow, Mr. E. Patton, Much Wenlock, Salop.
Walker, Mr. J., Northleach
Walker, J., Esq., Holmer, Hereford.
Walker, Mr. J., Knightwick, Worcester.
Wathen, Mr., Bridge Court, Hereford.

SUBSCRIBERS.

Wenlock, The Right Hon. Lord, Wenlock, Salop.
Wheeler, E. V., Esq., Kyrewood House, Tenbury.
Wheeler, Mr. T., Wormhill, Eaton Bishop, Hereford.
White, Mr. J. Lyndors, Monmouth.
Whitehouse, J. H., Esq., Ipsley Court, Redditch.
Whittaker. Mr., Bratton, Wilts.
Wicksted, C., Esq., Shakenhurst, Bewdley.
Wigmore, Mr. J., Bickerton Court, Dymock.
Williams, Mr. J., Kingsland, Leominster.
Williams, Mr., Culmington, Salop.
Williams, Mr. R., Pencelley Castle, Brecon.
Williams, Mr., Yazor. Hereford.
Williams, Mr. R. L., Velyn-nwyd, Brecon.
Williams, Mr. Maes-y-rony, Glasbury, Hay.
Williams, Mr. T., New House Farm, Bromfield, Salop.
Woolley, Mr. T., Weston, Pembridge, Leominster.
Worcestershire Agricultural Society.
Wright, E., Esq., Halston Hall, Oswestry.

Y.

Yeomans, Mr. H., Canon Pyon, Hereford.
Yeomans, Mr. W., Stretton Court, Hereford.

INDEX TO THE PLATES.

                                                                                       PAGE.

SIR OLIVER THE SECOND (1733), bred by Mr. T. Rea, the property of Mr. J. H. Arkwright. Winner of a first prize at the Worcester meeting of the Royal Agricultural Society of England; first at the Bristol meeting of the Bath and West of England Society; besides local prizes.

ADFORTON (1839), bred by Mr. W. Tudge, the property of Mr. T. Edwards. Winner of a third prize at the Battersea meeting, and first at the Worcester meeting of the Royal Agricultural Society of England, besides local prizes 2

COMMODORE (2472), bred by and the property of Mr. T. Duckham. Winner of first prizes at the Newcastle and Plymouth meetings of the Royal Agricultural Society of England and other prizes (see pedigree)... 26

DINEDOR (2497), bred by and the property of Mr. J. R. Paramore. Winner of a first prize at the Plymouth meeting of the Royal Agricultural Society of England, and at the Hereford meeting of the Bath and West of England Society 32

SIR THOMAS (2228), bred by Mr. T. Roberts, the property of Mr. J. Monkhouse. Winner of a first prize at the Worcester meeting of the Royal Agricultural Society of England, and other prizes. See pedigree, vol. 5, page 90 100

TAMBARINE (2254), bred by Lord Bateman, the property of Mr. W. Taylor. Winner of first prizes at the Worcester and Newcastle meetings of the Royal Agricultural Society of England; first at the Exeter meeting of the Bath and West of England Society; second at their Hereford meeting; and other prizes (see pedigree, vol. 5 p. 96) ... 106

FAIRY QUEEN (dam Fairy, vol. 5 p. 151), bred by and the property of Mr. J. Monkhouse. Winner of a first prize at the Newcastle meeting of the Royal Agricultural Society of England, and second at their Plymouth meeting; also a third prize at the Hereford meeting of the Bath and West of England Society; and a first, as one of a pair, at the meeting of the Herefordshire Agricultural Society, 1864.............. 131

INDEX TO THE PLATES.

	PAGE.
CLEMENTINE (dam Columbine), bred by and the property of Mr. J. Monkhouse. Winner of a first prize at the Worcester meeting of the Royal Agricultural Society of England	163
DUCHESS OF BEDFORD THE SECOND, bred by Mr. T. Roberts, the property of Mr. J. Baldwin. Winner of first prizes at the Newcastle and Plymouth meetings of the Royal Agricultural Society of England, and other prizes (see pedigree)	189
LADY ASHFORD (bred by Mr. W. Tudge, the property of Mr. J. Baldwin. Winner of a first prize at the Worcester meeting of the Royal Agricultural Society of England, and other prizes (see pedigree)	229
PRINCESS MARY (dam Maude, sire Deception, 2491) bred by and the property of Major-General the Hon. A. N. Hood. Winner of a first prize at the Plymouth meeting of the Royal Agricultural Society of England	254
MISS HASTINGS THE SECOND, bred by Mr. T. Roberts, the property of Mr. J. Baldwin. Winner of first prizes at the Worcester, Newcastle, and Plymouth meetings of the Royal Agricultural Society of England, and other prizes (see pedigree)	261
SPANGLE THE SECOND, bred by Mr. J. Rea, the property of Mr. J. Baldwin. Winner of first prizes at the Worcester and Newcastle meetings of the Royal Agricultural Society of England	321

APPENDIX.

HEREFORD OX, bred by Mr. T. L. Meire, fed by Mr. Heath. Winner of a first prize and the gold medal as the best fat ox or steer of any breed exhibited at the meeting of the Smithfield Club, 1863.

HEREFORD OX, bred by Mr. Tully, Huntington, Hereford, fed by Mr. Westcar, Aylesbury, Bucks. Winner of the first prize as the best fat ox or steer of any breed at the meeting of the Smithfield Club, 1799 7

The following reading is on the original picture:—

THE PRIZE OX, purchased by Mr. Chapman, Fleet Market, for £100. Height, 6ft. 7in.; girth, 10ft. 4in.; length, 9ft.; weight, 247st. 3lb.; tallow, 36st. 4lb.; offal, 29st, of 8lb. per stone. Published Feb. 1, 1800.

THE HERD BOOK
OF
HEREFORD CATTLE.

VOL. VI.

BULLS.

(2365) **A 1 THE SECOND.**

Red with white face, calved in the year 1861; bred by Mr.
Y. Cooke, Moreton House, Hereford, the property of Mr. Tudge
Ox House, Shobdon, Leominster; got by A 1 (1478), dam
(Young Countess) by Sportsman (1395), g.d. (Countess) by Sir
David (349), g.g.d. (Lovely) by Colossus (591), g.g.g.d. (Nain)
by Holmer (1616), g.g.g.g.d. (Countess) by Royal (331),
g.g.g.g.g.d. (Countess) by a son of Fitzfavourite (441).

(2366) **AAR.**

Red with white face, calved December 2, 1863; bred by and
the property of Mr. P. W. Bowen, Shrawardine Castle, Salop;
got by Claret (1177), dam (Bedstone Prettymaid the Second)
by Bedstone (2411), g.d. (Bedstone Prettymaid) by Conrad
(1183), g.g.d. (Prettymaid) by Lottery the Second (408), g.g.g.d.
(Prettymaid) by Dinedor (395).

BULLS.

(2367) ABBOT.
Red with white face, calved October 3, 1862; bred by and the property of Mr. J. P. Apperley, Fownhope, Hereford; got by Coroner (1555), dam (Snowdrop) by Wonder (1458), g.d. (Snowberry) by Wonder (420), g.g.d. (Snowball) by Young Kingsland (1464), g.g.g.d. (Miss Seeward) by Hector (181).

(2368) ABERHONDDU THE SECOND.
Red with white face, calved March 12, 1864; bred by and the property of Mr. John Ricketts, Trebarried, Bronllys, Hay; got by Trebarried (2833), dam (Gaylass) by Aberhonddu (903), g.d. (Spot) by Robin (1053), g.g.d (Fairmaid) by Chancellor.

(2369) ABERNANT.
Red with white face, calved in the month of February, 1863; bred by Mr. Rees Keene, Pencraig, Caerleon, the property of Mr. B. Leonard, Abernant, Caerleon, Monmouth; got by Pencraig (2671), dam (Nelly) by Prince Albert (2168), g.d. (Spot) by Young David (2325), g.g.d. (Old Spot) by Foxhall (2520).

(2370) ACHILLES.
Red with white face, calved September 16, 1863; bred by and the property of Mr. T. Olver, Penhallow, Grampound, Cornwall; got by Conservative (1930), dam (Alma) by Great Eastern (1598), g.d. (Annie) by Duke of Cornwall (1569).

(2371) ADFORTON THE SECOND.
Red with white face, calved August 2, 1863; bred by Mr. Thos. Edwards, Wintercott, Leominster, the property of Mr. Preece, Blangovenny, Llanvihangel, Abergavenny; got by Adforton (1839), dam (Rosebud the Second) by Sir Newton (1731), g.d. (Rosebud) by Croft (937), g.g.d (Daisy) by Stretford (1749), g.g.g.d. (Dainty) by Coningsby the Second (1552).

BULLS.

(2372) ADFORTON THE THIRD.

Red with white face, calved August 14, 1863; bred by Mr. Thomas Edwards, Wintercott, Leominster, the property of Mr. Godfrey, Brierley, Leominster; got by Adforton (1839), dam (Lady the Second) by Croft (937), g.d. (Lady) by Paddock (773), g.g.d (Lovely) by Coningsby the Second (1552).

(2373) ADFORTON THE FOURTH.

Red with white face, calved August 27, 1863; bred by Mr. Thomas Edwards, Wintercott, Leominster, the property of Mr. Lowe, Petchfield, Ludlow; got by Adforton (1839, dam (Blowdy), by Sir Newton (1731), g.d. (Pink the Second) by Wellington (1113), g.g.d. (Pink) by Stretford (1749), g.g.g.d. (Lady) by Coningsby the Second (1552).

(2374) ADMIRAL.

Red with white face, calved May 10, 1864, bred by and the property of Mr. B. Rogers, the Grove, Pembridge, Leominster; got by North Star (2138), dam (Curly the Fourth) by The Grove (1764), g.d. (Curly the Third) by Severus (1062), g.g.d. (Curly the Second) by Young Royal (1470), g.g.g.d. (Curly) by Prince (251).

(2375) ADMIRAL.

Red with white face, calved January 25, 1863; bred by Mr. F. W. Stone, Moreton Lodge, Guelph, Canada West, the property of Mr. J. Merryman, Hayfields, Cockeysville, Maryland, America; got by Patriot (2150), dam (Gentle) by Carlisle (923), g.d. (Lady) by The Knight (185), g.g.d. — by Monarch (504), g.g.g.d. — bred by the late Mr. J. Turner, Noke, Leominster.

Admiral was a winner of the Second Prize in his class at the Kingston Meeting of the Upper Canada Provincial Exhibition.

BULLS.

(2376) ADVENTURER.

Red with white face, calved July 7, 1864, bred by and the property of Mr. John E. Hewer, jun., Vern House, Hereford; got by Avenger (1855), dam (Miss Mary) by Doctor (1964), g.d. (Lady Mary) by Governor (464), g.g.d. (Lucy) by Lot (364), g.g.g.d. (Fanny) by Young Sovereign (506.)

(2377) AGITATOR.

Red with white face, calved October 15, 1863; bred by and the property of Mr. J. H. Arkwright, Hampton Court, Leominster; got by Dan O'Connell (1952), dam (Spot) by Riff Raff (1052) g.d. — by Quicksilver the Second, g.g.d. — by Jupiter (1289), g.g.g.d — by Reliance (278).

(2378) AJAX.

Red with white face, calved July 14, 1863, bred by Mr. Bourn, Mawley Town Farm, Cleobury Mortimer; got by Cardinal (1526), dam (Ada) by Attingham (911), g.d. (Princess), g.g.d. (Brecon) by Young Hope (343), g.g.g.d. (Duchess the Second) by Young Byron (832), g.g.g.g.d. (Duchess) by Chance (348).

(2379) ALABAMA.

Red with white face, calved September 4, 1864; bred by and the property of Mr. Charles Britten, Woodhouse, Leominster; got by Logic (2079), dam (Alice Maude) by Napoleon the Third (1019), g.d. (Victoria) by Hope (439) g.g.d. (Countess) by Young Chance (449).

(2380) ALBERT.

Red with white face, calved September 10, 1861; bred by the late Lord Berwick, Cronkhill, Salop, the property of Mr. Elsmere, Berrington, Salop; got by Severn (1382), dam (Silver) by Emperor (221).

BULLS.

(2381) ALBERT VICTOR.

Red with white face, calved January 14, 1864, bred by and the property of Mr. W. Newbery, Fernhill, Kenilworth; got by Comus (2477), dam (Lofty) by Attingham (911), g.d. (Grey Oak Apple) by Tom Thumb (243), g.g.d. (Oak Apple) by Commerce (354), g.g.g.d. (Strawberry) by a son of Guinea, bred by the late Mr. Jeffries, The Grove.

(2382) ALBERT VICTOR.

Red with white face, calved September 1, 1861; bred by Mr. Sobey, Penhallow, Grampound, Cornwall, the property of Mr. Martyn, Laddock, Cornwall; got by Conservative (1931), dam (Rosebud) by Pembridge (721), g.d. (Moss Rose) by Prince Dangerous (362), g.g d. (Fanny) by The Sheriff (356), g.g.g.d. (Cherry) by Crabstock (303), g.g.g.g.d. (Duchess) by Sovereign (404.)

(2383) ALDERMAN.

Red with white face, calved November 20, 1862; bred by and the property of Mr. T. Elsmere, Berrington, Salop; got by Franky (1243), dam (Silver) by Emperor (221).

(2384) ALEXANDER.

Red with white face, calved September 11, 1861; bred by Mr. John Monkhouse, The Stow, Hereford, the property of Mr. David Rogers, The Rhodd, Presteign; got by Chieftain (930), dam (Haughty) by Madoc (899), g.d. (Lofty) by Phantom (1035), g.g.d. (Stately) by Sir Andrew (183), g.g.g.d. (Stately) by a son of Sovereign (404)

BULLS.

(2385) ALFRED THE GREAT.

Red with white face, calved January 2, 1863; bred by and the property of Messrs. J. and W. Vaughan, Lawton, Leominster; got by Chelmsford (1915), dam (Crinoline) by Purifier (1364), g.d. (Curly) by Garrick (1248), g.g.d. (Moss Rose) by Hope (411), g.g.g.d (Old Moss Rose) by Old Wellington (507).

(2386) ALIGATOR.

Red with white face, calved in the month of October, 1864; bred by and the property of Mr. J. H. Arkwright, Hampton Court, Leominster; got by Dan O'Connell (1952), dam (Spot) by Riff Raff (1052), g.d. — by Quicksilver the Second, g.g.d. — by Jupiter (1289), g.g.g.d. by Reliance (278).

(2387) ALLURER.

Red with white face, calved December 27, 1863; bred by and the property of Mr. John E. Hewer, jun., Vern House, Hereford; got by Sir John (2225), dam (Clara) by Mameluke (1307), g.d. (Matilda) by Magnum Bonum (1303), g.g.d. (Purton Lass) by Royal Prince (528), g.g.g.d. (Moss Rose) by Hope (411).

(2388) AMBASSADOR.

Red with white face, calved July 17, 1863; bred by Mr. Philip Turner, The Leen, Pembridge, Leominster, the property of Mr. Hawkins, Rye Court, Ledbury; got by Bolingbroke (1883), dam (Ladybird) by Bertram (1513), g.d. (Countess) by Silurian (1064), g.g.d. (Princess) by Andrew the Second (619), g.g.g.d. (Brenda) by Viscount (816), g.g.g.g.d. (Rarity) by Cupid (1950).

Countess was a winner of the First Prize in her class at the meeting of the Smithfield Club, 1862.

BULLS.

(2389) ANDREW.

Red with white face, calved December 27, 1864; bred by and the property of Mr. T. Elsmere, Berrington, Salop; got by Albert (2380), dam (Silver) by Emperor (221).

(2390) ARGENTINE.

Red with white face, calved September 9, 1862; bred by and the property of Mr. Richard S. Fetherstonhaugh, Rockview, Killucan, Ireland; got by Cropper (1559), dam (Adelaide) by Attingham (911), g.d. (Silver) by Emperor (221).

(2391) ARTFUL.

Red with white face, calved August 19, 1861; bred by the late Mr. Thos. Rea, Westonbury, Leominster, the property of Mr. Davies, Abersewin, Brecon; got by England's Glory (1983), dam (Spot the Second) by Cholstrey (217), g.d. (Spot by Hope) (439), g.g.d. (Spot) by Primate (204), g.g.g.d. — by Forester (112).

(2392) ARTLESS.

Red with white face, calved March 9, 1864; bred by the late Mr. Rea, Westonbury, Leominster, the property of Lord Wenlock, Bourton Grange, Wenlock; got by Artful (2391), dam (Primrose) by Glendower (898), g.d. (Primrose) by Cholstrey (217), g.g.d. (Primrose) by Gallant (239).

(2393) AVON.

Red with white face, calved February 9, 1863; bred by and the property of Mr. Wm. Powell, Eglwysnunydd, Taibach, Glamorgan; got by General (1251), dam (Coquette) by Van Tromp (1440), g.d. (Lofty the Second) by Governor (464), g.g.d. (Lovely) by Hope (411), g.g.g.d. (Old Lofty) by Original the

First (455), g.g.g.g.d. (Blossom) by Old Wellington (507), g.g.g.g.g.d. (Old Blossom) bred by Mr. W. Hewer, The Hardwicke, g.g.g.g.g.g.d. (Old Lofty) by Silver (540).

(2394) BANJO.

Red with white face, calved September, 14, 1862; bred by and the property of Mr. Philip Ballard, Leighton Court, Bromyard; got by Tambarine (2254), dam (Leighton Pigeon) by Young Sir David (1137), g.d. (Thingehill Pigeon) by Reform (508), g.g.d. (Hampton Pigeon) by Young Sovereign (506), g.g.g.d. (Sylph) by Chance (355).

(2395) BANTER.

Red with white face, calved January 30, 1861; bred by Mr T. S. Bradstock, Cobrey Park, Ross, the property of Mr. Robert Keene, Elberton, Bristol; got by Melon (2111), dam (Eva) by Uncle Tom (1108), g.d. (Poston) by Foxwhelp (2522), g.g.d. (Daisy) by Son of Sovereign (404).

(2396). BANTING.

Red with white face, calved August 19, 1864; bred by and the property of Mr. T. Olver, Penhallow, Grampound, Cornwall; got by Zippor (2354), dam (Alma) by Great Eastern (1598), g.d. (Annie) by Duke of Cornwall (1569).

(2397) BARGAIN.

Red with white face, calved December 5, 1860; bred by Mr. James P. Apperley, Fownhope, the property of Mr. E. Jones, The Moat, Knighton; got by Baron of Noke (1862), dam Miss Julia) by Sir David (349).

(2398) BARLING.

Red with white face, calved September 24, 1864; bred by Mr. T. S. Bradstock, Cobrey Park, Ross; got by Young Rambler (2335), dam (Comely) by Daniel (1201), g.d. (Cocky) by Longitude, g.g.d. (Prettymaid) by Sovereign the Third, g g.g.d. (Eaton) by Lottery the Second (408).

(2399) BARON.

Red with white face, calved October 14, 1862; bred by Mr. Thos. Sheriff, Coxall, Ludlow, the property of Mr. E. Taylor, Whitton, Leintwardine, Salop; got by Sir Colin (2216), dam (Princess Royal) by Coxall (1196), g.d. — by Brilliant (1518), g g d. — by Confidence (367), g.g.g.d — by Emperor (221).

(2400) BARON.

Red with white face, calved February 22, 1864; bred by and the property of Mr. Joseph Meire, Berrington, Salop; got by Franky (1243), dam (Sabrina) by Severn (1382), g.d. (Beauty) by Attingham (911), g.g.d. (Silver) by Emperor (221).

(2401) BARON.

Red with white face, calved October 25, 1864; bred by and the property of Mr. E. Wright, Halston Hall, Oswestry; got by Hero (2039), dam (Lady Noble) by Young Byron (832) g.d. (Miss Noble) by Noble (543), g.g.d. (Favourite) by Son of Sovereign (404), g.g.g.d. (Damsel) by Young Wellington (505).

(2402) BARON GROVE.

Red with white face, calved in the month of November, 1858; bred by Mr. Edwards, Brampton Bryan, the property of Mr. William Connop, Street Court, Leominster; got by The Grove (1764), dam — by Stanage (1742), g.d. — by Old Court (306).

BULLS.

(2403) BARRISTER.

Red with white face, calved March 5, 1862; bred by Mr. Richard Shirley, Baucott Munslow, Church Stretton, the property of Mr. Price, Dodmore, Ludlow; got by Pilot (1036), dam (Miss Susan) by Baucott (1507), g.d. (Old Dilwyn) by Sir Andrew (183).

(2404) BARTER.

Red with white face, calved October 29, 1864; bred by and the property of Mr. T. S. Bradstock, Cobrey Park, Ross; got by Young Rambler (2335), dam (Fairy) by Daniel (1201), g.d. (Freckle) by Deluge (1210), g.g.d. (Blossom) by Young Royal (1468).

(2405) BATHURST.

Red with white face, calved August 1, 1862; bred by Mr. T. S. Bradstock, Cobrey Park, Ross; got by Young Rambler (2335), dam (Comely) by Daniel (1201), g.d. (Cocky) by Longitude, g.g.d. (Prettymaid) by Sovereign the Third, g.g.g.d (Eaton) by Lottery the Second (408).

(2406) BATTENHALL.

Red with white face, calved July 27, 1862; bred by Mr. Thomas Roberts, Ivington Bury, Leominster, the property of Mr. Edwards, Haywood, Hereford; got by Sir Thomas (2228), dam (Duchess) by King James (978), g.d. (Pyot) by Andrew the Second (619), g.g.d. (Old Pyot) by Prince by Dayhouse (299), g.g.g.d. bred by the late Mrs. Davies, Croft Castle.

Battenhall was a winner of the First Prize in his class at the Worcester Meeting of the Royal Agricultural Society of England.

BULLS.

(2407)   BEARTON.

Red with white face, calved January 14, 1863; bred by Mr. George Bray, jun., Henwood, Dilwyn, Leominster, the property of Mr. John Worrall, Llanwarda, Llangadock, Carmarthen; got by Valentine (2288), dam (Chance) by Haven (1610), g.d. — by Pilot (1689), g.g.d. — by Weobley (2308), g.g.g.d. — by Wistaston.

(2408)   BEAU.

Red with white face, calved in the month of April, 1863; bred by and the property of Mr. William Yeomans, Stretton Court, Hereford; got by Monaughty (2118), dam (Dinah) by Defence (1207).

(2409)   BEAUTY'S GOLDEN HORN.

Red with white face, calved May 24, 1862; bred by Mr. Wm. Perry, St. Oswald, Cholstrey, Leominster, the property of Mr. John Perry, Much Cowarne, Bromyard; got by Young Salisbury (2336), dam (Beauty) by Noble Boy (1337), g.d. (Bury the Third) by Noble Boy (751), g.g.d. (Gloucester) by Marden (564), g.g.g.d. (Bury the Second) by Goldfinder (383).

(2410)   BEAUTY'S WORCESTER.

Red with white face, calved July 5, 1863; bred by Mr. Wm. Perry, St. Oswald, Cholstrey, Leominster, the property of Mr. John Cave, Wallend, Monkland; got by Valentine (2288), dam (Beauty) by Noble Boy (1337), g.d. (Bury the Third) by Noble Boy (751), g.g.d. (Gloucester) by Marden (564), g.g.g.d. (Bury the Second) by Goldfinder (383).

BULLS.

(2411) BEDSTONE.

Red with white face, calved in the month of March, 1847; bred and used by Mr. William Stedman, Bedstone Hall, Salop; got by Young Emperor (1811), dam (Fat Rumps) by Dinedor (395), g.d. (Violet) by Waterloo (49), g.g.d. (Victoria), bred by the late Mr. Knight, Downton Castle.

(2412) BELAMITE.

Red with white face, calved January 20, 1862; bred by Mr. T. S. Bradstock, Cobrey Park, Ross, the property of Mr. Baker, Llanvihangel Court, Monmouth; got by Young Rambler (2335), dam (Fancy) by Daniel (1201), g.d. (Freckle) by Deluge (1210), g.g.d. (Blossom) by Young Royal (1468).

(2413) BENDIGO.

Red with white face, calved December 7, 1863; bred by and the property of Mr. Wm. Lane, Compton Casey, Andoversford; got by Hardy (2027), dam (Broad) by Planet (1690).

(2414) BERRINGTON.

Red with white face, calved October 29, 1863; bred by and the property of Mr. Thomas Roberts, Ivington Bury; got by Sir Thomas (2228), dam (Lady Ashton) by Ashton (1500), g.d. — by The Knight (185).

(2415) BERRINGTON.

Red with white face, calved September 29, 1863; bred by Mr. Prosser, Honeybourne Grounds, Worcester, the property of Mr. C. Ward, Saintbury Grounds, Worcester; got by The Jew (2266), dam (Beauty) by Berrington (435).

(2416) **BERTIE.**

Red with white face, calved July 21, 1864; bred by Mr. James Bourn, Mawley Town Farm, Cleobury Mortimer, the property of Mr. W. B. Peren, Compton, South Petherton; got by Cardinal (1526), dam (Peony) by Wigmore (1800), g.d. (Pansy) by Kyrewood (2062), g.g.d. (Pink the Third) by The Duke (2265), g.g.g.d. (Pink) by Silurian (1386), g.g.g.g.d. (Old Pink) by The Bishop (2260).

(2417) **BERTRAM.**

Red with white face, calved September 25, 1863; bred by and the property of Mr. C. H. Hinckesman, The Poles, Ludlow; got by Berwick (1874), dam (Birthday) by Butler (921), g.d. (Jessie) by Byron (559).

(2418) **BERWICK THE SECOND.**

Red with white face, calved October 12, 1864; bred by and the property of Mr. C. H. Hinckesman, The Poles, Ludlow; got by Berwick (1874), dam (Church House) by Andrew the Second (619), g.d. Church House by Cotmore (376).

(2419) **BILBOA.**

Red with white face, calved August 25, 1864; bred by and the property of Lieut.-Col. Feilden, Dulas Court, Hereford; got by Vincent (2858), dam (Young Fairmaid) by Mameluke (1307), g.d. (Old Fairmaid) by Garrick (1248), g.g.d. (Old Fan) by Defiance (416), g.g.g.d. — by Young Sovereign (506).

(2420) **BILLINGSLY.**

Red with white face, calved December 1, 1862; bred by and the property of Mr. J. Prosser, Honeybourne Grounds, Worcester; got by The Jew (2266), dam (Woodmaid) by Medalist (1009).

(2420A) BIRD'S EYE.
Red with white face, calved January 7, 1861; bred by and the property of Mr. James W. James, Mappowder, Blandford, Dorset; got by Statesman (1744), dam (Ellen) by Chance (2452).

(2421) BIRTHDAY.
Red with white face, calved April 15, 1863; bred by and the property of Mr. Wm. Beaumand, Vron End, Clun, Salop; got by Newcastle (2651), dam (Lady Montgomery) by Titterell (1096), g.d. (Old Browny) by The Baron (704).

(2422) BISHOPSTONE.
Red with white face, calved in the month of May, 1862; bred by and the property of Mr. William Bennett, North Cerney, Cirencester; got by Murphy (1331), dam (Countess) by Warrior (2305), g.d. (Cowslip) by Champion (1906), g.g.d. — by Fitz-favourite (441).

(2423) BLONDIN.
Red with white face, calved October 30, 1861; bred by Mr. Charles Nott, Bury House, Wigmore, the property of Mr. Broadhurst, Tenbury, Worcestershire; got by Truelove (2840), dam — by Trusty Ben — g.d. — by Mortimer (896), g g.d. — by Son of Confidence (367), g.g.g.d. — by Old Court (306), g.g.g.g.d. — by Cotmore the Second (1191).

(2424) BLUCHER.
Red with white face, calved September 5, 1862; bred by and the property of Mr. Wm. Tudge, Adforton, Ludlow; got by Sir Colin (2216), dam (Countess) by The Doctor (1083), g.d. (Duchess) by Stanage (1741), g.g.d. (Laurel) by Nelson (1021), g.g.g.d. (Lily the Third) by Turpin (300), g.g.g.g.d. (Lily the Second) by a Tully Bull.

BULLS.

(2425) BLUE CAP.

Red with white face, calved March 20, 1863; bred by and the property of Mr. William Bennett, North Cerney, Cirencester; got by Murphy (1331), dam (Miss Muslin) by Meteor (1319), g.d. (Grace) by Governor (464), g.g.d. (Victoria) by Prince (524), g.g.g.d. (Old Moss Rose) by Old Wellington (507).

(2426) BOCHYM.

Red with white face, calved January 28, 1863; bred by Mr. R. Davey, M.P., Polsue House, Grampound, Cornwall, the property of Mr. W. Horton Davey, Bochym, Helstone, Cornwall; got by Penhallow (2154), dam (Graceful), by Big Ben (1875), g.d. (Cherry), bred by the late Earl St. Germans.

(2427) BOLD BOY.

Red with white face, calved in the month of February, 1864; bred by and the property of Mr. Rees Keene, Pencraig, Caerleon, Monmouth; got by Cholstrey the Second (1919), dam (Cherry) by Young Chance, g.d. (Jenny), by Young David (2325), g.g.d. (Rose) by Foxhall (2520).

Bold Boy, with his sire and dam, were winners of the Second Prize in their class at the Tredegar Show, 1864.

(2428) BOLD DAVID THE SECOND.

Red with white face, calved December 1, 1863; bred by and the property of Mr. John Jones, Llwyn-y-gaer, Raglan, Monmouth; got by Bold David (1881), dam (Gager) by Dolphin (2500), g.d. (Gager) by a bull bred by the late Mr. Morris, Stockton.

Bold David the Second was a winner of the first prize in his class at the meeting of the Abergavenny Agricultural Society, 1864.

(2429) BONIFACE.

Red with white face, calved April 23, 1863; bred by Mr. Richard Shirley, Baucott, Munslow, Church Stretton, Salop, the property of Mr. Price, Moss Hill, Eardisland, Leominster; got by Pilot (1036), dam (Miss Page) by Baucott 1507), g.d (Old Miss Page) by Knockerell (1630), g.d. — by Holdgate (1615).

(2430) BRECON.

Red with white face, calved June 15, 1864; bred by Mr. B. Rogers, The Grove, Pembridge, the property of Mr. Downes, Brynich, Brecon; got by North Star (2138), dam (Daisy) by Severus (1062), g.d. (Prettymaid the Second) by Young Royal (1470) g.g.d (Prettymaid) by Prince (251), g.g.g.d. (Curly the Fourth) by Charity the Second (1535).

(2431) BRITAIN.

Red with white face, calved August 9, 1860; bred by Mr. Price, Pembridge, Leominster, the property of Mr. J. Wathen, Bridge Court, Hereford; got by Salisbury (2204), dam (Blossom) by Pembridge (721), g.d. (Countess) by Sir David (349), g.g.d. (Snowdrop) by Forester (398), g.g.g.d. (Red Rose) by Crabstock (303).

(2432) BROMFIELD.

Red with white face, calved September 10, 1864; bred by and the property of Mr. Hinckesman, The Poles, Ludlow; got by Berwick (1874), dam (Larkspur) by The Friar (1085), g.d, (Lady) by Old Jim (1026), g.g.d. (Duchess) by Nelson (1021), g.g.g.d. bred by Mr. W. Jellicoe, by Lundyfoot (88), g.g.g.g.d. bred by Mr. B. Tomkins.

BULLS.

(2433)   BROWN WILLIE.

Red with white face, calved June 12, 1859; bred by and the property of Mr. G. Lobb, Lawhitton, Launceston; got by Young (Orleton) 1476, dam (Giantess) by Rory O'More (1711), g.d. (Old Giantess) by Defiance (416), g.g.d. — bred by the late Earl St. Germans.

Brown Willie was a winner of the Second Prize in his class at the Liskeard and the Truro Meetings of the Royal Cornwall Agricultural Society, and First at their Saltash Meeting. He was also a winner of First Prizes at Launceston, 1859 and 1863; at Tavistock, 1862 and 1863; and at Liskeard, 1864.

(2434)   BUCKINGHAM.

Red with white face, calved June 2, 1863; bred by and the property of Mr. James Bennett, Ingestone, Ross; got by Castor (1900), dam (Dainty) by Breinton (1155), g.d. (Dainty) by Grove (1268), g.g.d. (Dainty) by Lottery the Second (408).

(2435)   BUCKMAN.

Red with white face, calved July 21, 1864; bred by and the property of Mr. T. S. Bradstock, Cobrey Park, Ross; got by Young Rambler (2335), dam (Eva) by Uncle Tom (1108), g.d. (Poston) by Foxwhelp (2522), g.g.d. (Daisy) by Son of Sovereign (404).

(2436)   BUCKSTONE.

Red with white face, calved July 30, 1862; bred by the late Mr. John Rogers, Stockton, Presteign, the property of Messrs. R. and T. Lewis, Stapleton Castle, Presteign; got by Delight (1564), dam (Conningsby) by Zealous (2345), g.d. — by The Count (1760), g.g.d. — by Royal (331), g.g.g d. — by Conningsby (718).

BULLS.

(2437) BULWER.
Red with white face, calved March 20, 1862; bred by and the property of Mr. William Bennett, North Cerney, Cirencester; got by Murphy (1331), dam (Miss Trueman) by the Duke (493), g.d. (Folly) by Tobias (487), g.g.d. by Sovereign (404).

(2438) BURNSIDE.
Red with white face, calved January 3, 1864; bred by and the property of Mr. T. Olver, Penhallow, Grampound, Cornwall; got by Zippor (2354), dam (Beatrice) by Earl Derby (1979), g.d. (Beauty) by Young Walford (1820).

(2439) BUTLER.
Red with white face, calved September 26, 1864; bred by and the property of Mr. P. Ballard, Leighton Court; got by Banjo (2394), dam (Leighton Pigeon) by Young Sir David (1137), g.d. (Thingehill) Pigeon by Reform (508), g.g.d. (Hampton Pigeon) by Young Sovereign (506), g.g.g.d. (Sylph) by Chance (355).

(2440) BYRON.
Red with white face, calved December 5, 1863; bred by and the property of Mr. William Perry, St. Oswald, Cholstrey, Leominster; got by Witchend the Second (2315), dam (Prettymaid) by Noble Boy (751), g.d. (Change) by Monkland (552), g.g.d. — by Goldfinder (383), g.g.g.d. — bred by Mr. Tomkins.

(2441) CANADIAN CHIEF.
Red with white face, calved November 9, 1863; bred by and the property of Mr. F. W. Stone, Moreton Lodge, Guelph, Canada West; got by Sailor (2200), dam (Peach) by Albert Edward

(859), g.d. (Cherry the Thirteenth) by Walford (871), g.g.d. (Red Cherry) by Tom Thumb (243), g.g.g.d. (Cherry the Fifth) by Cholstrey (868), g.g.g.g.d. (Cherry the Fourth) by Green's Grey Bull (850A), g.g.g.g.g.d. (Cherry the Third) by Chancellor (156), g.g.g.g.g.g.d. (Cherry the Second) by Thickset (1769).

Canadian Chief was a winner of the First Prize in his class at the Hamilton Meeting of the Canadian Agricultural Society, 1864.

(2442) CAPTAIN.

Red with white face, calved July 13, 1864; bred by Mr. F. W. Stone, Moreton Lodge, Guelph, Canada West, the property of Mr. T. Aston, Elyria Ohio, United States, America; got by Sailor (2200), dam (Baroness the Second) by Patriot (2150), g.d. (Baroness) by Carlisle (923), g.g.d. (Little Beauty) by Andrew the Second (619), g.g.g.d. (Dainty) by Vulcan (1446), g.g.g g.d. — bred by the late Mr. Turner, Noke, Leominster.

Captain was a winner of the Third Prize in his class at the Hamilton Meeting of the Canadian Agricultural Society, 1864.

(2443) CAPTAIN.

Red with white face, calved October 8, 1861; bred by and the property of Mr. Charles Wicksted, Shakenhurst, Bewdley; got by General (1251), dam (Penelope) by Chancellor (929), g.d. (Perfection) by Trader (1101).

(2444) CAPTAIN PERRY.

Red with white face, calved in the month of May 1862; bred by Mr. J. Perry, Much Cowarne, Bromyard, the property of Mr. J. P. Apperley, Hall Court, Much Marcle; got by Noble Boy (1337), dam (Peggy) by Monkland (552), g.d. (Primrose) by Noke.

(2445) CARACTACUS.

Red with white face, calved in the year, 1859; bred by Mr. Vaughan, Cholstrey, Leominster, the property of Mr. T. Vaughan, Lawton, Leominster; got by Plunder (1038), dam (Damsel) by Emperor (373), g.d. — by Cholstrey (217).

(2446) CARAUSIUS.

Red with white face, calved December 20, 1856; bred by and the property of Mr. W. H. Oatley, Wroxeter, Salop; got by Rustic (863), dam (Faustina) by Uriconium (598), g.d. (Fausta) by Bryony (599), g.g.d. Princess Royal.

(2447) CARDIFF.

Red with white face, calved December 18, 1863; bred by and the property of Mr. William Powell, Eglwysnunydd, Taibach, Glamorganshire; got by General (1251), dam (Gaylass the Third) by Grateful (1260), g.d. (Gaylass) by Young Emperor (1811), g.g.d. — by Venison (1441), g.g.g.d. — by Lottery the Second (408), g.g.g.g.d. — by Hector (535), g.g.g.g.g.d. — by Son of Waterloo (49).

(2448) CARLISLE THE SECOND.

Red with white face, calved August 9, 1863; bred by Mr. P. R. Jackson, Blackbrook, Skenfrith, Monmouth, the property of Mr. Evans, Skenfrith, Monmouth; got by Carlisle (923), dam (Miss Gay the Second) by Murphy (1331), g.d. (Miss Gay) by Gaylad (400), g.g.d. — by Berrington (435), g.g.g.d. by Dewshall (358).

BULLS.

(2449) CARLTON,

Red with white face, calved November 4, 1862; bred by and the property of Mr. T. Olver, Penhallow, Grampound, Cornwall; got by Sir Hugh (2223), dam (Cheerful) by Great Eastern (1598), g.d. (Patience) by Colossus (591), g.g.d. (Cheerful) by Invincible (592), g.g.g.d. (Cherry) by Reform (508).

(2450) CERVANTES.

Red with white face, calved June 17, 1864; bred by and the property of Mr. Philip Turner, The Leen, Pembridge, Leominster; got by Bolingbroke (1883), dam (Celia) by Veracity (1443), g.d. (Princess) by Andrew the Second (619), g.g.d. (Brenda) by Vicount (816), g.g.g.d. (Rarity) by Cupid (1950).

(2451) CHAMPAGNE.

Red with white face, calved July 9, 1863, bred by Mr. William Bennett, North Cerney, Cirencester, the property of Mr. John Haynes, Llanrothall, Monmouth; got by Trueman (2841), dam (Pretty Maid) by Mameluke (1307), g.d. (Countess) by Warrior (2305) g.g.d. — by Berrington (435).

(2452) CHANCE.

Red with white face, calved in the year 1847; bred by the late Mr. Stephens, Sheephouse, Hay; sold to the late Mr. James, Mappowder, Blandford, Dorset.

(2453) CHANCELLOR.

Red with white face, calved July 10, 1863; bred by and the property of Mr. Thomas Williams, Newhouse, Bromfield, Salop; got by Tugford (2849), dam (Piety) by a Son of Sir Walter (352), g.d. (Old Piety) by Mercury (361), g.g.d. — by Corfton (1188), g.g.g.d. — by Doubtful (1971).

(2454) CHARITY.

Red with white face, calved August 15, 1862, bred by Mr. Thomas Wheeler, jun., Wormhill, Eaton Bishop, Hereford, the property of Mr. George Matthews, Tibberton, Hereford; got by Troubadour (1780), dam (Duchess) by Muley (1330), g.d. (Prettymaid) by Young Royal (1136), g.g.d. (Prettymaid) by Reform (508).

(2455) CHARLES THE FIRST.

Red with white face, calved August 14, 1863; bred by Mr. Charles Britten, Woodhouse, Leominster, the property of Mr. T. Powell, Castle Froome; got by Bolingbroke (1883), dam (Young Blossom) by Petchfield (2676), g.d. (Blossom) by Wroxeter (386), g g.d. — bred by Mr. Stubbs, Woofferton.

Charles the First was a winner of the First Prize in his class at the meeting of the Ludlow Agricultural Society, 1864.

(2456) CHARLES THE SECOND.

Red with white face, calved August 6, 1864; bred by and the property of Mr. C. Britten, Woodhouse, Leominster; got by Logic (2079), dam (Young Blossom) by Petchfield (2676), g.d. (Blossom) by Wroxeter (386), g.g.d. — bred by Mr. Stubbs, Woofferton.

(2457) CHIEFTAIN THE THIRD.

Red with white face, calved September 9, 1862; bred by and the property of Mr. William Stallard, Brockhampton, Ross; got by Chieftain the Second (1917), dam (Gwenny the Second) by Chieftain (930), g.d. (Gwenny) by Regent (891), g.g.d. (Venus the Fifth) by Albert (330), g.g.g.d. (Winifred) by

Monaughty (220), g.g.g.g.d. (Venus the Fourth) by Duke (304) g.g.g.g.g.d. (Venus the Third) by Regulator (360), g.g.g.g.g.g.d. (Venus the Second) by Noble (238), g.g.g.g.g.g.g.d. (Venus) by Crabstock (303).

Chieftain the Third was a winner of the First Prize in his class at the meeting of the Ross Agricultural Society, 1864.

(2458)   CHIEFTAIN THE FOURTH.

Red with white face, calved February 1, 1864; bred by and the property of Mr. J. A. Hollings, Hillend, Hereford; got by Chieftain the Second (1917), dam (Rose of Weston) by St. Clement (2201), g.d. (Rose the Fifth) by Voltigeur (1445), g.g.d. (Rose the Fourth) by Byron (380), g.g.g d (Rose the Third) by Herald (2037), g.g.g.g.d (Rose the Second) by Cornet (1933) g.g.g.g.g.d. (Rose the First) by Young Waterloo (2341).

Chieftain the Fourth, with his sire, and dam, were winners of the First Prize in their class at the meeting of the Herefordshire Agricultural Society, 1864.

(2459)   CHOLSTREY GOLDFINDER.

Red with white face, calved December 7, 1862; bred by Mr. William Perry, St. Oswald, Cholstrey, Leominster, the property of Mr. James Gregg, Fencote Abbey, Leominster; got by Cowarne (1942), dam (Silver the Second) by Goldfinder the Second (959), g.d. (Silver) by Marden (564), g.g.d. (Princess) by Albert (330), g.g.g.d. (Princely) by Goldfinder (388).

(2460)   CHRYSALIS.

Red with white face, calved August 24, 1862; bred by and the property of Mr. Lewis Loyd, Monks' Orchard, Addington, Surrey; got by Sir Colin (1390), dam (Butterfly) by the Doctor (1083), g.d. (Redrose) by Orleton (901), g.g.d. (Redrose) by Nelson (1021), g.g.g.d. (Redrose) by Turpin (300).

(2461) CLARENCE.

Red with white face, calved January 21, 1864; bred by and the propery of Mr. P. R. Jackson, Blackbrook, Skenfrith, Monmouth; got by Carlisle (923), dam (Miss Chance) by Young Sir David (1818), g.d. (Cherry) by Young Walford (1820).

(2462) CLARET.

Red with white face, calved January 10, 1863; bred by and the property of Mr. Thomas Thomas, St. Hilary, Cowbridge, Glamorgan; got by Goldfinder the Second (959), dam (Fancy) by Young Royal (1467).

Claret was a winner of the First Prize in his class at the meetings of the Carmarthenshire and Cowbridge Agricultural Society, 1863 and 1864.

(2463) CLAUDIO.

Red with white face, calved July 28, 1862; bred by Mr. Philip Turner, the Leen, Pembridge, Leominster, the property of Mr. Bridgewater, Porthamel, Breconshire; got by Bolingbroke (1883), dam (Nonpariel) by Bertram (1513), g.d. (Exquisite) by Sir David (349), g.g.d. (Nell Gwynne) by The Knight (185), g.g.g.d. (Belle) by Sir Walter (352), g.g.g.g.d. (Myrtle) by Commerce (354).

(2464) CLIFFORD.

Red with white face, calved August 12, 1862; bred by Mr. Philip Ballard, Leighton Court, Bromyard, the property of Mr. James Gwillim, Lower Court, Clifford, Hay; got by Leighton (2069), dam (Lovely) by Temperance (1405), g.d. (Spot) by Grand Turk (1595).

BULLS.

(2465)   COCKLEFORD.
Red with white face, calved January 23, 1864; bred by and the property of Mr. J. M. Read, Elkstone, Cheltenham; got by Colesborne (2467), dam (Symphony) by Sebastopol (1381), g.d. (Pigeon) by The Sheriff (356), g.g.d. — by a celebrated bull bred by the late Mr. Price, of Ryall

(2466)   COLENSO.
Red with white face, calved July 10, 1864; bred by Mr. J. Bourn, Mawley Town Farm, Cleobury Mortimer, the property of Mr. Vincent Gosford, Tanylan, Holywell, North Wales; got by Sir Colin (1389), dam (Prudence) by Cardinal (1526) g.d. (Prettymaid) by Kyrewood (2062), g.g.d. (Pink the Third) by The Duke (2265).

(2467)   COLESBORNE.
Red with white face, calved February 22, 1862; bred by and the property of Mr. J. M. Read, Elkstone, Cheltenham; got by Caliban (1163), dam (Washington) by Carlisle (923), g.d. (Winsome) by Vension the Second (1442), g.g.d. (Whiteback) by Jeffries (587), g.g.g.d. (Old Whiteback) bred by Lord Radnor.

(2468)   COMBERTON.
Red with white face, calved September, 20, 1862; bred by the late Mr. Lowe, Petchfield, Ludlow, the property of Mr. Henry Mason, Comberton, Ludlow; got by Sultan (2795), dam (Beauty) by The Sultan (1419).

(2469)   COMET.
Red with white face, calved July 18, 1863; bred by and the property of Mr. Thomas Edwards, Wintercott, Leominster; got

by Sir William (2233), dam (Lively) by Leominster (1634), g.d. (Lady the Second) by Croft (937), g.g.d. (Lady) by Paddock (773), g.g.g.d. (Lovely) by Coningsby the Second (1552)

(2470) COMMERCE.

Red with white face, calved January 19, 1864; bred by Mr. Naylor, Leighton Hall, Montgomeryshire, the property of Miss Thomas, Llwynmaddoc, Montgomeryshire; got by Salisbury (2204), dam (Prudence) by Silvester (797), g.d. (Blowdy) by Big Ben (248), g.g.d. (Mottle) by Prince (251), g.g.g.d. (Beauty) by Claret (253), g.g.g.g.d. (Spot) by Trump (490).

(2471) COMMODORE.

Red with white face, calved September 26, 1863; bred by and the property of Mr. Olver, Penhallow, Grampound, Cornwall; got by Conservative (1931), dam (Ring Dove), by Young Walford (1820), g.d. — bred by the late Mr. Longmore. Buckton, Salop.

(2472) COMMODORE.

Red with white face, calved August 8, 1862; bred by and the property of Mr. T. Duckham, Baysham Court, Ross; got by Castor (1900), dam (Carlisle) by Albert Edward (859), g.d. (Silver) by Emperor (221).

Commodore was a winner of a Third Prize in his class at the meeting of the Herefordshire Agricultural Society, 1863; First at the Bristol and Hereford meetings of the Bath and West of England Society; First at the Newcastle meeting of the Royal Agricultural Society of England; First at the Coventry meeting of the Warwickshire Agricultural Society; First at Hereford; First at Tredegar, together with the Extra Prize, for the best bull of any breed exhibited, and Third at Ludlow, 1864.

BULLS.

(2473)   COMMODORE.

Red with white face, calved January 31, 1863; bred by and the property of Mr. F. W. Stone, Moreton Lodge, Guelph, Canada West; got by Patriot (2150), dam (Vesta the Second) by Shobdon (1725), g d. (Vesta) by Carlisle (923), g.g.d (Lady) by The Knight (185), g.g.g.d. — by Monarch (504), g.g.g.g.d. — bred by the late Mr. J. Turner, Noke, Leominster.

Commodore was a winner of the First Prize in his class at the Kington meeting of the Upper Canada Provincial Exhibition.

(2474)   COMPTON.

Red with white face, calved May 17, 1862; bred by the Rev. Archer Clive, Whitfield, Hereford, the property of Mr. W. B. Peren, Compton, South Petherton; got by Bertram (1513), dam (Wanton) by Mameluke (1307).

Compton was a winner of the Second Prize in his class at the meetings of the Yeovil and Crewkerne Agricultural Societies, 1864.

(2475)   COMUS.

Red with white face, calved in the month of December, 1863; bred by Mr. B. Rogers, The Grove, Pembridge Leominster, the property of Mr. Jones, Downton, Kington; got by Bolingbroke (1883), dam (Silver the Fifth) by Severus (1062), g.d. (Silver the Fourth) by Young Royal (1470), g.g.d. (Silver the Third) by Gaylad the Second (1589), g.g g.d. (Silver the Second) by Portrait (272).

(2476)   COMUS.

Red with white face, calved November 9, 1864; bred by and the property of Mr. T. Powell, the Bage, Madley, Hereford; got by Interest (2046), dam (Moss Rose) by Courtier (1194), g.d.

(Rosabella) by Mameluke (1307), g.g.d. (Rossabella) by Pope (527), g.g.g.d. (Old Silver) by Old Wellington (507), g.g.g.g.d. (Beauty) by Sovereign (404), g.g.g.g.g.d. (Old Gentle) by Chance (355).

(2477) COMUS.

Red with white face, calved August 20, 1859; bred by Mr. W. Tudge, Ashford, Ludlow; the property of Mr. W. Newbery, Fern Hill, Kenilworth; got by The Doctor (1083), dam (Victoria) by Orleton (901), g.d. (Spot) by Turpin (300).

(2478) CONFEDERATE.

Red with white face, calved October 12, 1863; bred by Mr. E. Tanner, Hopton Castle, Ludlow, the property of Mr. John Gittins, Hinton House, Stottesden; got by the Doctor (1083), dam (Margery) by son of Buckton (1891), g.d. (Bountiful) by Young Walford — g.g.d. (Beauty) by Northampton (600).

(2479) CONFEDERATE.

Red with white face, calved August 1, 1863; bred by and the property of Mr. Joseph Meire, Berrington, Salop; got by Friendship (1995), dam (Comfort) by Lawyer the Second (2066), g.d. (Comely) by Lawyer (627), g.g.d. (Old Comely) by Speculation (387), g.g.g.d. (Countess) by Young Waxy (451).

(2480) CONSTANT.

Red with white face, calved February 16, 1863; bred by and the property of Mr. W. Powell, Eglwysnunydd, Taibach, Glamorgan; got by General (1251), dam (Vesta) by Carlisle (923), g.d. (Lady) by the Knight (185), g.g.d. — Monarch (504), g.g.g.d. — bred by the late Mr. Turner, of Noke Court.

BULLS.

(2481)   CORNER COP.

Red with white face, calved in the year 1852; bred by Mr. Vaughan, Cholstrey, Leominster, the property of Mr. Bedford, Corner Cop, Leominster; got by Goldfinder the Second (959), dam bred by the late Mr. Jeffries, The Grove.

(2482)   CORONET THE SECOND.

Red with white face, calved February 29, 1864; bred by and the property of Mr. John Sparkman, Little Marcle, Ledbury; got by Coronet (1936), dam (Prettymaid) by General (1251), g.d. (Redrose) by Chance (355), g.g.d. (Rosebud) by Young Wellington (505), g.g.g.d. (Old Redrose) by Waxy (403), g.g.g.g.d. (Prettymaid) by Old Wellington (507), g.g.g.g.g.d. — by Silver (540).

(2483)   CORPORAL.

Red with white face, calved in the month of December, 1863; bred by and the property of Mr. B. Rogers, The Grove, Pembridge, Leominster; got by Rifleman (2189), dam (Cherry) by Sampson (1061), g.d. (Cherry) by Red Ben (768).

(2484)   COUND.

Red with white face, calved December 4, 1862; bred by and the property of Mr. John Richards, Cound, Shrewsbury; got by Pirate (2158), dam (Countess) by the Knight (185).

(2485)   CROMWELL.

Red with white face, calved June 16, 1863; bred by and the property of Mr. Naylor, Leighton Hall, Montgomeryshire; got by Salisbury (2204), dam (Skylark) by Tom of Lincoln (1099), g.d. (Cress) by Silvester (797), g.g.d. (Lily) by Young Persian, g.g.g.d. (Greystock) by Young Charity, g.g.g.g.d. — by a bull bred by the late Mr. Tully, g.g.g.g.g.d. — by Blood Royal bull.

BULLS.

(2486) CRONKHILL.

Red with white face, calved June 29, 1863; bred by and the property of Mr. Naylor, Leighton Hall, Montgomeryshire; got by Blondin (1880), dam (Apple Blossom) by Attingbam (911), g.d. (Grey Oak Apple) by Tom Thumb (243), g g.d. (Oak Apple) by Commerce (354), g.g.g.d. (Strawberry) bred by the late Mr. E. Jefferies, by a son of Guinea.

(2487) CRONKHILL.

Red with white face, calved July 15, 1864; bred by and the property of the Hon. and Rev. H. Noel Hill, Cronkhill, Salop; got by Conqueror (1929), dam (Polyanthus) by Albert Edward (859), g.d. (Primrose) by Walford (871), g.g.d. (Young Nutty) by Tom Thumb (243), g.g.g.d. (Nutty the Third) by the Count (351), g g.g.g.d (Nutty the Second) by Young Trueboy (1475), g.g.g.g.g.d. (Nutty the First) by Cholstrey (868).

(2488) CYRUS THE SECOND.

Red with white face, calved May 29, 1862, bred by Mr. Wm. Beaumand, Vron End, Clun, Salop, the property of Mr. T. Vaughan, Felhampton, Ludlow; got by Cyrus (1199), dam (Lady Montgomery) by Titterell (1096), g.d. (Old Browny) by The Baron (704).

(2489) DANDY.

Red with white face, calved August 6, 1864; bred by and the property of Mr. John Monkhouse, The Stow, Hereford; got by Chieftain (930), dam (Vanity) by Madoc (899), g.d. (Lofty) by Phantom (1035), g.g.d. (Stately) by Sir Andrew (183).

BULLS.

(2490) DEACON.

Red with white face, calved May 10, 1864; bred by and the property of Mr. W. Tudge, Adforton, Ludlow; got by Pilot (2156), dam (Dora) by Carbonel (1525), g.d. (Dainty) by The Doctor (1083), g.g.d. (Dainty) by Orleton (901), g.g.g.d. (Pretty Maid) by Nelson (1021).

(2491) DECEPTION.

Red with white face, calved July 18, 1862; bred by the late Mr. James Rea, Monaughty, Knighton, the property of Major-General the Hon. A. N. Hood, Cumberland Lodge, Windsor; got by Sir Benjamin (1387), dam (Nonsuch) by Wellington (1112), g.d. (Fairlass) by Chieftain (930), g.g.d. (Fairmaid the Third) by Cholstrey (217), g.g.g.d. (Fairmaid the Second) by Gallant (239).

(2492) DEFENCE.

Red with white face, calved January 2, 1862; bred by and the property of Mr. Thomas Sheriff, Coxall, Ludlow; got by Defender (1956), dam (Bloomer) by Brilliant (1518), g.d. (Ruby) by Dinedor (395), g.g.d. (Gaylass) by Emperor (221), g.g.g.d. (Lily of the Valley) by Byron (440).

(2493) DEFIANCE.

Red with white face, calved July 1, 1862; bred by and the property of Mr. P. R. Jackson, Blackbrook, Skenfrith, Monmouth; got by Carlisle (923), dam (Miss Gay the Second) by Murphy (1331), g.d. (Miss Gay) by Gaylad (400), g.g.d. — by Berrington (435), g.g.g.d. — by Dewshall (358).

BULLS.

(2494) DEMETRIUS.

Red with white face, calved May 12, 1862; bred by Mr. Philip Turner, The Leen, Pembridge, the property of Mr. Edward Russell, The Showers, Kingsland; got by Bolingbroke (1883), dam (Tulip) by Felix (953), g.d. (Moss Rose) by Silurian (1064), g.g.d. (Ada) by Andrew the Second (619), g.g.g.d. (Johanna) by Albert Edward (754), g.g.g.g.d. (Lady) by Sir Walter (352), g.g.g.g.g.d. (Peeress) by Viscount (816).

(2495) DICK THE DUSTMAN.

Red with mottle face, calved August 21, 1862; bred by and the property of Mr. W. R. Grose, Penpont, Wadebridge, Cornwall; got by Conservative (1931), dam (Lovely) by Pembridge (721), g.d. (Luck's All) by Prince Dangerous (362), g.g.d. (Mottle) by The Sheriff (356), g.g.g.d. (Lady) by Forester (398).

(2496) DILIGENCE.

Red with white face, calved November 11, 1864; bred by Mr. James Smith, Ridby Farm, Much Dewchurch, the property of Mr. Thomas Wheeler, Wormhill Farm, Hereford; got by Battenhall (2406), dam (Countess) by Governor the Second (2018), g.d. — by Young Sovoreign (506).

(2497) DINEDOR (A TWIN).

Red with white face, calved October 15, 1863; bred by and the property of Mr. J. R. Paramore, Dinedor Court, Hereford; got by The Jew (2266), dam (Young Countess) by Carlisle (923), g.d. (Countess) by The Duke (493).

Dinedor was a winner of the First Prize in his class at the Hereford meeting of the Bath and West of England Society, 1865.

BULLS.

(2498) DISCOUNT.

Red with white face, calved December 25, 1864; bred by and the property of Mr. B. Rogers, The Grove, Pembridge, Leominster; got by Grove the Second (2556), dam (Myrtle) by Dutiful (1978,) g.d. (Musk Rose) by Mameluke (1307,) g.gd. (Marchioness) by Young Sovereign (379), g.g.g.d (Miss Cotmore the Second), by Lottery the Second (408), g g.g.g.d. Miss Cotmore) by Cotmore (376), g.g.g.g.g.d. — by Conqueror (412).

(2499) DISCOVERY.

Red with white face, calved July 20, 1862; bred by and the property of Mr. T. S. Bradstock, Cobrey Park, Ross; got by Geologist (2012), dam (Eva), by Uncle Tom (1108), g.d. (Poston) by Foxwhelp (2522), g g.d. (Daisy) by son of Sovereign (404.)

(2500) DOLPHIN.

Red with white face; bred by the late Mr. David Williams, Newton, Brecon; used by Mr. James Jones, Hollow Farm, Hereford, and Mr. Watkins, Penrose, Monmouth.

(2501) DOMINIE SAMPSON.

Red with white face, calved September 1, 1863; bred by and the property of Mr. William Tudge, Adforton, Ludlow; got by Pilot (2156(, dam (Camilla) by Kyrewood — g.d. (Comely), by Orleton (901), g.g.d. (Young) Spot by Nelson (1021), g.g.g.d. (Spot) by Turpin (300), g.g.g.g.d. (Cherry) by a Tully Bull,

BULLS.

(2502) DON JUAN.

Red with white face, calved July 24, 1862; bred by and the property of Mr. Edward Bowen, Corfton, Ludlow; got by Abdel Kader (1837), dam (Lady Wiseman) by Cardinal Wiseman (1168), g.d. — by Governor (464), g.g.d. — by Mercury (361), g.g.g.d. — by Corfton (1188), g.g.g.g.d. — bred by the late Lord Rodney.

(2503) DON QUIXOTE.

Red with white face, calved September 8, 1864; bred by and the property of Mr. Philip Turner, The Leen, Pembridge, Leominster, got by Bolingbroke (1883), dam (Sylph) by Felix (953), g.d. (Cherry) by Tom of Lincoln (1099), g.g.d. (Nymph) by Andrew the Second (619), g.g.g.d. (Sylph) by Viscount (816,(g.g.g.g.d. (Sylph,) by Old Court the Second (1341).

(2504) DOT.

Red with white face, calved August 21, 1864; bred by and the property of Mr. S. W. Urwick, Leinthall, Ludlow; got by Severus the Second (2747), dam (Prettymaid) by Young Royal (1469), g.d. (Whitehorn) by Newton (1667), g.g.d. — by Sir David (349).

(2505) DOUGLAS.

Red with white face, calved November 10, 1863; bred by and the property of Mr. William Tudge, Adforton, Ludlow; got by Pilot (2156), dam (Dainty) by The Doctor (1083,) g.d. (Dainty by Orleton (901), g.g.d. (Prettymaid) by Nelson (1021), g.g.g.d. (Prettymaid) by Turpin (300).

Douglas was a winner of the Third Prize in his class at the Hereford meeting of the Bath and West of England Society.

BULLS.

(2506) DUKE.

Red with white face, calved June 12, 1863; bred by and the property of Mr. John Richards, Cound, Salop; got by Pirate (2158), dam (Duchess) by The Knight (185), g.d. — bred by Mr. Rammell.

2507) DUKE OF BEDFORD.

Red with white face, calved September 27, 1862; bred by Mr. Thomas Roberts, Ivington Bury, Leominster, the property of Mr. Turner, Eardisley Park, Hereford; got by Sir Thomas (2228), dam (Duchess of Bedford) by Arthur Napoleon (910,) g.d — bred by Mr. Vaughan, Cholstrey.

Duke of Bedford was a winner of the Second Prize in his class at the meetings of the Ludlow and Leominster Agricultural Societies, 1863.

(2508) DUKE OF WELLINGTON.

Red with white face, calved January 7, 1860; bred by and the property of Mr. Samuel Gilliland, Brook Hall, Londonderry; got by Widgeon (1792), dam (Silk) g.d. Original by Original the First (455).

Duke of Wellington was a winner of the First Prize in his class at the meeting of the North West of Ireland Agricultural Society, 1863.

(2509) EARL DERBY.

Red with white face, calved June 25, 1858; bred by Mr. Edward Price, Court House, Pembridge, Leominster, sold to Mr. Strafford, for exportation to Australia; got by Goldfinder the Second (959), dam (Symmetry) by Sir David (349,) g.d. (Curly) by Prince Dangerous (362), g.g.d. (Countess) by Sheriff (356), g.g.g.d. (Tidy) by Forester (398), g.g.g.g.d. (Silk) by Crabstock)303.)

BULLS.

(2510) EARL DERBY THE SECOND.

Red with white face, calved August 12, 1860; bred by Mr. E. Price, Court House, Pembridge, Leominster, the property of Mr. Thomas Woolley, Weston Court, Pembridge; got by Earl Derby (2509), dam (Young Windsor) by Magnet (823), g.d. (Windsor) by Pembridge (721), g.g.d. (Symmetry) by Sir David (349), g.g.g.d. (Wegtail) by Prince Dangerous (362), g.g.g.g.d. (Morocco) by the Sheriff (365).

(2511) EARL OF MONMOUTH.

Red with white face, calved November 3, 1864; bred by and the property of Mr. P. R. Jackson, Blackbrook, Skenfrith, Monmouth; got by Florence (1991), dam (Silver) by Treasurer (1105), g.d. (Glow-worm) by Young Conrad (2322), g.g.d. (Blossom) by Caracticus (659), g.g.g.d. (Blossom) by Old Court (306), g.g.g.g.d. — bred by Mr. Rea, Monaughty.

(2512) EMERALD.

Red with white face, calved June 7, 1862; bred by Mr. T. Duckham, Baysham Court, Ross, the property of Mr. T. Smith, Bodenham, Much Marcle, Ledbury; got by Garibaldi (2003), dam (Ruby) by Breinton (1155), g.d. (Eywood) by Cotmore the Second (1191), g.g.d. — bred by the late Earl of Oxford.

(2513) FAIRFAX.

Red with white face, calved August 5, 1864; bred by and the property of Mr. R. Davey, M.P., Polsue House, Grampound, Cornwall; got by Zippor (2354), dam (Fairmaid) by Big Ben (1875), g.d. (White Rose,) bred by the late Earl St. Germans.

(2514) FALSTAFF.

Red with white face, calved July 16, 1862; bred by Mr. Philip Turner, The Leen, Pembridge, the property of Mr. W. Rolls, The Hendre, Monmouth; got by Bolingbroke (1883,) dam (Daisy) by Felix (953), g.d. (Rosebud) by Andrew the Sceond (619), g.g.d. (Daisy) by Marmion (763), g.g.g.d. (Desdemona) by Sir Walter (352).

(2515) FERRY MAN.

Red with white face, calved April 14, 1863; bred by and the property of Mr. Richard Shirley, Baucott. Munslow, Church Stretton, Salop; got by Pilot (1036), dam (Nutty the Seeond) by Baucott (1507), g.d. (Old Nutty) by Dulluggan (759), g.g.d. — by The Count (2263).

(2516) FIREBRAND.

Red with white face, calved October 6, 1863; bred by Lord Wenlock, Bourton Grange, Wenlock, the property of Mr. E. Davies, Patton, Much Wenlock; got by Dreadnought (1973), dam (Star) by Kinlet (1293), g.d. (Perfection) by Bedstone (2411), g.g.d. — by Perfection (538), g.g.g.d. (Violet) by Dinedor (395, g.g.g.g.d. — by Trojan (542), g.g.g.g.g.d. — by son of Waterloo.

(2517) FIREMAN.

Red with white face, calved December 2, 1864; bred by and the property of Lord Wenlock, Bourton Grange,, Wenlock; got by Dreadnonght (1973), dam (Star) by Kinlet (1293), g.d. (Perfection) by Bedstone (2411), g.g.d. — by Perfection (538,) g.g.g.d. (Violet) by Dinedor (395), g.g.g.g.d. — by Trojan (542), g.g g.g.g.d. — by son of Warterloo.

(2518) FITZ PILOT.

Red with white face, calved September 4, 1863; bred by and the property of Mr. Richard Hill, Orleton Court, Ludlow; got by Pilot (1036), dam (Evangeline) by Chanticleer (1173), g.d. (Eva) by The Knight (185).

(2519) FLAME.

Red with white face, calved November 28, 1861; bred by Mr. T. L. Meire, Cound Arbour, Salop, the property of Mr. Bromley, Lea Hall, Salop; got by Franky (1243), dam (Ruby) by Cound (1193), g.d. (Princess) by Uckington (2286), g.g.d. (Splendour) by Lawyer (627), g.g.g.d. (Slut) by Dinedor (395).

(2520) FOXHALL.

Red with white face, bred by the late Mr. Edwards, Foxhall, Ross, and sold to the late Mr. Rees Keene, Pencraig, Caerleon, Monmouth.

(2521) FOXLEY'S SIR BENJAMIN.

Red with white face, calved February, 1, 1863; bred by the late Mr. James Rea, Monaughty, Knighton, the property of the Rev. G. H. Davenport, Foxley, Hereford; got by Sir Benjamin (1387), dam (Grace Darling) by Pilot (1037), g.d. (Grace) by Young Conrad (2322), g.g.d. (Lively the Second) by Barrister (658), g.g.g.d. (Lively) by Gallant (239).

(2522) FOXWHELP.

Red with white face, calved August 13, 1851; bred by Mr. T. C. Yeld, The Broome, Leominster, the property of Mr. Thomas Dew, Poston Court, Ross; got by The Knight (185), dam (Lovely) by Big Ben (248), g.d. (Strapper) g.g.d. — by Old Court (306), g.g.g.d..— bred by the late Mr. Rea, Monaughty.

BULLS.

(2523) FRANCISO.

Red with white face, calved November 26, 1864; bred by and the property of Mr. T. Duckham, Baysham Court, Ross; got by Franky (1243), dam (Silver) by Colossus (591), g.d. (Sylph) by Pope (527), g.g.d. (Eywood) by Cotmore the Second (1191), g.g.g.d. — bred by the late Earl of Oxford.

(2524) FRED.

Red with white face, calved February 26, 1864; bred by and the property of Capt. Peploe, Garnstone, Weobley; got by Leo (2070), dam (Winifred the Second) by Regent (891), g.d. (Clara) by Caractacus (659), g.g.d. (Winifred) by Monanghty (220). g g.g.d. (Venus the Fourth) by Duke (304).

(2525) GAIETY.

Red with white face, calved November 18, 1862; bred by Mr. T. L. Meire, Cound Arbour, Salop, the property of Mr. J. Meire, Brockton House, Salop; got by Harnage (2028), dam (Nosey) by Walford (871), g.d. (Stately) by Layman (767) g.g.d. (Blossom) by Speculation (387).

(2526) GAINFUL.

Red with white face, calved April 15, 1862; bred by Mr. Benjamin Hawkins, Orleton, Ludlow, the property of Mr. Cheatham, Billingsly, Bridgenorth; got by The Grove (1764), dam (Symmetry) by Merry Andrew (1011), g.d. (Silver) by Northampton (600), g.g.d. (Silver) by Young Chance (449), g.g.g.d. — bred by the late Mr. Thomas Jeffries, The Grove.

(2527) GALBA.

Red with white face, calved March 15, 1864; bred by and the property of Mr. W. H. Oatley, Wroxeter, Salop; got by Franky (1243), dam (Agrippina) by Julius Cæsar (2054), g.d. (Long Horns) by Surprise (779), g.g d. (Princess Royal) by Grandson of Sovereign.

(2528) GAYLEN.

Red with white face, calved October 19, 1863; bred by Mr. Henry Gibbons Hampton Bishop, Hereford, the property of Mr. J. Guilding, Bushley Park, Tewkesbury; got by Shamrock the Second (2210), dam (Rose) by Medallist (1009), g.d. (Rose) by The Admiral (1078), g.g.d. (Blossom) by Young Gaylad (1163).

(2529) GAMBLER.

Red with white face, calved September 15, 1864; bred by and the property of Mr. Henry Gibbons, Hampton Bishop, Hereford; got by Shamrock the Second (2210), dam (Strapper) by Defence the Second (1208), g.d. (Strapper) by The Admiral (1078), g.g.d. (Old Strapper) by Young Gaylad (1463).

(2530). GAMESTER.

Red with white face, calved April 28, 1863; bred by Mr. Benjamin Hawkins, Orleton, Ludlow, the property of Mr. Thos. Hawkins, Sugwas Court, Hereford; got by The Grove (1764), dam (Symmetry) by Merry Andrew (1011), g.d. (Silver) by Northampton (600), g.g.d. (Silver) by Young Chance (449), g.g.g.d. — bred by the late Mr. Thos. Jeffries, The Grove.

BULLS.

(2531) GARIBALDI.

Red with white face, calved October 19, 1863; bred by Mr. T. Roberts, Ivington Bury, Leominster, the property of Mr. J. Bosley, Lower Lyde, Hereford; got by Sir Thomas (2228), dam (Wildrose the Second) by Master Butterfly (1313), g.d. (Moss Rose) by Uncle Tom (1108), g.g.d. (Yellow Rose) by Rattler (802).

(2532) GARRICK JUNIOR.

Red with white face, bred by and the property of the late Mr. C. Bulmer, Holmer, Hereford; got by Garrick (1248), dam (Criss), bred by the late Mr. Jones, Tarrington.

(2533) GARRICK THE SECOND.

Red with white face, calved in the month of March, 1859; bred by Mr. John Jones, Lower Hill, Leominster, the property of Mr. J. G. Alford, The Thorne, Leominster; got by Garrick, (1248). dam — bred by the late Mr. Walker, Wormsley.

(2534) GARRICK THE THIRD.

Red with white face, calved July 11, 1864; bred by and the property of Mr. Thomas Roberts, Ivington Bury, Leominster; got by Garrick the Second (2533), dam (Lady Ann) by Arthur Napoleon (910), g.d. (Lady Jane) by Cholstrey (217).

(2535) GARRON.

Red with white face, calved December 26, 1862; bred by and the property of Mr. Edward Drinkwater, Treribble, Llangarren, Ross; got by Noble Boy, (1337), dam (Blossom) by a bull bred by Mr. Williams, Kingsland, Leominster.

BULLS.

(2536) GARWAY.
Red with white face, calved in the year 1859; bred by Mr. Walter Prosser, Garway Court, the property of Mr. John Partridge, Bishop's Wood, Ross; got by the Duke (1410), dam — bred by Mr. Paul Prosser, Garway Court.

(2537) GARWAY THE SECOND.
Red with white face, calved May 30, 1864; bred by and the property of Mr. J. Partridge, Bishop's Wood, Ross; got by Garway (2536), dam (Miss Noble) by Cardinal Wiseman (1168), g.d. (Noble) by Young Byron (832), g.g.d. (Miss Noble) by Noble (543), g.g.g.d. (Favourite) by Sovereign (404), g.g.g.g.d. (Damsel) by Young Wellington (505).

(2538) GAY BOY.
Red with white face, calved May 22, 1864; bred by and the property of Mr. B. Rogers, The Grove, Pembridge, Leominster; got by North Star (2138), dam (Young Gay) by Claret (1921), g.d. (Gay) by Young Royal (1470), g.g.d. (Damsel) by Gaylad the Second (1589), g.g.g.d. (Curly) by Charity the Second (1535).

(1539) GENERAL.
Red with white face, calved November 19, 1864; bred by and the property of Mr. J. R. Paramore, Dinedor Court, Hereford; got by The General (2817), dam (Blue Bell) by Brecon (918), g.d. (Pretty), bred by the late Mr. Davies, Tarrington.

(2540) GENERAL GARIBALDI.
Red with white face, calved October 19, 1863; bred by and the property of Mr. Thomas Roberts, Ivington Bury, Leominster; got by Sir Thomas, (2228), dam (Wildrose the Second)

by Master Butterfly (1313), g.d. (Moss Rose) by Uncle Tom (1108), g.g.d. (Yellow Rose) by Rattler (802), g.g.g.d. (Wildrose) by Chevalier (799).

(2541) GENERAL LEE.
Red with white face, calved March 14, 1864; bred by Mr. Warren Evans, Llandowlas, Usk, the property of Mr. Jones, Castletown, Newport; got by Monaughty (2117), dam (Damsel the Second) by Oakley (1673), g.d. (Damsel) by Stockton (2243), g.g.d. (Fashion) by Stumpfoot (2245).

(2542) GERALD.
Red with white face, calved August 9, 1864; bred by and the property of Mr. Henry Gibbons, Hampton Bishop, Hereford; got by Shamrock the Second (2210), dam (Daisy) by The Admiral (1078), g.d. (Beauty) by Young Gaylad (1463), g.g.d. (Old Beauty) by Zephyr (1826).

(2543) GIFT THE SECOND.
Red with white face, calved August 12, 1864; bred by and the property of Lieut-Col. Feilden, Dulas Court, Hereford; got by Gift (1254), dam (Pretty Maid) by Zouave (2359), g.d. (Handsome) by Newton (1022), g.g.d. — by Monacrh (219), g.g.g.d. — by Gallant (239).

(2544) GIGANTIC.
Red with white face, calved April 20, 1863; bred by and the property of Mr. Richard Shirley, Baucott, Munslow, Church Stretton: got by Pilot, (1036), dam (Giantess) by Marlow (2104), g.d. (Tastey) by Knockerell (1630), g.g.d. (Nutty) by Dolluggan (759), g.g.g.d — by The Count, (2263).

(2545) GIN.

Red with white face, calved in the month of July, 1863; bred by and the property of Mr. J. O. G. Pollock, Mountainstow, Navan, Ireland; got by Master Willie (2637), dam (Juniper) by Walford (871), g.d. (Rebecca) by Governor (464), g.g.d. (Old Prettymaid) by Young Sovereign (1472), g.g.g.d. — by White nob (345), g.g.g.g.d. — by Young Wellington (505).

(2546) GLADSTONE.

Red with white face, calved September 11, 1863; bred by and the property of Mr. Henry Gibbons, Hampton Bishop, Hereford got by Shamrock the Second (2210), dam (Pyat) by The Admiral (1078), g.d. (Beauty) by Young Gaylad (1463), g.g.d. (Beauty) by Zephyr (1826),

(2547) GLADSTONE.

Red with white face, calved July 2, 1862; bred by and the property of Mr. Naylor, Leighton Hall, Montgomeryshire; got by Salisbury (2204), dam (Miriam) by Silvester (797), g.d. (Lovely) by Big Ben (248), g.g.d. (Strapper) bred by Mr. Rea, Monaughty, g.g.g.d. — by Old Court (306).

(2548) GLEAM.

Red with white face, calved December 10, 1862; bred by and the property of Mr. T. L. Meire, Cound Arbour, Salop; got by Franky (1243), dam (Rose) by Cound (1193), g.d. (Princess) by Uckington (2286), g.g.d. (Splendour) by Lawyer (627), g.g.d. (Slut) by Dinedor (395).

BULLS.

(2549).   GLENDOWER.

Red with white face, calved October 22, 1864; bred by and the property of Mr. Philip Turner, The Leen, Pembridge, Leominster; got by Grove the Second (2556), dam (Sal) by Sir David (349), g.d. (Gaudy) by Defiance (1209), g.g.d. (Beauty) by Old Court (306), g.g.g.d. — bred by Mr. Child, Wigmore Grange.

(2550)   GLENDOWER.

Red with white face, calved September 2, 1862; bred by Mr. Henry Gibbons, Hampton Bishop, Hereford, the property of Mr. Blenkin, Lower House, Stanton-on-Wye; got by Shamrock the Second (2210), dam (Spot) by The Admiral (1078), g.d. (Old Spot) by Young Gaylad (1463).

(2551)   GLOSSY.

Red with white face, calved March 27, 1863; bred by Mr. Richard Shirley, Baucott, Munslow, Church Stretton; the property of Mr. Hamer, Newcastle, Clun, Salop; got by (Pilot) (1036), dam (Silky Mottle Face) by Marlow (2104), g.d. (Mottle Silky) by Knockerell (1630), g.g.d. (Silky) by Dollugan (759), g.g.g.d. (Tidy) by The Count, (2263).

(2552)   GOLDEN CHAIN.

Red with white face, calved in the month of January 1864; bred by the late Mr. H. E. Powell, Great Brampton, Madley, the Property of Mr. J. H. Whitehouse. Ipsley Court, Redditch; got by Vincent (2858), dam (Laburnum) by Musssulman (1333), g.d. (Lady) by Young Sovereign (379), g.g.d. (Miss Cotmore the Second) by Lottery the Second (408), g.g.g d. (Miss Cotmore) by Cotmore (376), g.g.g.g.d. — by Conqueror (412).

BULLS.

(2553)   GRAHAM.

Red with white face, calved September 16, 1864; bred by and the property of Mr. Henry Gibbons, Hampton Bishop, Hereford; got by Shamrock the Second (2210), dam (Lily) by The Admiral (1078), g.d. (Lily) by Young Gaylad (1463), g.g.d. (Lily) by Zephyr (1826).

(2554)   GRANDEE.

Red with white face, calved August 5, 1864; bred by and the property of Mr. John Monkhouse, The Stowe Hereford; got by Chieftain (930), dam (Grand Duchess) by Madoc (899), g.d. (Duchess) by Young Hope (343), g.g.d. (Duchess) by Chance (348), g.g.g.d. — bred by the late Mr. D. Williams, Newton, Brecon.

(2555)   GRATEFUL.

Red with white face, calved August 28, 1863; bred by and the property of Mr. Henry Gibbons, Hampton Bishop, Hereford; got by Shamrock the Second (2210), dam (Daisy) by The Admiral (1078), g.d. (Beauty) by Young Gaylad (1463), g.g.d. (Beauty) by Zephyr (1826).

Grateful was a winner of the Second Prize in his class at the Hereford meeting of the Bath and West of England Society.

(2556)   GROVE THE SECOND.

Red with white face, calved August 10, 1862; bred by Mr. B. Rogers, The Grove, Pembridge, Leominster, the property of Mr. John Rogers, Allt-yr-ynys, Abergavenny; got by Bolingbroke (1883), dam (Silver the Fifth) by Severus (1062), g.d. (Silver the Fourth) by Young Royal (1470), g.g.d. (Silver the Third) by Gaylad the Second (1589), g.g.g.d. (Silver the Second) by Portrait (372).

BULLS.

(2557) HALLINGWOOD.

Red with white face, calved March 6, 1862; bred by Mr. John Gay Attwater, Ablington, Amesbury, Wilts, the property of Mr. Joshua Whitaker, Bratton, Westbury, Wilts; got by Casey (1527), dam (Blossom the Second) by Tyro (1786), g.d. (Blossom) by Hospodar (1621), g.g.d. — by Whittington (1797).

Hallingwood was a winner of the Second Prize in his class at the meeting of the Melksham Agricultural Society, 1863.

(2558) HAMPTON LAD.

Red with white face, calved November 30, 1862; bred by and the property of Mr. Henry Gibbons, Hampton Bishop, Hereford; got by Shamrock the Second (2210), dam (Blossom) by The Admiral (1078), g.d. (Lily) by Young Gaylad (1463), g.g.d. (Lily) by Zephyr (1826) g.g.g.d. (Silk) by a son of Dewshall (358).

(2559) HANSA.

Red with white face, calved August 3, 1863; bred by Mr. James Taylor, Stretford Court, Leominster, the property of Mr. James Connop, Noke Court, Leominster; got by Pleasant (2679), dam (Empress the Eighth), by St. Oswald (1378), g.d. (Empress the Second) by Orleton (2144), g.g.d. (Empress) bred by the late Mr. Bowen, Monkland.

(2560) HAPPY JACK.

Red with white face, calved July 2, 1863; bred by and the property of Mr. John E. Hewer, jun., Vern House, Hereford; got by Random Jack (2181), dam (Jenny Lind) by Van Tromp (1440), g.d. (Curly) by Garrick (1248), g.g.d. (Moss Rose) by Hope (411), g.g.g.d. (Old Moss Rose) by Old Wellington (507).

(2561) HAPPY LAND.

Red with white face, calved February 6, 1859; bred by Mr. E. Price, Court House, Pembridge, Leominster, the property of Mr. J. W. James, Mappowder, Blandford, Dorset; got by Goldfinder the Second (959), dam (Beauty) by Pembridge (721) g.d. (Blowdy) by Sir David (349), g.g.d. (Wagtail) by Prince Dangerous (362), g.g.g.d. (Partridge) by Crabstock (303).

Happy Land was a winner of the First Prize in his class at the meeting of the Sturminster Agricultural Society, 1860.

(2562) HARLEY.

Red with white face, calved September 21, 1862; bred by and the property of Mr. William Hall, Ashton, Leominster; got by Ashton, (1500), dam (Rose), by Young Cotmore (601), g.d. (Pigeon) by Young Cotmore (601), g.g.d. (Fillpail) by Favourite.

(2563) HAVELOCK.

Red with white face, calved September 8, 1857; bred by the late Mr. J. Rea, Monaughty, Knighton, the property of the Misses Abley, Norton, Presteign; got by Grenadier (961), dam (Lively the Second) by Barrister (658), g.d. (Lively) by Gallant (239) g.g.d. (Pert) by Old Court)306).

(2564) HERCULES.

Red with white face, calved November 12, 1862; bred by Mr. T. Olver, Penhallow, Grampound, Cornwall, the property of Messrs. Clark, Julian St. Ewe, Cornwall; got by Penhallow (2154), dam (Honey) by Earl Derby (1979), g.d. (Honeysuckle) by Duke of Cornwall (1569), g.g.d. (Honeysuckle) by Woodbine (1120).

BULLS.

(2565) HERCULES.

Red with white face, calved December 25, 1864; bred by and the property of Mr. B. Rogers, The Grove, Pembridge, Leominster; got by Grove the Second (2556), dam (Daisy the Second) by The Grove (1764), g.d. (Prettymaid the Second) by Young Royal (1470), g.g.d (Prettymaid) by Prince (251), g.g.g.d. (Curly) by Charity the Second (1535).

(2566) HERDSMAN.

Red with white face, calved January 30, 1864; bred by and the property of Mr. W. Lane, Compton Casey, Cheltenham; got by Hardy (2027), dam (Silk) by Hospodar (1621,) g.d. (Silky) by Planet (1690), g.g.d. (Cockhorn) by Perfection (1685).

(2567) HEREFORD.

Red with white face, calved June 15, 1864; bred by Mr. B. Rogers, The Grove, Pembridge, the property of Mr. J. Wigmore, Bickerton Court, Ledbury; got by Bolingbroke, (1883), dam (Fairmaid the Second) by Mowley by Madley (1301), g.d. (Fairmaid) by Gaylad the Second (1589), g.g.d. (Fairmaid) by Prince (251).

(2568) HERMIT.

Red with white face, calved January 20, 1864, bred by and the property of Mr. J. P. Apperley, Fownhope, Hereford; got by Volunteer (2862), dam (Snowdrop) by Wonder (1458), g.d. (Snowberry) by Wonder (420), g g.d. (Snowball) by Young Kingsland (1464), g.g.g.d. (Miss Seeward) by Hector (181).

BULLS.

(2569)   HERO OF THE WEST.

Red with white face, calved in the month of July 1863; bred by and the property of Mr. Thomas Morris, Therrow, Llyswen, Hay; got by Don Salisbury (1969), dam Miss (Byron the Third) by Telegraph (1404), g·d. (Miss Byron) by Young Byron (832), g.g.d. (Miss Hope) by Young Hope (343), g.g.g.d. (Beauty the Second) by Counsellor (422), g.g.g.g.d. (Prettymaid) by White Nob (345), g.g.g.g.g.d. (Lovely) by Charity the Second (516).

(2570)   HESIOD.

Red with white face, calved December 25, 1864; bred by and the property of Mr. J. Baldwin, Luddington, Stratford-on-Avon; got by Battersea (1865), dam (Miss Hastings the Second) by Sir Thomas (2228), g.d. (Lady Hastings) by Master Butterfly (1313), g.g.d. (Prima Donna) by King James (978), g.g.g.d. (Long Horns) by Andrew the Second (619), g.g.g.g.d. (Pigeon) by Prince by Dayhouse (299).

(2571)   HIGH SHERIFF.

Red with white face, calved December 5, 1864; bred by and the property of Mr. T. Rogers, Coxall, Ludlow; got by Grove the Second (1764), dam (Violet) by Malcolm (1305), g.d bred by Mr. Mason, Yatton, Aymestry.

(2572)   HOLLY.

Red with white face, calved November 29, 1862; bred by Mr. Richard S. Fetherstonhaugh, Rockview, Killucan, Ireland, the property of the Rev. John Fetherstonhaugh, Griffenstown House, Killucan, Ireland; got by Leominster (2071), dam (Heliotrope) by Attingham (911), g.d. (Grey Dove) by Wonder (420), g.g.d. (Dove) by Ashley Moor (791), g.g.g.d. (Pigeon) by Ashley Moor White Bull (870), g.g.g.g.d. (Damsel) by Cholstrey (868).

BULLS.

(2573)   HOPE.

Red with white face, calved August 21, 1863; bred by Mr. H. R. Evans, jun., Swanstone Court, Leominster, the property of Mr. W. Evans, Llandowlas, Usk.; got by Chatham (1914), dam (Nora) by Rambler (1046), g.d. (Stately) by Swanstone (1072), g.g.d. (Juno) by Emperor (373), g.g.g.d. (Countess) by Coningsby (718), g.g.g.g.d. (Lovely) by Young Trueboy (1475).

(2574)   HOPEFUL.

Red with white face, calved March 10, 1864; bred by and the property of Mr. Thomas, Roberts, Ivington Bury, Leominster; got by Sir Thomas (2228), dam (Hope the Second) by Master Butterfly (1313), g.d. (Hope) by King James (978), g.g d. (Rose of the Valley) by Coningsby (718). g g.g.d. — Goldfinder (383.)

(2575)   HOPTON FAVORITE.

Red with white face, calved October 29, 1864; bred by and the property of Mr. E. Tanner, Hopton Castle, Clun, Salop; got by Doctor (1964,) dam (Mulberry) by Northampton (600), g.d. Oakley.

(2576)   IRONMASTER.

Red with white face, calved July 20, 1863; bred by Mr. T. S. Bradstock, Cobrey Park, Ross, the property of Mr. T. Allaway, Lydney; got by Young Rambler (2335), dam (Margaret) by Daniel (1201) g.d. (Stately) by Deluge (1210), g.g.d. (Lofty) by Young Gaylad (1463).

(2577)   JERRY.

Red with white face, calved August 16, 1864; bred by and the property of Mr. Powell, The Bage, Madley, Hereford; got by

Interest (2046), dam (Gaylass) by Troubadour (1780), g.d. (Governess) by Governor the Second (2018), g.g.d (Fairmaid) by Royal (331), g.g.g.d. Old Fairmaid.

(2578) JOHN BULL.

Red with white face, calved September 6, 1863; bred by and the property of Mr. Philip Ballard, Leighton Court, Bromyard; got by Leighton (2069), dam (Young Prettymaid) by Temperance (1405), g.d. (Prettymaid) by Grand Turk (1595).

(2579) JOHN BARLEYCORN.

Red with white face, calved August 7, 1863; bred by and the property of Mr. Philip Ballard, Leighton Court, Bromyard; got by Leighton (2069), dam (Lovely) by Temperance (1405), g.d. (Spot) by Grand Turk (1595).

(2580) JOLLY BOY.

Red with white face, calved in the month of September, 1862; bred by Mr. J. O. G. Pollock. Mountainstowe, Navan, Ireland, the property of Sir William Burrington, Bart., Glenotall Castle, Limerick, Ireland; got by Sir Robert (2227), dam (Jessamine) by Attingham (911), g d. (Becky) by Young Byron (832), g.g.d. (Rebecca) by Governor (464).

(2581) JOLLY MILLER THE THIRD.

Red with white face, calved in the month of May, 1855; bred by and the property of Mr. Samuel Gilliland, Brook Hall, Londonderry, Ireland; got by Jolly Miller the Second, dam Red Rose.

Jolly Miller the Third was a winner of First Prizes at the meetings of the North West of Ireland and Londonderry Agricultural Societies, in the years 1856, 1857, and 1858; also a First Prize at the meeting of the Royal Agricultural Society of Ireland, 1858.

(2582) JOLLY MILLER THE FOURTH.

Red with white face, calved in the month of May, 1856; bred by and the property of Mr. Samuel Gilliland, Brook Hall, Londonderry, Ireland; got by Jolly Miller the Second, dam Red Rose the First.

Jolly Miller the Fourth was a winner of First Prizes at the meetings of the North West and the Londonderry Agricultural Societies, 1859.

(2583) JOLLY MILLER THE FIFTH.

Red with white face, calved in the month of June 1857; bred by and the property of Mr. Samuel Gilliland, Brook Hall, Londonderry, Ireland; got by Jolly Miller the Third (2581), dam (Lady Hereford).

(2584) JOLLY MILLER THE TENTH.

Red with white face, calved January 4, 1861; bred by and the property of Mr. Samuel Gilliland, Brook Hall, Londonderry, Ireland; got by Jolly Miller the Fourth (2582), dam (Silk) by Widgeon (1792), g.d. — by Original the First (455), g.g.d. — bred by Mr. Berrow, Allensmore, Hereford.

Jolly Miller the Tenth was a winner of First Prizes at the Sligo meeting of the Royal Agricultural Society of Ireland, and at the meetings of the North West and Londonderry Agricultural Societies, 1864.

(2585) JOLLY MILLER THE ELEVENTH.

Red with white face, calved in the month of May, 1861; bred by and the property of Mr. Samnel Gilliland, Brook Hall, Londonderry, Ireland; got by Jolly Miller the Fourth (2582), dam (Hereford Lass).

BULLS.

(2586) JOLLY MILLER THE THIRTEENTH.

Red with white face, calved January 10, 1862; bred by and the property of Mr. Samuel Gilliland, Brook Hall, Londonderry, Ireland; got by Jolly Miller the Fourth (2582), dam (Beauty) by Widgeon (1792), g.d. — by Original the First (455), g.g.d. — bred by Mr. Berrow, Allensmore, Hereford.

(2587) JOLLY MILLER THE FOURTEENTH.

Red with white face, calved May 20, 1864; bred by and the property of Mr. Samuel Gilliland, Brook Hall, Londonderry, Ireland; got by Jolly Miller the Tenth (2584), dam (Beauty) by Widgeon (1792), g.d. — by Original the First (455), g.g.d. — bred by Mr. Berrow, Allensmore, Hereford.

(2588) JOLLY MILLER THE FIFTEENTH.

Red with white face, calved May 20, 1864; bred by and the property of Mr. Samuel Gilliland, Brook Hall, Londonderry, Ireland; got by Jolly Miller the Tenth (2584), dam (Pale Face).

(2589) JULIUS CÆSAR.

Red with white face, calved September 7, 1864; bred by and the property of Mr. Philip Turner, The Leen, Pembridge, Leominster; got by Grove the Second (2556), dam (Fairmaid) by Felix (953), g.d. (Venus) by Andrew the Second (619), g.g.d. (Lady) by Sir Walter (352), g.g.g.d. (Peeress) by Viscount (816).

(2590) JUSTICE.

Red with white face, calved in the month of July, 1862; bred by Mr. B. Rogers, The Grove, Pembridge; the property of Mr.

Wilde, Michaelchurch, Kington; got by Bolingbroke (1883), dam (Spot the Second) by Severus (1062), g.d. (Spot) by Old Court the Second (2140), g.g.d. (Spot) by a bull bred by the late Mr. Jeffries, The Grove.

(2591)   KEARSAGE.

Red with white face, calved September 4, 1863; bred by Mr. J. Rogers, Allt-yr-ynys, Abergavenny, the property of Mr. A. Dowle, Bernithen Court, Ross; got by Zouavite (2364), dam (Pigeon) by The Count (1760), g.d. (Pigeon) by Royal (331), g.g.d. (Pigeon) by Prince (251), g.g.g.d. (Pigeon) by Sovereign (404).

(2592)   KENILWORTH DUKE.

Red with white face, calved March 24, 1863; bred by Mr. John Palmer, Hampton-on-the-Hill, Warwick, the property of Mr. W. Newbery, Fern Hill, Kenilworth; got by Sir Edmund Lyons, (2219), dam (Caroline) by Cardinal Wiseman (1168), g.d. (Hampton Lass) by Mark (424), g.g.d. (Miss Hampton) by Garrick (1248), g.g.g.d. (Lady Hampton) by Reform (508).

(2593)   KING CHARLES.

Red with white face, calved September 20, 1858; bred by Mr. Nott, The Bury House, Wigmore, the property of Mr. Edy, The Frith, Ledbury; got by a Son of Plunder (1038), dam (Miss Radnor) by Mortimer (896), g.d. — by Son of Confidence (367), g.g.d. — by Cotmore the Second (1191), g.g.g.d. by Dayhouse.

(2594) KING CHARLES THE SECOND.

Red with white face, calved August 13, 1863; bred by and the property of Mr. Thomas Smith, Bodenham, Much Marcle, Ledbury; got by King Charles (2593), dam — bred by Earl Somers, g.d. — bred by Mr. Mason.

(2595) LEOTARD.

Red with white face, calved September 19, 1864; bred by and the property of Mr. T. Olver, Penhallow, Grampound, Cornwall; got by Zippor (2354), dam (Lady) by Conservative (1931), g.d. (Strawberry) by Attingham (911), g.g.d. (Young Oak Apple) by Tom Thumb (243), g.g.g.d. (Oak Apple) by Commerce (354), g.g.g.g.d. (Strawberry) bred by the late Mr. Jeffries, The Grove.

(2596) LIBERAL.

Red with white face, calved in the month of July, 1862; bred by Mr. B. Rogers, The Grove, Pembridge, the property of Mr. Rogers, The Homm, Weobley; got by Bolingbroke (1883), dam (Daisy) by Severus (1062), g.d. (Prettymaid the Second) by Young Royal (1470) g.g.d. (Prettymaid), by Prince (251), g.g.g.d. (Curly the Third) by Charity the Second (1535).

(2597) LICHFIELD.

Red with white face, calved August 20, 1862; bred by Mr. Naylor, Leighton Hall, Montgomeryshire, the property of Colonel Dyott, Freeford, Lichfield; got by Salisbury (2204), dam (Apple Blossom) by Attingham (911), g.d. (Grey Oak Apple) by Tom Thumb (243), g.g.d. (Oak Apple) by Commerce (354), g.g.g.d. (Strawberry) bred by the late Mr. E. Jeffries, by a Son of Guinea.

BULLS.

(2598)   LINCOLN.

Red with white face, calved in the month of August, 1862; bred by Mr. B. Rogers, The Grove, Pembridge, the property of Mr. John Downes, Flintsham, Kington; got by Bolingbroke (1883), dam (Curly) by Severus (1062), g.d. (Curly) by Gaylad the Second (1589), g.g.d. (Curly) by Young Royal (1470), g.g.g.d. (Curly) by Portrait (372).

(2599)   LION.

Red with white face, calved April 26, 1863; bred by and the property of Mr. Thomas Lewis, Newchurch, Kinnersley, Hereford; got by Wellington (1113), dam (Beauty) by Cardinal Wiseman (1168), g.d. (Beauty) by Monarch (1655), g.g.d. (Beauty) by Curate (1560).

(2600)   LION.

Red with white face, calved July 7, 1862; bred by and the property of Mr. E. Wright, Halston Hall, Oswestry; got by Magnet the Second (989), dam (Lioness) by Carlisle (923), g.d. (Lofty) by Albert Edward (859).

Lion was a winnner of the Second Prize in his class at the Newcastle meeting of the Royal Agricultural Society of England.

(2601)   LION THE SECOND.

Red with white face, calved November 24, 1863; bred by and the property of Mr. E. Wright, Halston Hall, Oswestry; got by Magnet the Second (989), dam (Lioness) by Carlisle (923), g.d. (Lofty) by Albert Edward (859).

(2602) LITTLE JOHN.

Red with white face, calved August 22, 1864; bred by and the property of Mr. F. W. Stone, Moreton Lodge, Guelph, Canada West; got by Sailor (2200), dam (Nelly) by Carlisle (923), g.d. (Peeress) by Monarch (504), g.g.d. — by St. Germans (227), g.g.g.d. — bred by the late Mr. Turner, Noke, Leominster.

(2603) LITTLE BEN.

Red with white face, calved October 2, 1864; bred by and the property of Mr. P. Ballard, Leighton Court, Bromyard; got by Banjo (2394), dam (Lovely) by Telltale (1757), g.d. (Spot) bred by the late Mr. Ballard.

(2604) LLANARTH.

Red with white face, calved January 5, 1862; bred by and the property of Mr. T. Edwards, Llanarth, Raglan, Monmouth; got by Chancellor (1172), dam (Snowdrop) by Newton (344), g.d. (Snowdrop) by Enterprise (948), g.g.d. (Silver) by Charity the Second (516), g.g.g.d. (Damsel) by White Nob (345).

(2605) LLANDILO.

Red with white face, calved August 15, 1863; bred by Mr. P. R. Jackson, Blackbrook, Skenfrith, Monmouth, the property of Mr. Powell, Llandilo, Monmouth; got by Carlisle (923) dam (Violet) by Young Sir David (1818), g.d. (Rosebud) by Young Walford (1820).

(2606) LLANROTHALL.

Red with white face, calved in the month of October, 1864; bred by and the property of Mr. John Haynes, Llanrothall, Monmouth; got by Garibaldi (2008), dam (Damsel) by Pope (527), g.d. (Old Damsel) by a Garway bull.

BULLS.

(2607) LLANTILLIO.

Red with white face, calved March 23, 1863; bred by Mr. W. Powell, White House, Llantillio, Monmouth; the property of Mr. John Wigmore, Bickerton, Much Marcle, Ledbury; got by King David (2060), dam (Victoria) by Captain David (1897), g.d. (Damsel) by The Nugget (1091).

(2608) LLOWES.

Red with white face, calved in the month of August, 1857; bred by Mr. Edward Williams, Llowes Court, Hay, the property of Mr. John Monkhouse, The Stow, Hereford; got by Radnor (1366), dam (Beauty) by Glasbury (709), g.d. (Alice) by Quicksilver. (353).

(2609) LLYWELLYN.

Red with white face, calved September 17, 1863; bred by and the property of Mr. Naylor, Leighton Hall, Montgomeryshire; got by Salisbury (2204), dam (Patience) by Silvester (797), g.d. (Blowdy) by Big Ben (248), g.g.d. (Mottle) by Prince (251), g.g.g.d. (Beauty) by Claret (253), g.g.g.g.d. (Spot) by Trump (490).

(2610) LONGBOW.

Red with white face, calved January 10, 1864; bred by and the property of Mr. P. J. Kearney, Miltown House, Clonmellon, Ireland; got by Sir Benjamin (1387), dam (Lofty the Second) by Treasurer (1105), g.d. (Lofty) by Regent (891), g.g d. (Lofty) by Barrister (658), g.g.g.d. (Lofty) by Mask (307), g.g.g.g.d. — by Duke (304).

BULLS.

(2611) LONGITUDE.

Red with white face, calved in the month of May, 1863; bred by Mr. John Jones, Llwyn-y-gaer, Raglan, Monmouth, the property of Mr. Peter Marfell, Cwm Farm, Clytha, Monmouth; got by Chancellor (1172), dam (Lucy Long) by Patron the Third (2151), g.d. (Long Page) by a Son of Garway (2009), g.g.d. — by Chance's Son (1911).

(2612) LORD BATEMAN.

Red with white face, calved December 19, 1861; bred by the late Mr. C. Bulmer, Holmer, Hereford; got by Lord Hereford (2617) dam (Shut) by Carlisle (923), g.d. (Strapper) by Monarch (504), g.g.d. — bred by Lord Bateman, Shobdon Court.

(2613) LORD CLIFDEN.

Red with white face, calved July 20, 1862; bred by Messrs. T. & W. Vaughan, Lawton, Leominster, the property of Mr. Davies, Chanston, Hereford; got by Lord Wellington (2094), dam (Fairmaid) by Plunder (1038), g.d. (Daisy) by Emperor (373), g.g.d. (Letton) by Charity (375).

Lord Clifden was a winner of the First Prize in his class at the meetings of the Ludlow and the Leominster Agricultural Societies, 1863.

(2614) LORD CLYDE.

Red with white face, calved August 27, 1863; bred by Mr. T. Duckham, Baysham Court, Ross, the property of Mr. J. Fry, Woodgate, Wellington, Somerset; got by Victory (2296), dam (Carlisle) by Albert Edward (859), g.d. (Silver) by Emperor (221).

BULLS.

(2615) **LORD CLYDE.**

Red with white face, calved August 18, 1862; bred by Captain Savery, Hardwick Lodge, Chepstow, the property of Mr. Henry Clay, Piercefield Park, Chepstow; got by Hatfield (2030), dam (Curious) by Partner (1031), g.d. (Chance) by Corporal (1190), g.g.d. (Dainty) by Withington (1457). g.g.g.d. (Violet) by Yorick (1461).

(2616) **LORD HASTINGS.**

Red with white face, calved September 15, 1863; bred by Mr. Roberts, Ivington Bury, Leominster, the property of Mr. T. Morris, Therrow, Llyswen, Hay; got by Sir Thomas (2228), dam (Lady Hastings) by Master Butterfly (1313), g.d. (Prima Donna) by King James (978), g.g.d. (Long Horns) by Andrew the Second (619), g.g.g.d. (Pigeon) by Prince by Dayhouse (299).

Lord Hastings was a winner of the First Prize in his class at the meeting of the Leominster Agricultural Society, 1864.

(2617) **LORD HEREFORD.**

Red with white face, calved January 11, 1856; bred by the late Mr. C. Bulmer, Holmer, Hereford; got by Cardinal Wiseman (1168), dam (Gentle) by Governor (464), g.d. (Moss Rose) by Hope (411), g.g.d. (Fanny) by Defiance (416), g.g.g.d. (Old Fanny) by Fitzfavourite (441).

(2618) **LORD LINCOLN.**

Red with white face, calved August 13, 1863; bred by Mr. T. S. Bradstock, Cobrey Park, Ross, Herefordshire, the property of Mr. T. Cadle, Longcroft, Gloucester; got by Young Rambler

(2335), dam (Laura) by the Jew (2266), g d. (Margaret) by Daniel (1201), g g.d. (Stately) by Deluge (1210), g.g.g.d (Lofty) by Young Gaylad (1463).

(2619)   LORD LUCY.

Red with white face, calved September 9, 1863; bred by Mr. Thomas Roberts, Ivington Bury, Leominster, the property of Capt. Power, Hill Court, Ross; got by Sir Thomas (2228), dam (Lady Lucy) by Arthur Napoleon (910), g.d. (Lady Jane) by Cholstrey (217).

(2620)   LORD NOEL HILL.

Red with white face, calved August 29, 1862; bred by and the property of Mr. R. W. Reynell, Killynan, Killucan, Ireland; got by Cropper (1559), dam (Cronkhill Lofty) by Lottery the Second (408), g.d. (Empress) by Dinedor (395).

(2621)   LORD OF THE ISLES.

Red with white face, calved August 25, 1863; bred by and the property of Mr. Richard Hill, Orleton Court, Ludlow; got by Pilot (1036), dam (Jenny Lind) by Chance, g.d. (Patch) by Zealous (2348), g.g.d. (Patty) by Lottery (410).

(2622)   LORD OF THE MANOR.

Red with white face, calved November 26, 1862; bred by Mr. William Tudge, Adforton, Ludlow, the property of Mr. Green Price, M.P., Norton Manor, Presteign; got by Sir Colin (2216), dam (Dainty) by The Doctor (1083), g.d. (Dainty) by Orleton (901), g.g.d. (Pretty Maid) by Nelson (1021), g.g.g.d. (Pretty Maid) by Turpin (300).

BULLS.

(2623)   LORD RODNEY.

Red with white face, calved November 4, 1864; bred by and the property of Mr. Thomas Roberts, Ivington Bury, Leominster; got by Sir Thomas (2228), dam (Lady Ashton) by Ashton (1500), g.d. —— by The Knight (185).

(2624)   LORD STANLEY.

Red with white face, calved February 8, 1864; bred by and the property of Mr. J. T. Pinches, Hardwick, Pembridge, Leominster; got by Earl Derby the Second (2510), dam (Lady the Third) by Burton (1159), g.d. (Lady) by Cholstrey, g.g.d. (Venus) by a bull bred by the late Mr. Jeffries.

(2625)   LOTTERY.

Red with white face, calved August 6, 1863; bred by and the property of Mr. J. Lewis, Milton, Pembridge, Leominster; got by Zeno (1825), dam (Beauty the Third) by Gratitude (1261), g.d. (Beauty the Second) by Monarch (1655), g.g.d. (Beauty) by Young Sovereign,

(2626)   LOWER HILL.

Red with white face, calved in the month of October, 1864; bred by Mr. William Yeomans, Stretton Court, Hereford, the property of Mr. John Haynes, Llanrothall, Monmouth; got by Young Mameluke (2331), dam (Young Duchess) by The Priest (1416), g.d. (Duchess) by Capsicum (697), g.g.d. (Old Duchess) by Peter Simple (342), g.g.d. (Old Lofty) by Lottery (410).

(2627) MACARONI.

Red with white face, calved August 10, 1864; bred by and the property of Mr. Edward Bowen, Corfton, Ludlow; got by Oxenbold (2145), dam (Lady Wiseman) by Cardinal Wiseman (1168), g.d. — by Governor (464), g.g.d. — by Mercury (361), g.g.g.d. — by Corfton (1188), g.g.g.g.d. — bred by the late Lord Rodney.

(2628) MACKNEY.

Red with white face, calved October 15, 1863; bred by and the property of Messrs. T. and W. Vaughan, Lawton, Leominster, got by Cholstrey The Second (1919), dam (Crinoline) by Purifier (1364), g.d. (Curly) by Garrick (1248), g.g.d. (Moss Rose) by Hope (411), g.g.g.d. (Old Moss Rose) by Old Wellington (507).

(2629) MAJOR.

Red with white face, calved in the month of July, 1863; bred by Mr. B. Rogers, The Grove, Pembridge, Leominster, the property of Mr. Hall, Ashton, Leominster; got by Interest (2046), dam (Daisy) by Severus (1062), g.d. Prettymaid the Second) by Young Royal (1470), g.g.d. (Prettymaid) by Prince (251), g.g.g.d. (Curly The Third) by Charity The Second (1535).

(2630) MAJOR.

Red with white face, calved October 3, 1862; bred by Mr. Naylor, Leighton Hall, Montgomeryshire, the property of Mr. William Davies, Guildsfield, Welshpool; got by Salisbury (2204), dam (Delight) by Silvester (797), g.d. (Daisy Queen) by The Knight (185), g.g.d. (Toby Pigeon) by Big Ben (248), g.g.g.d. (Duchess) by Tobias (487), g.g.g.g.d. (Duchess) by Sovereign (404), g.g.g.g.g.d. — bred by the late Mr. Turner. Noke.

(2631) MALLARD.
Red with white face, calved July 31, 1862; bred by Mr. Thomas Powell, Madley, Hereford, the property of Mr. William Mumford, Credenhill, Hereford; got by Harlequin (2031), dam (Plum) by Greengage (1266), g.d. (Pigeon), by a Son of Madley (1301), g.g.d. (Blossom) by Waxy (403), g.g.g.d. (Old Pleasant) by Defiance (416), g.g.g.g.d. (Stately) by Byron (440), g.g.g.g.g.d. (Old Stately) by Huntington (539).

(2632) MAMELUKE THE SECOND,
Red with white face, calved in the month of July 1860; bred by and the property of Mr. John Hewer, Vern House, Hereford; got by Mameluke (1307), dam (Prettymaid) by General (1251), g.d. (Red Rose) by Chance (355), g.g.d. (Rosebud) by Young Wellington (505), g.g.g.d. (Old Red Rose) by Waxy (403).

(2633) MARION.
Red with white face, calved July 3, 1863; bred by and the property of John Merryman, Hayfields, Cockeysville, Maryland, America; got by Curly (801), dam (Milton) by Wonder (420), g.d. (Old Milton) by Milton, bred by the late Mr. J. Longmore, Orleton, Ludlow.

(2634) MARS.
Red with white face, calved August 1, 1863; bred by and the property of Mr. Naylor, Leighton Hall, Montgomeryshire; got by Blondin (1880), dam (Crocus) by Sir David (349), g.d.

(Verbena) by Attingham (911). g.g.d. (Young Rebecca) by Young Hope (343), g.g.g.d. (Rebecca) by Governor (464), g.g.g.g.d. (Old Prettymaid) by Young Sovereign (1472), g.g.g.g.g.d. — by White Nob (345), g.g.g.g.g.g.d. — by Young Wellington (505).

(2635) MARS.

Red with white face, calved August 20, 1864; bred by Mr. B. Rogers, The Grove, Pembridge, the property of Mr. Pitt, Chadnor Court, Leominster; got by North Star (2138), dam (Strawberry) by Severus (1062), g.d. (Strawberry) by Young Royal (1470), g.g.d. (Miss Grove the Second) by Gaylad the Second (1589), g.g.g.d. (Miss Grove) by Defiance (1209).

(2636) MASTER BENJAMIN.

Red with white face, calved April 5, 1863; bred by and the property of Mr. R. H. Ridler, Gattertop, Leominster; got by Sir Benjamin (1387), dam (Rosebud) by the Count (1760), g.d. — by Young Royal, g.g.d. — by Young Albert, g.g.g.d. — by Portrait (272), g.g.g.g.d. — by Sovereign (404).

(2637) MASTER WILLIE.

Red with white face, calved in the month of January 1861; bred by Mr. Evan Bowen, Ens Down House, Shrewsbury, the property of Mr. J. O. G. Pollock, Mountainstowe, Navan, Ireland; got by Alliance (1490), dam (Merry Legs) by Chieftain (930), g.d. (Silver) by Quicksilver (353), g.g.d. — by a Son of Waterloo.

Master Willie was a winner of the First Prize in his class at the meeting of the Royal Dublin Agricultural Society, 1863.

(2638)　　　　MATCHLESS.

Red with white face, calved June 13, 1863; bred by and the property of Mr. W. B. Peren, Compton, South Petherton; got by Bertram (1513), dam (Wanton) by Mameluke (1307).

(2639)　　　　MIDDLETON.

Red with white face, calved March 26, 1863; bred by Mr. G. Tipton, Easton Farm, Middleton-on-the-Hill, Tenbury, the property of Mr. Whitehouse, Ipsley Court, Redditch; got by Baron Grove (2402), dam — by Herdsman (967).

(2640)　　　　MILTON.

Red with white face, calved May 4, 1863; bred by and the property of Mr. J. Lewis, Milton, Pembridge, Leominster; got by Zeno (1825), dam (Beauty the Fourth) by Golding (1257), g.d. (Beauty the Third) by Gratitude (1261), g.g.d. (Beauty the Second) by Monarch (1655), g g.g.d. (Beauty) by Young Sovereign.

(2641)　　　　MIRACLE.

Red with white face, calved January 12, 1863; bred by Mr. John Smith, Sevenhampton, Andoversford, the property of Mr. Morris, Stanley, Pontlarge, Winchcomb; got by Saint Michael (1718), dam (Matchless) by Phantom (1035), g.d. (Old Stately).

(2642)　　　　MONARCH.

Red with white face, calved April 27, 1864; bred by and the property of Mr. J. H. Whitehouse, Ipsley Court, Redditch; got by Vincent (2858), dam (Mulberry) by Musselman (1333), g.d. (Mayoress) by Prince (524), g.g.d. (Marchioness) by Young Sovereign (379), g.g.g.d. (Miss Cotmore the Second) by Lottery the Second (408), g.g.g.g.d. (Miss Cotmore) by Cotmore (376), g.g.g.g.g.d. — by Conqueror (412).

BULLS.

(2643) **MONAUGHTY.**

Red with white face, calved February 4, 1863; bred by the late Mr. James Rea, Monaughty, Knighton, the property of Mr. Jackson, Blackbrook, Skenfrith, Monmouth; got by Garland 2006), dam (Diana the Second) by Wellington (1112), g.d. (Diana) by Grenadier (961), g.g.d. (Spot the Second) by Cholstrey (217), g.g.g.d. (Spot) by Hope (439).

(2644) **MORMON.**

Red with white face, calved April 10, 1864; bred by and the property of Mr. P. W. Bowen, Shrawardine Castle, Salop; got by Claret (1177), dam (Merrylegs) by Chieftain (930), g.d. (Silver) by Quicksilver (353), g.g.d. — by Waterloo (99).

(2645) **MOUNTAIN CHIEF.**

Red with white face, calved August 10, 1862; bred by and the property of Mr. John J. Stone, Scyborwen, Llantrissent, Monmouth; got by Franky (1243), dam (Agatha) by Attingham (911), g.d. (Silver) by Emperor (221).

(2646) **MOUNTAINEER.**

Red with white face, calved October 21, 1863; bred by Mr. Edward Price, Court, House, Pembridge, Leominster; the property of Mr. J. O. G. Pollock, Mountainstowe, Navan, Ireland; got by Earl Derby the Second (2510), dam (Violet the Second) by Goldfinder the Second (959), g.d. (Violet) by Magnet) (823), g.g.d. (Violet) by Forester (398), g.g.g.d. (Countess) by Sir David (349).

(2647) MY LORD.

Red with white face, calved March 4, 1862; bred by and the property of Mr. J. Palmer, Hampton-on-the-Hill, Warwick; got by Abdel Kader (1837), dam (Curly) by Garrick (1248) g.d. (Moss Rose) by Hope (411), g.g.d. (Old Moss Rose) by Old Wellington (507), g.g.g.d. (Fanny) by Fitz Favourite (441),

(2648) NABOB.

Red with white face, calved April 12, 1862; bred by Mr. H. H. Chattock, Solihull, the property of Mr. William Lort, King's Norton, Birmingham; got by Colonel (1924), dam (Colleen Bawn) by Carlisle (923), g.d. (Graceful) by Alma (1144), g.g.d. (Primrose) by Duke (304).

(2649) NAMELESS.

Red with white face, calved in the year 1860; bred by Mr. W. H. Oatley, Wroxeter, Salop; got by Augustus Cæsar (1854), dam (Faustina) by Uriconium (598), g.d. (Fausta) by Bryony (599), g.g.d. (Princess Royal) by Emigrant (1980), g.g.g.d. (Sovereign Cow) by Sovereign (404).

(2650) NERO.

Red with white face, calved January 15, 1862; bred by and the property of Mr. W. H. Oatley, Wroxeter, Salop; got by Tiberius Cæsar (2272), dam (Agripina) by Julius Cæsar (2054), g.d. (Long Horns) by Surprise (779), g.g.d. (Princess Royal) by Emigrant (1980), g.g.g.d. — by Sovereign (404)

(2651) NEWCASTLE.

Red with white face, calved July 5, 1860; bred by and the property of Mr. John Hewer, Vern House, Hereford; got by Mameluke (1307), dam (Hampton Lass) by Mark (424), g.d.

(Miss Hampton) by Garrick (1248), g.g.d. (Lady Hampton) by Reform (508), g.g.g.d. (Moss Rose) by Hope (411).

(2652) NOBLEMAN.

Red with white face, calved in the month of July 1864; bred by Mr. J. B. Green, Marlow, Leintwardine, the property of Mr. Middleton, Chapel Lawn, Bucknell, Salop; got by Zeal (2342), dam (Daisy) by Vanguard (1109), g.d. (Damsel) by Beefy Ben (1869), g.g.d. (Dainty) by Cholstrey (217), g.g.g.d. (Dove) by Zest of Oxford (2352).

(2653) NOBLE VAN TROMP.

Red with white face, calved August 20, 1863; bred by and the property of Mr. William Perry, Cholstrey, Leominster; got by Witchend the Second (2315), dam (Young Pigeon) by Van Tromp (2291), g.d. (Pigeon) by Newton (1023), g.g d. (Miss Hewer) by Noble Boy (751), g.g.g.d. (Chadnor) by Turpin (365).

(2654) NONPAREIL.

Red with white face, calved September 29, 1864; bred by and the property of Mr. R. H. Capper, The Northgate, Ross; got by Garibaldi (2008), dam (Nonsuch) by Wellington (1112), g.d. (Fairlass) by Chieftain (930), g.g.d. (Fairmaid) by Cholstrey (217), g.g.g.d. (Fairmaid) by Gallant (239), g.g.g.g.d. (Fairmaid) by Portrait (372).

(2655) NORTHGATE.

Red with white face, calved August 17, 1862; bred by and the property of Mr. R. H. Capper, The Northgate, Ross; got by Lord Wellington (2094), dam (Margaret) by Attingham (911), g.d. (Dorcas the Third) by Tom Thumb (243), g.g.d. (Dorcas the

Second) by Wonder (420), g.g.g.d. (Dorcas) by Ashley Moor White Bull (870), g.g.g.g.d. (Old Damsel) by Coleman's Bull (1547), g.g.g.g.g.d. (Old Daisy) by Thickset (1769).

(2656) OAK APPLE.

Red with white face, calved in the month of September, 1864; bred by and the property of Mr. Thomas Morris, Therrow, Lyswen, Hay; got by Prince Imperial (2171), dam (Miss Byron the Second) by Prior (1359), g.d. (Miss Byron) by Young Byron (832), g.g.d. (Miss Hope) by Young Hope (343), g.g.g.d. (Beauty the Second) by Counsellor (422), g.g.g.g.d. (Prettymaid) by White Nob (345).

(2657) OAKLEY.

Red with white face, calved October 21, 1864; bred by and the property of Mr. C. H. Hinckesman, The Poles, Ludlow; got by Salopian (2739), dam (Jessie) by Byron (559), g.d. — bred by the late Mr. Tully.

(2658) OLIVER CROMWELL.

Red with white face, calved July 10, 1863; bred by Mr. Thomas Roberts, Ivington Bury, Leominster, the property of Mr. G. Gardiner, West Hope, Canon Pyon, Hereford; got by Oliver the Second (1733), dam (Lady Ann) by Arthur Napoleon (910), g.d. (Lady Jane) by Cholstrey) (217).

(2659) ORCUS.

Red with white face, calved December 1, 1864; bred by and the property of Mr. J. M. Read, Elkstone, Cheltenham; got by Colesborne (2467), dam (Theora) by Sebastopol (1381), g.d. (Cherry the Seventh) by Hotspur (855), g.g.d. (Cherry the Fifth) by Cholstrey (868), g.g.g.d. (Cherry the Fourth) by

Green's Grey Bull (850A), g.g.g.g.d. (Cherry the Third) by Chancellor (156), g.g.g.g.g.d. (Cherry the Second) by Thickset (1769), g.g.g.g.g.g.d. (Cherry) bred by the late Mr. Knight, Downton Castle.

(2660) ORIGINAL.

Red with white face, calved August 22, 1862; bred by and the property of Mr. Michael Ockey, Thruxton, Hereford; got by Grateful (1596), dam (Lofty) by Berrington (435), g.d. (Red Rose) by Governor (464), g.g.d. (Primrose) by Matchless (415), g.g.g.d. (Primsose) by Reform (508).

(2661) ORLEANS.

Red with white face, calved December 1, 1862; bred by and the property of Mr. William Tudge, Adforton, Ludlow; got by Magnum Bonum (2097), dam (Maud) by The Doctor (1083), dam (Morella) by Young Walford (1820), g.g d. (Cherry) by Nelson (1021), g.g.g.d. (Cherry) by Turpin (300), g.g.g.g.d (Cherry) by a bull bred by the late Mr. Tully.

(2662) ORPHAN.

Red with white face, calved August 24, 1862; bred by and the property of Mr. R. H. Capper, The Northgate, Ross; got by Lord Wellington (2094), dam (Blossom) by Walford (871), g d, (Becky) by Young Byron (832), g.g.d. (Rebecca) by Governor (464), g.g.g.d. (Old Prettymaid) by Young Sovereign (1472), g.g.g.g.d. — by Whitenob (345), g.g.g.g.g.d. — by Young Wellington (505).

Orphan with Dainty and their offspring were winners of the Third Prize in their class at the Hereford meeting of the Bath and West of England Society.

BULLS.

(2663)   OTHO (TWIN).

Red with white face, calved February 27, 1863; bred by Mr. W. H. Oatley, Wroxeter, Salop, the property of Mr. Clayton, Wrockwardine, Salop; got by Claudius Cæsar (1922), dam (Agripina) by Julius Cæsar (2054), g.d. (Long Horns) by Surprise (779), g.g.d. (Princess Royal) by Emigrant (1980), g.g g.d. (Sovereign Cow) by Sovereign (404).

(2664)   PALMERSTON.

Red with white face, calved January 7, 1864; bred by Mr. B. Rogers, The Grove, Pembridge, the property of Mr. J. D. Allen, Pyt House, Tisbury, Wilts.; got by North Star (2138), dam (Young Prettymaid) by Severus (1062), g d. (Prettymaid the Second) by Young Royal (1470), g.g.d. (Prettymaid) by Prince (251), g.g.g d. (Curly the Fourth) by Charity the Second (1535).

(2665)   PARAGON.

Red with white face, calved August 19, 1863; bred by Mr. T. Powell, The Bage, Madley, Hereford, the property of Mr. Price, Court House, Pembridge, Leominster; got by Portly (2165), dam (Plum) by Greengage (1266), g.d. (Pigeon) by Son of Madley (1301), g.g.d. (Blossom) by Waxy (403), g.g.g.d. (Old Pleasant) by Defiance (416), g.g.g.g.d. (Stately) by Byron (440), g.g.g g.g.d. (Old Stately) by Huntington (539).

(2666)   PATENT.

Red with white face, calved August 10, 1863; bred by Mr. Thomas Morris, Therrow, Llyswen, Hay, the property of Mr. C. Partridge, Clyro Court, Hay; got by Don Salisbury (1969), dam (Lucy) by Newton (344), g.d. (Lucy) by Young Hope (343), g.g.d. (Lady) by Prince Llewellyn (713), g.g.g.d. (Lily) by Charity the Second (516), g.g.g.g.d. (Lily) by Whitenob (345).

BULLS.

(2667) PATENT.

Red with white face, calved July 18, 1863; bred by Mrs. Edwards, Dayhouse, Kingsland, Leominster, the property of Messrs. T. and W. Vaughan, Lawton, Leominster; got by a Son of Carlisle (923), dam (Miss Taylor) by Cholstrey (217).

(2668) PATRIARCH.

Red with white face, calved December 5, 1857; bred by and the property of Mr. James Williams, Culmington, Bromfield, Salop; got by Clipper (1178), dam (The Abbess) by Corfton (1188).

(2669) PATRON.

Red with white face, calved August 4, 1864; bred by and the property of Mr. Thomas Roberts, Ivington Bury, Leominster; got by Sir Thomas (2228), dam (Patroness) by Duke of Marlborough (1974), g.d. (Patroness) by Fairboy (617), g.g.d. (Patroness) by Original (216), g.g.g.d. (Perry) by Woodman (255).

Patron was a winner of the Third Prize in his class at the Hereford meeting of the Bath and West of England Society.

(2670) PEACEFUL.

Red with white face, calved in the month of November, 1854; bred by Mr. Vaughan, Cholstrey, Leominster, the property of the late Mr. T. Vaughan, Lawton, Leominster; got by Dayhouse, dam (Young Newton) by Cholstrey (217), g.d. — bred by the late Mr. D. Williams, of Newton, Brecon.

(2671) PENCRAIG.

Red with white face, calved in the month of January 1861; bred by the late Mr. Reese Keene, Pencraig, Caerleon, Monmouth, the property of Mr. George Williams, Caerlicken, Penhow, Newport; got by Odd Trick (1674) dam (Nancy) by Prince Albert (2168), g.d. (Gaylass) by Young David (2325), g.g.d. (Old Gaylass) by a bull bred by the late Mr. Edwards, Foxhall.

(2672)   PENMARSH.

Red with white face, calved October 20, 1862; bred by Mr. G. T. Forester, High Ercall, Wellington, Salop, the property of Mr. George Brookes, Walton, Wellington, Salop; got by Severn (1382), dam (Maid Marian) by Darling (1202), g.d. (Miller's Daughter) by Quicksilver (353), g.g.d. (Maid of the Mill) by Hope (439), g.g.g.d. — by Royal (331), g.g.g.g.d. — bred by the late Mr. Yeomans, Moreton, Hereford.

(2673)   PEREMPTORILY.

Red with white face, calved January 8, 1862; bred by and the property of Mr. J. M. Read, Elkstone, Cheltenham; got by Caliban (1163), dam (Beauty the Second) by Dodmore (1217A), g.d. (Beauty the First) by Taurus (577), g.g.d. (Lovely) by Conqneror (684), g.g.g.d. (Blowdy), bred by Mr. Stephens, Sheep House, Hay.

Peremptorily was a winner of the First Prize in his class at the Cirencester and Cheltenham meetings of the Gloucestershire Agricultural Society, 1863 and 1864, and a Second Prize at the Coventry meeting of the Warwickshire Agricultural Society, 1864.

(2674)   PERFECTION.

Red with white face, calved August 16, 1863; bred by Mr. Edward Price, Pembridge, Leominster, the property of Mr. R. S. Featherstonhaugh, Rockview, Killucan, Ireland; got by Earl Derby (2509), dam (Bombazin the Second) by Goldfinder the Second (959), g.d. (Bombazin) by Magnet (823), g.g.d. — by Prince Dangerous (362), g.g.g.d. — by The Sheriff (356), g.g.g.g.d. — by Crabstock (303).

(2675) PERFECTION.
Red with white face, calved January 23, 1863, bred by the late Mr. William Stokes, Llyd-y-dy-way, Cusop, Hay, the property of Mrs. Stokes, Llyd-y-dy-way, Cusop, Hay; got by Chieftain (930), dam (Sprightly) by Alma (1144), g.d. (Primrose) by Quicksilver (353), g.g.d. (Old Primrose) by Prince of Wales (1041).

(2676) PETCHFIELD.
Red with white face; bred by the late Mr. Lowe, Petchfield, Ludlow, the property of Mr. Henry Mason, Comberton, Ludlow; got by Newton (1667).

(2677) PETER THE GREAT.
Red with white face, calved June 8, 1863; bred by Mr. John Monkhouse, The Stow, Hereford, the property of Mr. H. Allen, The Priory, Clifford, Hay; got by Chieftain (930), dam (Catherine) by Madoc (899), g.d. (Lofty) by Phantom (1035), g.g.d. (Stately) by Sir Andrew (183), g.g.g.d. (Stately) by a Son of Sovereign (404).

(2678) PILOT.
Red with white face, calved September 14, 1863; bred by Mr. Thos. Powell, Bage, Madley, Hereford, the property of Mr. W. Jones, Canon Bridge, Hereford; got by Portly (2165), dam (Young Fairmaid) by Greengage (1266), g.d. (Fairmaid) by Royal (331), g.g.d. Old Fairmaid.

(2679). PLEASANT.
Red with white face, calved September 30, 1864; bred by Mr. T. Olver, Penhallow, Grampound, Cornwall, the property of Mr. T. Golding, Callington, Cornwall; got by Zippor (2354), dam (Rosa) by Great Eastern (1598), g.d. (Lady) by The Earl (1761), g.g.d. — bred by the late Earl St. Germans.

(2680) PLOUGHMAN.

Red with white face, calved April 22, 1863; bred by Mr. John Evans, Treberfa, Knighton, the property of Mr. H. Marston, Monkhall, Bridgenorth; got by Agriculturist (1842), dam (Gaylass) by Newton (1022), g.d. (Bounty) by Monarch (219), g.g.d. (Sal) by Turpin (300).

(2681) PLUNDER THE SECOND.

Red with white face, calved in the month of October, 1859; bred by Mr. William Vaughan, Cholstrey, Leominster, the property of Mr. Walter Greenhouse, Harbour, Kingsland; got by Plunder (1038), dam (Silk) by Emperor (373), g.d. (Newton) bred by the late Mr. D. Williams, Newton, Brecon.

(2682) POLSUE.

Red with white face, calved February 28, 1862; bred by Mr. T. Duckham, Baysham Court, Ross, the property of Messrs. S. and R. Davey, Tywarntrayle Farm, St. Agnes, Cornwall; got by Castor (1900), dam (Prettymaid) by Colossus (591), g.d. (Beauty) by Young Royal (1136), g g.d. (Prettymaid) by Mercury (1317).

(2683) POMPEY.

Red with white face, calved October 2, 1864; bred by and the property of Mr. James Taylor, Stretford Court, Leominster; got by Trustful (2845), dam (Melody) by Chieftain (930), g.d. (Eva) by Regent (891), g.g.d. (Spot) by Caractacus (659), g.g.g.d. (Spot) by Hope (439).

BULLS.

(2684) POOL WHARF.

Red with white face, calved in the month of July 1862; bred by Mr. John Haynes, Llanrothall, Monmouth, the property of Mr. Jones, Pool Wharf, Dewchurch, Hereford; got by Abdel Kader (1637), dam (Marianne) by Meteor (1319), g.d. (Promise) by Lot (364), g.g.d. (Fanny) by Young Sovereign (506) g.g.g.d. (Old Fanny) by Fitzfavourite. (441).

(2685) PORTRAIT.

Red with white face, calved September 12, 1858; bred by Mr. Edward Williams, Llowes Court, Hay, the property of Mr. Edward Drinkwater, Treribble, Llangarren, Ross; got by Tipperary Boy (1424), dam (Miss Chester) by Protection (794), g.d. (Broady) by Prince (333).

(2686) PREMIER.

Red with white face, calved in the month of September, 1864; bred by and the property of Mr. B. Rogers, The Grove, Pembridge; got by Grove the Second (2550), dam (Miss Stanton) by The Grove (1764), g.d. (Miss Stanton) by Severus (1062), g.g.d. (Miss Stanton) by Gaylad the Second (1589), g.g.g.d. (Miss Stanton) by Portrait (372).

(2687) PRESIDENT.

Red with white face, calved August 20, 1863; bred by and the property of Mr. William Tudge, Adforton, Ludlow; got by Pilot (2156), dam (Gaylass the Third) by the Doctor (1083), g.d. (Gaylass the Second) by Stanage (1741), g.g.d. (Gaylass) by Turpin (300), g.g.g.d. (Comely) by Tully Bull.

BULLS.

(2688) PRESTON.

Red with white face, calved December 10, 1852; bred by Mr. T. Powell, The Bage, Madley, Hereford, the property of Mr. John Davies, Preston-on-Wye, Hereford; got by Madley, (1301), dam (Fairmaid) by Royal (331), g.d. (Old Fairmaid.)

(2689) PRIME MINISTER.

Red with white face, calved August 3, 1863; bred by and the property of Mr. T. Roberts, Ivington Bury, Leominster; got by Sir Thomas (2228), dam (Prima Donna) by King James (978) g.d. (Long Horns) by Andrew the Second (619).

(2690) PRINCE.

Red with white face, calved March 18, 1864; bred by Mr. F. W. Stone, Moreton Lodge, Guelph, Canada West, the property of Mr. G. Clarke, East Springfield, Otsego Co, New York, U.S., America, got by Patriot (2150), dam (Princess) by Carlisle (923), g.d. (Peeress) by Monarch (504), g g.d. — by St. Germans (227), g.g.g.d. bred by the late Mr. Turner, Noke, Leominster.

Prince was a winner of the Second Prize in his class at the Hamilton meeting of the Canadian Agricultural Society, 1864.

(2691) PRINCE.

Red with white face, calved July 2, 1862; bred by and the property of Mr. E. Wright, Halston Hall, Oswestry; got by Magnet the Second (989), dam (Miss Noble) by Brecon (918), g.d. (Lady Noble) by Young Byron (832), g.g.d. (Miss Noble) by Noble (543), g g g.d. (Favourite) by a Son of Sovereign (404), g.g.g.g.d. (Damsel) by Young Wellington (505).

(2692) PRINCE ALBERT.

Red with white face, calved December 28, 1861; bred by and the property of Mr. J. Merryman, Hayfields, Cockeysville, Maryland, America; got by Blenheim (1879), dam (Cora) by Cronkhill, (867), g.d. (Milton) by Wonder (420), g.g.d. (Old Milton) by Milton, bred by the late Mr. J. Longmore, Orleton, Ludlow.

(2693) PRINCE ARTHUR.

Red with white face, calved November 4, 1864; bred by and the property of Major-General the Hon. A. N. Hood, Cumberland Lodge, Windsor; got by Garibaldi (2007), dam (Victoria) by Brecon (918), g.d.) Gwenllian) by Young Dewshall (1125).

(2694) PRINCE ARTHUR.

Red with white face, calved August 30, 1864; bred by and the property of Mr. S. W. Urwick, Leinthall, Ludlow; got by Severus the Second (2747), dam (Whitehorn) by Newton (1667), g.d. — by Sir David (349).

(2695) PRINCE ARTHUR.

Red with white face, calved November 24, 1862; bred by Mr. T. Duckham, Baysham Court, Ross, the property of Mr. Probert, Abbey Dore, Hereford; got by Castor (1900), dam (Delight) by Pope (527), g.d. (Eywood) by Cotmore the Second (1191), g.g.d. — bred by the late Earl of Oxford.

Prince Arthur was a winner of the Second Prize in his class at the Cirencester meeting of the Gloucestershire Agricultural Society, 1863.

BULLS.

(2696) **PRINCE EDWARD.**
Red with white face, calved September 1, 1864; bred by Mr. S. W. Urwick, Leinthall, Ludlow, the property of Mr. Mountford, Brockton, Ludlow; got by Severus the Second (2747), dam (Beauty) by Tatteridge (1755), g.d. (Whitehorn) by Newton (1667), g.g.d. — by Sir David (349).

(2697) **PRINCE LEOPOLD.**
Red with white face, calved September 27, 1864; bred by and the property of Major-General the Hon. A. N. Hood, Cumberland Lodge, Windsor; got by Deception (2491), dam (Venus) by Brecon (918), g.d. (Zoe) by Young Dewshall (1125).

(2698) **PRINCE ROYAL.**
Red with white face, calved August 1, 1864; bred by and the property of Mr. Thomas Powell, Bage, Madley, Hereford; got by Portly (2165), dam (Fairmaid the Second) by Greengage (1266), g.d. (Fairmaid) by Royal (331), g.g.d. Old Fairmaid.

(2699) **PRINCE ROYAL.**
Red with white face, calved August 14, 1864; bred by and the property of Mr. R. H. Capper, The Northgate, Ross; got by Garibaldi (2008), dam (Princess Royal) by Coxall (1196), g.d. — by Brilliant (1518), g.g.d. — by Confidence (367), g.g.g.d. by Emperor (221).

(2700) **PRINCE'S PIPPIN.**
Red with white face, calved May 19, 1862; bred by Mr. S. C. Good, Aston Court, Tenbury; the property of Mr. Richard Roberts, Burrington, Ludlow; got by Trumpet (2843), dam (Broady) by Walford (871), g.d. — Mr. S. Meire's bull, g.g.d. — by Old Court (306), g.g.g.d. — bred by the late Mr. Roberts, Little Heath, Salop.

(2701) PRINCE VICTOR.

Red with white face, calved June 24, 1863; bred by and the property of Mr. P. W. Bowen, Shrawardine Castle, Salop; got by Claret (1177), dam (Virgin) by Uncle Tom (1107), g.d. (Young Primrose) by Chieftain (930), g.g.d. (Primrose) by Quicksilver (353), g.g.g.d. (Lovely) by Bright, g.g.g.g.d. (Lovely) by a Son of Waterloo (99).

(2702) PRIZE FIGHTER.

Red with white face, calved November 14, 1863; bred by and property of Mr. Thomas Roberts, Ivington Bury, Leominster; got by Lord Warwick (2093), dam (Prize Daisy) by Sir Benjamin (1387), g.d. (Prize Flower) by Arthur Napoleon (910), g.g.d. (Long Horns) by Andrew the Second (619), g.g.g.d. (Pigeon) by Prince, by Dayhouse (299).

(2703) PRIZEMAN.

Red with white face, calved October 10, 1863; bred by his Grace the Duke of Bedford, Woburn Park, Beds., the property of the Rev. C. J. Campion Westoning, Woburn, Beds.; got by Carbonel (1525), dam (Prize Flower) by Magnet (823), g.d. (Queen) by Forester (398), g.g.d. (Blowdy) by Hope (439), g.g.g.d. (Beauty) by Sovereign (404).

(2704) PROFITABLE.

Red with white face, calved June 6, 1862; bred by and the property of Mr. G. Pitt, Chadnor, Leominster; got by Dilwyn (1961), dam (Perfection) by Plunder (1038), g.d. (Brandy) by Little Teddy (983), g.g.d. (Trumpet) by Northampton (600), g.g.g.d, (Rosebud) by Cookeridge (1187).

BULLS.

(2705) PROMISE.

Red with white face, calved November 19, 1864; bred by Mr. T. Rea, Westonbury, Leominster, the property of Mr. J. H. Whitehouse, Ipsley Court, Redditch; got by Zeno (1825), dam (Primrose the Second) by Sir Benjamin (1387), g.d. (Primrose) by Glendower (898), g.g.d. (Primrose) by Cholstrey (217), g.g.g.d. (Primrose) by Gallant (239).

(2706) PROTECTOR.

Red with white face, calved September 3, 1863; bred by and the property of Mr. Warren Evans, Llandowlais, Usk; got by Monaughty (2117), dam (Snowdrop) by Symmetry (1073), g.d. (Duchess) by Gaylad (400), g.g.d. (Fillpail) by Llanarth (2077), g.g.g.d. (Beauty) by Stumpfoot (2245).

(2707) RADICAL.

Red with white face, calved June 25, 1864; bred by and the property of Mr. Naylor, Leighton Hall, Montgomeryshire; got by Salisbury (2204), dam (Annie) by Silvester (797), g.d. (Elinor) by The Knight (185), g.g.d. (Lady Elinor) by Big Ben (248), g.g.g.d. (Butterfly) by Prince (251), g.g.g.g.d. (Nell) by a Son of Sir Andrew (183).

(2708) RADNOR.

Red with white face, calved May 24, 1864; bred by Mr. B. Rogers, The Grove, Pembridge, the property of Mr. E. Jenkins, The Grove House, Presteign; got by North Star (2138), dam (Silver) by The Grove (1764), g.d. (Silver) by Severus (1062), g.g.d. (Silver) by Young Royal (1470), g.g.g.d. (Silver) by Gaylad the Second (1589).

BULLS.

(2709)   RAGMAN.

Red with white face, calved December 6, 1864; bred by and the property of Mr. J. Jones, Llwyn-y-gaer, Raglan; got by Trusty (2846), dam (Laura) by Chancellor (1172), g.d. (Lalla) by Patron (803), g g.d. (Lily).

(2710)   RAINBOW.

Red with white face, calved January 8, 1863; bred by and the property of Mr. T. L. Meire, Cound Arbour, Salop; got by Franky (1243), dam (Bloom) by Cound (1193), dam (Princess) by Uckington (2286), g.g.d. (Splendour) by Lawyer (627), g.g.g.d. (Slut) by Dinedor (395).

(2711)   RAMBLER.

Red with white face, calved August 19, 1864; bred by and the property of Mr. T. Olver, Penhallow, Grampound, Cornwall; got by Zippor (2354), dam (Alma) by Great Eastern (1598), g.d. (Annie) by Duke of Cornwall (1569).

(2712)   RATTLING JACK.

Red with white face, calved April 18, 1863; bred by and the property of Mr. Richard Shirley, Baucott, Munslow, Church Stretton; got by Pilot (1036), dam (Nut Bush) by Baucott (1507) g.d. (Nutty) by Sir Henry (1730), g.g.d. (Big Nutty) by Byron (559), g.g.g.d. — by Tully Bull.

(2713)   RECKLESS.

Red with white face, calved November 24, 1863; bred by Mr. James Taylor, Stretford Court, Leominster, the property of Mr. J. Meredith, The Heldre, Welshpool; got by Trustful (2845), dam (Fancy Leominster) by King John (830), g.d. (Countess) bred by the late Mr. Bowen, Monkland.

BULLS.

(2714) RECTOR.

Red with white face, calved June 4, 1863; bred by Mr. John Haynes, Llanrothall, Monmouth, the property of Mr. Tippins, Park Grace Diew, Monmouth; got by The Priest (1416), dam (Marianne) by Meteor (1319), g.d. (Primrose) by Lot (364), g.g.d. (Fanny) by Young Sovereign (506), g.g.g.d. (Old Fanny) by Fitzfavourite (441).

(2715) RED MAN.

Red with white face, calved December 20, 1861; bred by Mr. John Jones, Llwyn-y-gaer, Raglan, Monmouth, the property of Mr. John Stone, Tarlton, Cirencester; got by Chancellor (1172), dam (Lovely) by The Noke (2268), g.d. (Lovely) by Chance's Son (1911).

Red Man was a winner of the First Prize in his class at the meeting of the Monmouthshire Agricultural Society, 1862.

(2716) REINDEER.

Red with white face, calved May 13, 1863; bred by Mr. Philip Turner, The Leen, Pembridge, Leominster, the property of Mr. Farr, Pennyworlod, Brecon; got by Bolingbroke (1883), dam (Sylph) by Felix (953), g.d. (Cherry) by Tom of Lincoln (1099), g.g.d. (Nymph) by Andrew the Second (619), g.g g.d. (Sylph) by Viscount (816).

(2717) REINDEER.

Red with white face, calved in the month of April, 1862; bred by and the property of Mr. J. O. G. Pollock, Mountainstown, Navan, Ireland; got by Sir Robert (2227), dam (Rose of the Valley) by Grenadier (961), g.d. (Dahlia) by Cholstrey (217), g.g.d. (Dahlia) by Monarch (219), g.g.g.d. (Dahlia) by Portrait (372), g.g.g.g.d. (Dahlia) by Sovereign (404).

Reindeer was a winner of the First Prize in his class at the Dublin Royal Agricultural Society; Spring Show, 1863.

BULLS.

(2718) RELIANCE.

Red with white face, calved in the month of September, 1864; bred by and the property of Mr. John Jones, Llwyn-y-gaer, Raglan; got by Red Man (2715), dam (Long the Second) by Chancellor (1172), g.d. (Lucy Long) by Patron the Third (2151), g.g.d. (Long) by a Son of Garway (2009), g.g.g.d. — by Chance's Son (1911).

(2719) RENOWN *alias* METEOR.

Red with white face, calved November 11, 1860; bred by Lord Berwick, Cronkhill, Salop, the property of Mr. J. B. Green, Marlow, Leintwardine, Salop; got by Severn (1382), dam (Star) by Albert Edward (859), g.d. (Stately) by Walford (871), g.g.d. (Miss Byron) by Baron (418), g.g.g.d. (Little Beauty) bred by Mr. J. Hewer.

(2720) REPEALER.

Red with white face, calved September 1, 1863; bred by Mr. T. Duckham, Baysham Court, Ross, the property of Mr. T. Price, The Helm, Ewyas, Harold Hereford; got by Victory (2296), dam (Ruby) by Breinton (1155), g.d. (Eywood) by Cotmore the Second (1191), g.g.d. — bred by the late Earl of Oxford.

(2721) REPEALER.

Red with white face, calved May 2, 1862; bred by Mr. J. B. Green, Marlow, Leintwardine, the property of Mr. G. Smythies, Marlow, Leintwardine; got by Sir Benjamin (1387), dam — by Vanguard (1109), g.d. — by Beefy Ben (1869), g.g.d. — by Cholstrey (217), g.g.g.d. by Zest of Oxford (2352).

(2722) REVELLER.

Red with white face, calved in the year 1846; bred by Mr. W. Racster, Thingehill Court, Hereford; got by Half Sovereign (964), dam (Old Stockton cow) bred by the late Mr. J. Morris, Stocktonbury.

(2723) REVELLER.

Red with white face, calved September 3, 1864; bred by and the property of Mr. T. Olver, Penhallow, Grampound, Cornwall; got by Zippor (2354), dam (Blanche) by Young Sir David (1818), g.d. (Beauty) by Young Walford (1820); g.g.d. — bred by Mr. T. Longmore, Buckton, Salop.

(2724) RIFLEMAN.

Red with white face, calved July 3, 1862; bred by and the property of Mr. Naylor, Leighton Hall, Montgomeryshire; got by Volunteer (2299), dam (Stella) by Tom of Lincoln (1099), g.d. (Young Beauty) by Uriconium (598), g.g.d. (Beauty) by Hampden (1603), g.g.g.d. (Venus) by Big Ben (248), g.g g.g.d. (Twist) by Prince (251), — (Gipsy) by Trump (490).

(2725) ROBIN HOOD.

Red with white face, calved August 22, 1864; bred by and the property of Mr. F. W. Stone, Moreton Lodge, Guelph, Canada West; got by Sailor (2200), dam (Nelly) by Carlisle (923), g.d. (Peeress) by Monarch (504), g.g.d. — by St. Germans (227), g.g g.d. — bred by the late Mr. Turner, Noke, Leominster.

(2726) ROB ROY.

Red with white face, calved July 24, 1864; bred by and the property of Mr. T. Duckham, Baysham Court, Ross; got by Cato (1902), dam (Carlisle) by Albert Edward (859), g.d. (Silver) by Emperor (221).

BULLS.

(2727) RODEN.

Red with white face, calved June 27, 1863; bred by Mr. G. T. Forester, High Ercall. Wellington, Salop, the property of Mr. Edwards, Cressage, Salop; got by Severn (1382), dam (Rarity) by Darling (1202), g.d. (Result) by Governor (464), g.g.d. (Maid of the Mill) by Hope (439), g.g.g.d. — by Royal (331), g.g.g.g.d. — bred by the late Mr. Yeomans, Moreton, Hereford.

(2728) ROLLO.

Red with white face, calved July 13, 1864; bred by and the property of Mr. J. O. G. Pollock, Mountainstown, Navan, Ireland; got by Master Willie (2637), dam (Rose of the Valley) by Grenadier (961), g.d. (Dahlia) by Cholstrey (217), g.g.d. (Dahlia) by Monarch (219), g.g.g.d. (Dahlia) by Portrait (372), g.g.g.g.d (Dahlia) by Sovereign (404.

(2729) ROUTER.

Red with white face, calved in the year 1860; bred by Mr. G. Lobb, Lawhitton, Launceston, Cornwall, the property of Mr. Holly, Okehampton; got by Young Orleton (1476), dam (Crafty) by Protector the Second (1362), g.d. (Round Horns) by Youngster (2899), g.g.d. — bred by Earl St. Germans.

(2730) ROVER.

Red with white face, bred by Mr. J. Hewer, Vern House, Marden, Hereford, the property of Mr. Colebatch, Rosemaund, Bromyard; got by Blucher by Paragon (1343), dam (Princess the Second) by Gift (1254), g.d. (Princess) by Lottery the Second (408), g.g.d. (Victoria) by Prince (524), g.g.g.d. (Old Moss Rose) by Old Wellington (507).

BULLS.

(2731) ROYAL.

Red with white face, calved November 13, 1864; bred by and the property of Mr. J. R. Paramore, Dinedor Court, Hereford; got by Portly (2165), dam (Young Countess) by Carlisle (923), g.d. (Countess) by The Duke (493),

(2732) ROYALIST.

Red with white face, calved August 4, 1864; bred by Mr. B. Rogers, The Grove, Pembridge. Leominster, the property of Mr. Williams, Brecon; got by Grove the Second (2556), dam (Spot) by Severus (1062), g.d. (Spot) by Old Court the Second (2140), g.g.d. (Spot) by a bull bred by the late Mr. Jeffries, The Grove.

(2733) ROYALTY.

Red with white face, calved December 2, 1863; bred by and the property of Mr. S. W. Urwick, Leinthall, Ludlow; got by Tatnor (1754), dam (Big Bess) by Young Royal (1469).

(2734) ROYAL OAK.

Red with white face, calved May 29, 1864; bred by and the property of Mr. John Baldwin, Luddington, Stratford-on-Avon; got by Battersea, (1865), dam (Pigeon) by Young Sir Andrew (1471), g.d. — by Sir Andrew (183).

(2735) RUFUS.

Red with white face, calved in the month of November, 1864; bred by and the property of Mr. T. Morris, Therrow, Llyswen, Hay; got by Prince Imperial (2171), dam (Red Cap) by Telegraph (1404), g.d. (Blossom) by Newton (344), g.g.d. (Silver) by White Nob (345), g.g.g.d. (Beauty) by Counsellor (422), g.g g.g.d. (Lovely) by Charity the Second (516).

BULLS.

(2736) **RUPEE.**

Red with white face, calved February 4, 1864; bred by and the property of Mr. W. Lort, The Cotteridge, King's Norton; got by Nabob (2648), dam (Loot) by Plunder (1038), g.d. — bred by Mr. Vaughan, of Cholstrey.

(2737) **SAINT ANTHONY.**

Red with white face, calved November 10, 1863; bred by Mr. J. Rogers, Allt-yr-ynys, Abergavenny, the property of Mr. D. Meredith, Cwm-y-oy, Abergavenny; got by Bolingbroke (1883), dam (Stately) by Mameluke (1307), g.d. (Stately) by The Count (1760), g.g.d (Stately) by Young Old Court, g.g.g.d. (Stately) by Charity (375), g.g.g.g.d. (Stately) by Sovereign (404).

(2738) **SALFORD.**

Red with white face, calved in the month of September, 1862; bred by Mr. P. R. Jackson, Black Brook, Skenfrith, Monmouth, the property of Mr. J. Farr, Pontrilas, Hereford; got by Carlisle (923), dam (Young Vick the Second) by Attingham (911), g.d. (Young Vick) by Wonder (420), g.g.d. (Victoria) by Hope (439), g.g.g.d. (Countess) by Young Chance (449).

(2739) **SALOPIAN.**

Red with white face, calved September 28, 1862; bred by Mr. Davis, Lady Meadow, Leominster, the property of Mr. C. H. Hinckesman, The Poles, Ludlow; got by Plato (2160), dam (Violet) by Corn Exchange (1935), g.d (Countess) by The Knight (185), g.g.d. — by Old Favourite (442).

(2740) **SAMBO.**

Red with white face, calved May 25, 1863; bred by and the property of Mr. Naylor, Leighton Hall, Montgomeryshire; got by

Salisbury (2204), dam (Pigeon) by Uriconium (598), g.d. (Toby Pigeon) by Big Ben (248), g.g.d. (Duchess) by Tobias (487), g.g.g.d. (Duchess) by Sovereign (404), g.g.g.g.d. — bred by the late Mr. Turner, Noke.

(2741)   SAN SEBSATIAN.

Red with white face, calved March 17th, 1864; bred by and the property of Mr. William Stallard, Brockhampton, Ross; got by Chieftain the Second (1917), dam (Gwenny the Fourth) by Sir Benjamin (1387), g.d. (Gwenny the Second) by Chieftain (930), g.g.d. (Gwenny) by Regent (891), g.g.g.d. (Venus the Fifth) by Albert (330), g.g.g.g.d. (Winifred) by Monaughty (220), — (Venus the Fourth) by Duke (304), — (Venus the Third) by Regulator (360), — (Venus the Second) by Noble (238), — (Venus) by Crabstock (303).

(2742)   SANTIAGO.

Red with white face, calved September 4, 1863; bred by Mr. George Pitt, Chadnor Court, Leominster, the property of Mr. James Taylor, Stretford Court, Leominster; got by San Jacinto (2209), dam (Spark) by Miliam (1321), g.d. (Spark) by Plunder (1038), g.g.d. (Highlass the Second) by Northampton (600), g.g.g.d (Highlass) by a Stockton Bull, g.g.g.g.d. (Fairlass) by Albert (330), (Fairmaid) by Young Favourite (460), (Damsel) bred by the late Mr. Morris, Stockton.

(2743)   SCYBORWEN.

Red with white face, calved October 2, 1864; bred by and the property of Mr. J. J. Stone, Scyborwen, Llantrissent, Monmouth; got by Mountain Chief (2645), dam (Jenny) by Attingham (911), g.d. (Barbara) by Albert Edward (859), g.g d. (Big Damsel) by The Count (351), g.g.g.d. (Prettymaid) by Cholstrey (868), g.g.g.g.d. (Old Damsel) by Coleman's Bull (1547), — (Old Daisy) by Chancellor (156), — (Cherry the Second) by Thickset (1769).

(2744)　　　SEIFTON.

Red with white face, calved June 12, 1863; bred by Mr. Thomas Williams, Newhouse, Bromfield, Salop, the property of Mr. E. Edwards, Bourton, Berrington, Salop; got by Tugford (2849), dam (Giantess) by Gift (1254), g.d. (Giantess) by Corfton (1188). g.g.d. — by Doubtful (1971).

(2745)　　　SEIGNIOR.

Red with white face, calved August 21, 1862; bred by Mr. William Stallard, Brockhampton, Ross; the property of Mr William Sexty, Mordiford, Hereford; got by Chieftain the Second (1917), dam (Queen's Gilliflower) by The Doctor (1083), g.d. (Dame's Violet) by Young Conrad (2322), g.g.d. (Dahlia) by Caractacus (659), g.g.g.d. (Dahlia) by Old Court (306).

(2746)　　　SERJEANT.

Red with white face, calved December 5, 1860; bred by Mr. J. P. Apperley, Fownhope, Hereford, the property of Mr. E. Jones, The Moat, Knighton; got by Baron of Noke (1862), dam (Miss Julia) by Sir David (349).

(2747)　　　SEVERUS THE SECOND.

Red with white face, calved January 22, 1860; bred by and the property of Mr. R. Roberts, Burrington, Ludlow; got by Severus (1062), dam (Nancy the Third) by Clungunford (869A), g.d. (Nancy the Second) by Old Court (306), g.g.d. — bred by the late Mr. Roberts, Little Heath, Salop.

BULLS.

(2748)　　　　SEVENHAMPTON.

Red with white face, calved April 10, 1862; bred by Mr. J. Smith, Sevenhampton, Cheltenham, the property of Mr. John Barton, Coln, Fairford, Gloucester; got by Cotswold (1938), dam (Charity) by Vanquish (1439), g.d. (Fatima) by Prince of Wales (1041), g.g d. (Young Faith) by The Baby (1079), g.g.g.d. (Faith) by Hope (439).

(2749)　　　　SHAMROCK.

Red with white face, calved September 24, 1864; bred by and the property of Mr. Naylor, Leighton Hall, Montgomeryshire; got by Salisbury (2204), dam (Yellow Beauty) by Bedstone (2411), g.d. (Young Violet) by Venison (1441), g.g.d. — by Dinedor (395).

(2750)　　　　SHAMROCK.

Red with white face, calved September 15, 1863; bred by Mr. Edward Price, Court House, Pembridge, Leominster, the property of Mr. Thomas Thomas, St. Hilary, Cowbridge; got by Earl Derby the Second (2510), dam (Rosebud the Third) by Salisbury (2204), g d. (Rosebud the Second) by Goldfinder the Second (959), g.g.d. (Rosebud) by Pembridge (721), g.g.g.d. (Moss Rose) by Prince Dangerous (362).

(2751)　　　　SHAKESPERE.

Red with white face, calved August 28, 1863; bred by and the property of Mr. Thomas Powell, Bage, Madley, Hereford; got by Portly (2165), dam (Governess the Second) by Greengage (1266), g.d. (Governess) by Governor the Second (2018), g.g.d. (Fairmaid) by Royal (331), g.g.g.d. (Old Fairmaid).

BULLS.

(2752) SHERIFF.

Red with white face, calved in the month of August, 1859; bred by Mr. J. H. Arkwright, Hampton Court, Leominster, the property of Mr. John Colebourn, Bodenham, Leominster; got by Mortimer (1328), dam — bred by Mr. Sheriff, Coxall, Ludlow.

(2753) SHROPSHIRE.

Red with white face, calved June 13, 1864; bred by and the property of Mr. Richard Shirley, Baucott, Munslow, Church Stretton; got by Pilot (1036), dam (Nutty) by Sir Henry (1730) g.d. (Big Nutty) by Byron (559), g.g.d. — by Tully Bull.

(2754) SHYLOCK.

Red with white face, calved July 6, 1863; bred by and the property of the Hon. and Rev. Noel Hill, Berrington, Salop: got by Van Tromp (2291), dam (Jewess) by Attingham (911), g.d. (Rebecca) by Governor (464), g.g.d. (Old Prettymaid) by Young Sovereign (1472), g.g.g.d. — by Whitenob (345).

(2755) SHYLOCK.

Red with white face, calved October 10, 1863; bred by Mr. Thomas Powell, Bage, Madley, Hereford, the property of Mr. T. Howells, Llancadle, Cowbridge, Glamorganshire; got by Portly (2165), dam (Pigeon the Second) by Delight (1564), g.d. (Pigeon) by a Son of Madley (1301), g.g.d. (Blossom) by Waxy (403), g.g.g.d. (Old Pleasant) by Defiance (416), g.g.g.g.d. (Stately) by Byron (440), — (Old Stately) by Huntington (539).

BULLS.

(2756) SILVER KING.

Red with white face, calved September, 2, 1864; bred by and the property of the Hon. and Rev. H. Noel Hill, Cronkhill, Salop; got by Albert (2380), dam (Gibsey) by Attingham (911), g.d. (Empress) by Venison (1441), g.g.d. (Empress) by Dinedor (395).

(2757) SIR BERTRAM.

Red with white face, calved January 26, 1863; bred by the late Mr. James Rea, Monaughty, Knighton, the property of Mr. Allies, Bromyard; got by Sir Benjamin (1387), dam (Ringdove the Second) by Pilot (1037), g.d. (Ringdove) by Grenadier (961), g.g.d. (Evergreen) by Regent (891), g.g.g.d. (Beauty) by Cholstrey (217).

(2758) SIR BENJAMIN THE SECOND.

Red with white face, calved December 23, 1864; bred by and the property of Mr. Thomas Rogers, Coxall, Ludlow; got by The Grove (1764), dam (Young Prettymaid) by Severus (1062), g.d. (Prettymaid the Second) by Young Royal (1470), g.g.d. (Prettymaid) by Prince (251), g.g.g.d. (Curly the Fourth) by Charity the Second (1535), g.g.g.g.d. (Curly the Third) by Portrait (372), (Curly the Second) by Sovereign the Second (1739).

(2759) SIR BENJAMIN THE SECOND.

Red with white face, calved November 10, 1863; bred by and the property of Mr. R. Harcourt Capper, The Northgate, Ross; got by Sir Benjamin (1387), dam (Duchess of York) by Balaclava (1505), g.d. (Duchess) by Regent (891), g.g.d.

(Duchess) by Barrister (658), g.g.g d. (Duchess) by Gallant (239), g.g.g.g.d. (Old Duchess) by Oldcourt (306), — by Regulator (360), — by Crabstock (303).

(2760) SIR CHARLES.

Red with white face, calved January 14, 1864; bred by Mr. Henry Higgins, Woolaston Grange, Chepstow, the property of Mr. Hawkins, Orleton, Ludlow; got by Sir Harry (2222), dam (Princess) by Sir David (349), g.d. (Milliner the Second) by Leamington (1631), g.g.d. (Milliner) by Gaylad (400), g.g.g.d. (Dressmaker) by Dinedor (395), g.g.g.g.d. Old Gay, bred by the late Mr. Powell, The Building.

(2761) SIR CUPIS BALL.

Red with white face, calved June 29, 1863; bred by the late Mr. James Rea, Monaughty Knighton, the property of Mr. P. J. Kearney, Miltown House, Clonmellon, Ireland; got by Sir Benjamin (1387), dam (Sultana) by Grenadier (961), g.d. (Gaylass) by Canute (890), g.g.d. (Gaylass) by Confidence (367), g.g.g.d. — by Charity (375).

(2762) SIR FRANK.

Red with white face, calved April 17, 1864; bred by the late Mr. Thomas Rea, Westonbury, Leominster, the property of Mr. William Taylor, Thingehill Court, Hereford; got by Sir Richard (1734), dam (Spangle the Second) by Wellington (1112), g.d. (Spangle) by Chieftain (930), g.g.d. (Young Venus) by Young Venus (217), g.g.g.d. (Venus) by Albert (330), g.g.g.g.d. (Countess) by Old Court (306).

(2763) SIR GEORGE.

Red with white face, calved August 12, 1862; bred by Mr. Thomas Roberts, Ivington Bury, Leominster, the property of Mr. Good, Felton, Bromyard; got by Sir Thomas (2228), dam (Duchess of Gloucester) by Arthur Napoleon (910), g.d. — bred by Mr. Vaughan, Cholstrey.

(2764) SIR GEORGE *alias* UMPIRE.

Red with white face, calved July 27, 1864; bred by and the property of Mr. Richard Hill, Orleton Court, Ludlow; got by Noble Boy (1337), dam (Lot) by Restorative (1369), g.d. (Patch) by a bull bred by Mr. Rea, g.g.d. (Patty) by Lottery (410),

Sir George was a winner of the Second Prize in his class at the Hereford Meeting of the Bath and West of England Society.

(2765) SIR GEORGE.

Red with white face, calved in the month of July, 1863; bred by Mr. Rogers, The Grove, Pembridge, Leominster. the property of Mr. J. Williams, Kingsland, Leominster; got by Interest (2046), dam (Prettymaid the Third) by The Grove (1764), g.d. (Young Prettymaid), by Severus (1062), g.g.d. (Prettymaid the Second) by Young Royal (1470), g.g.g.d (Prettymaid) by Prince (251).

(2766) SIR HARFORD.

Red with white face, calved June 30, 1863; bred by and the property of Mr. Charles Britten, The Woodhouse, Leominster; got by Agriculturist (1842), dam (Beauty) by a bull bred by Mr. Perry, Cholstrey, g.d. — bred by the late Mr. Bowen, Monkland, Leominster.

BULLS.

(2767) SIR HARRY.

Red with white face, calved January 22, 1862; bred by Mr. Henry Gibbons, Hampton Bishop, Hereford, the property of Mr. R. S. Featherstonhaugh, Rockview, Killucan, Ireland; got by Shamrock the Second (2210), dam (Strawberry) by The Admiral (1078), g.d. (Moap) by Zephyr (1826), g.g.d. (Old Moap) by a Son of Dewshall (358).

Sir Harry was a winner of the Second Prize in his class at the meeting of the Herefordshire Agricultural Society, 1863.

(2768) SIR JOHN.

Red with white face, calved July 15, 1862; bred by Mr. Thomas Robers, Ivington Bury, Leominster, the property of Mr. Williams, Abergavenny; got by Sir Thomas (2228), dam (Red Cap), by Croft (937), g.d. (Gipsy Queen) by King James (978), g.g.d. (Grey Gipsy) by North Star (758), g.g.g.d (Gipsy the Second) by Original (216), g.g.g.g.d. (Gipsy) by Woodman (255).

Sir John was a winner of the Third Prize in his class at the meeting of the Ludlow Agricultural Society, 1863.

(2769) SIR JOHN.

Red with white face, calved Semptember 19, 1863; bred by and the property of Mr. E. Wright, Halston Hall, Oswestry; got by Silver Horn (2213), dam (Sweet Meat) by Magnet 2nd (989).

Sir John was a winner of the First Prize in his class at the Newcastle and Plymouth meetings of the Royal Agricultural Society, England, 1864.

(2770) SIR JOHN.

Red with white face, calved February 8, 1864; bred by Mr. T. Rea, Westonbury, Leominster, the property of Mr. J. Burlton,

Luntley Court, Leominster; got by Sir Benjamin (1387), dam (Picture) by Grenadier (961), g.d. (Purity) by Regent (891), g.g.d. (Duchess) by Barrister (658), g.g g.d. (Duchess) by Gallant (239), g.g.g.g.d. — by Old Court (306).

(2771) SIR JOSEPH (A TWIN).
Red with white face, calved January 24, 1864; bred by Mr. J. B. Green, Marlow, Leintwardine, Salop, the property of Mr. Benjamin Hawkins, Orleton, Ludlow; got by Sir Benjamin (1387), dam (Governess) by Sovereign (404).

(2772) SIR JOSEPH.
Red with white face, calved November 6, 1863; bred by Mr. Thomas Roberts, Ivington Bury, Leominster, the property of Lord Bateman, Shobdon; got by Sir Thomas (2228), dam (Red Cap) by Croft (937), g.d. (Gipsy Queen) by King James (978), g.g.d. (Grey Gipsy) by North Star (758), g.g g.d (Gipsy the Second) by Original (216), g.g.g.g.d (Gipsy) by Woodman (255).

(2773) SIR OLIVER THE THIRD.
Red with white face, calved November 6, 1860; bred by Mr. Vaughan, Cholstrey, Leominster, the property of Mr. Holloway, Hampton Wafers, Leominster; got by Sir Oliver the Second (1733), dam (Lovely) by Cholstrey (217).

(2774) SIR PHILIP.
Red with white face, calved June 30, 1863; bred by Mr. J. B. Green, Marlow, Leintwardine, Salop, the property of Mr. Richard Shirley, Baucott, Munslow, Church Stretton, Salop; got by Zealot (2344), dam (Fairmaid) by Beefy Ben (1869), g.d. (Fancy) by Cholstrey (217), g.g.d. (Faithful) by Zest of Oxford (2352), g.g.g.d. — by Discount (339).

BULLS.

(2775)   SIR ROBERT.

Red with white face, calved June 15, 1863; bred by the late Mr. James Rea, Monaughty, Knighton, the property of Mr. R. Harcourt Capper, The Northgate, Ross; got by Sir Benjamin (1387), dam (Nonsuch) by Wellington (1112), g.d. (Fairlass) by Chieftain (930), g.g.d. (Fairmaid) by Cholstrey (217), g.g.g.d. (Fairmaid) by Gallant (239), g.g.g.g.d. (Fairmaid) by Portrait (372).

(2776)   SIR RICHARD.

Red with white face, calved September 8, 1863; bred by Mr. Jackson, Blackbrook, Skenfrith, Monmouth, the property of Mr. R. Harcourt Capper, The Northgate, Ross; got by Carlisle (923), dam (Cherry) by Young Walford (1820).

(2777)   SIR THOMAS.

Red with white face, calved December 21, 1864; bred by and the property of Mr. F. W. Stone, Moreton Lodge, Guelph, Canada West; got by Sailor (2200), dam (Vanquish) by Patriot (2150), g.d. (Verbena) by Carlisle (923), g.g.d. (Flower) by Radnor, g.g.g.d. (Old Fancy) bred by the late Mr. Galliers, Shobdon.

(2778)   SIR THOMAS THE SECOND.

Red with white face, calved June 23, 1864; bred by and the property of Mr. Thomas Roberts, Ivington Bury, Leominster; got by Sir Thomas (2228), dam (Young Pleasant) by Master Butterfly (1313), g.d. (Pleasant) by Andrew the Second (619), g.g.d. (Duchess) by Prince by Dayhouse (299).

BULLS.

(2779) SIR WILLIAM.

Red with white face, calved May 10, 1864; bred by and the property of Mr. Matthew Good, Felton, Bromyard; got by Young Monk (2332), dam (Rose) by Governor (464), g.d. (Lilac) by Berrington (423), g.g.d. (Knockrell) by Berrington (435).

(2780) SIR WILLIAM.

Red with white face, calved February 11, 1863; bred by the late Mr. T. Rea, Westonbury, Leominster, the property of Mr. G. Dent, Moreton Jefferies, Bromyard; got by Lord Nelson (2088), dam (Beautiful) by Sir Benjamin (1387), g.d. (Beauty the Fourth) by Young Conrad (2322), g.g.d. (Beauty the Third) by Nero (884), g.g.g.d. (Beauty the Second) by Albert (330), g.g.g.g.d. (Beauty) by Old Court (306).

(2781) SKENFRITH.

Red with white face, calved October 15, 1864; bred by and the property of Mr. P. R. Jackson, Blackbrook, Skenfrith, Monmouth; got by Defiance (2493), dam (Violet) by Young Sir David (1818), g.d. (Rosebud) by Young Walford (1820).

(2782) SNOWBALL.

White, calved October 1, 1862; bred by and the property of Mr. A. R. Boughton Knight, Downton Castle, Ludlow; got by Lord Grey (2085), dam Countess) by Tatteridge (1755), g.d. — by Marquis (992), g.g.d. — by Andrew the Second (619).

(2783) SOCRATES.

Red with white face, calved May 16, 1863; bred by Mr. Thomas Wheeler, Jun., Wormhill, Eaton Bishop, Hereford, the

BULLS.

property of Mr. William Badham, Pontypinna, Vowchurch, Hereford; got by Mentor (2112), dam (Beauty) by Medallist (1009), g.d. (Browny) by Governor (464).

(2784) SOL.

Red with white face, calved August 6, 1863; bred by Mr. William Stallard, Brockhampton, Ross, the property of Mr. Davies, Chipp's House, Leominster; got by Chieftain the Second (1917), dam (Luna) by Baron of Noke (1862), g.d. (Lofty) by Sir David (349), g.g.d. (Lofty) by Andrew the Second (619), g.g.g.d. — by Monarch (504).

(2785) SOOTHSAYER.

Red with white face, calved October 29, 1863; bred by and the property of Mr. William Stallard, Brockhampton, Ross; got by Chieftain the Second (1917), dam (Queen's Gilliflower) by The Doctor (1083), g.d (Dame Violet) by Young Conrad (2322), g.g.d. (Dahlia) by Caractacus (659), g.g.g.d. (Dahlia) by Old Court (306).

Soothsayer was a winner of the Second Prize in his class at the meeting of the Herefordshire Agricultural Society, and First at Ross, 1864.

(2786) SOVEREIGN.

Red with white face, calved June 12, 1863; bred by and the property of Mr. J. R. Paramore, Dinedor Court, Hereford; got by Bertram (1513), dam (Linnet) by Alma (1144), g.d. (Lark) by Royal (331).

(2787) SPARCHFORD.

Red with white face, calved November 27, 1864; bred by Mr. Edward Tanner, Ayntree, Ludlow, the property of Mr. Edward Tanner, Hopton Castle, Ludlow; got by Royal Butterfly (2196), dam (Pigeon) by Northampton (600).

BULLS.

(2788) **SPECULATION.**

Red with white face, calved July 5, 1864; bred by and the property of Mr. William Hall, Ashton, Leominster; got by Ashton (1500), dam (Rosebud) by Rodney (1373), g.d. (Stately) by Uncle Tom (1107), g.g.d. (Moss Rose) by Monkland (551), g.g.g.d. (Fat Rumps) by Young Cotmore (601), **g.g.g.g.d.** — bred by the late Mr. T. Jeffries.

(2789) **SPECULATION.**

Red with white face, calved September 9, 1864; bred by and the property of Mr. Thomas Rogers, Coxall, Ludlow; got by Grove (1764), dam (Silver) by Claret (1921), g.d. — by Emperor the Second (1572), g.g.d. — by Young Royal (1470), g.g.g.d. — by Gaylad the Second (1589), g.g.g.g.d — by Sovereign the Second (1739).

(2790) **STANWAY.**

Red with white face, calved September 11, 1864; bred by and the property of Mr. William Tudge, Adforton, Ludlow; got by Pilot (2156), dam (Darling) by Carbonel (1525), g.d. (Daisy) by The Doctor (1083), g.g.d. (Dainty) by Orleton (901), **g.g.g.d.** (Prettymaid) by Nelson (1021).

(2791) **STAR OF ENGLAND.**

Red with white face, calved June 4, 1864; bred by and the property of Mr. Richard Shirley, Baucott, Munslow, Church Stretton; got by Pilot (1036), dam (Silky Mottle Face) by Marlow (2104), g.d. (Mottle Silky) by Knockerell (1630), **g.g.d.** (Silky) by Dolluggan (759), g.g.g d. (Tidy) by The Count (2263).

(2792) STOCKWELL.

Red with white face, calved August 3, 1864; bred by and the property of Mr. Thomas Powell, Bage, Madley, Hereford; got by Interest (2046), dam (Rosabelle) by Mameluke (1307), g.d. (Rosabelle) by Pope (527), g.g.d. (Old Silver) by Old Wellington (507), g.g.g.d. (Beauty) by Sovereign (404), g.g.g.g.d. (Old Gentle) by Chance, bred by the late Mr. Hewer, The Hardwicke.

(2793) STOCKWELL.

Red with white face, calved July 16, 1862; bred by Mr. Philip Turner, The Leen, Pembridge, the property of Mr. W. D. Turner, The Lynch, Leominster; got by Bolingbroke (1883), dam (Bessie) by Silurian (1064), g.d. (Comely) by a Son of Confidence (367), g.g.d. (Gaudy) by Defiance (1209), g.g.g.d. (Beauty) by Old Court (306).

(2794) SUBALTERN.

Red with white face, calved March 5, 1864; bred by and the property of Mr. William Stallard, Brockhampton, Ross; got by Chieftain the Second (1917), dam (Gwenny the Second) by Chieftain (930), g.d. (Gwenny) by Regent (891), g.g.d. (Venus the Fifth) by Albert (330), g.g.g.d. (Winifred) by Monaughty (220), g.g.g.g.d. (Venus the Fourth) by Duke (304), — (Venus the Third) by Regulator (360), — (Venus the Second) by Noble (238), — (Venus) by Crabstock (303).

(2795) SULTAN.

Red with white face; bred by Mr. Vaughan, Cholstrey, Leominster, the property of Mr. Lowe, Petchfield, Leominster; got by Plunder (1038), dam (Lovely) by Cholstrey (217).

BULLS.

(2796) SULTAN.

Red with white face, calved August 12, 1862; bred by the late Mr. J. Rea, Monaughty, Knighton, the property of Mr. J. M. Read, Elkstone, Cheltenham; got by Sir Benjamin (1387), dam (Sultana) by Grenadier (961), g.d. (Gaylass) by Canute (890), g.g.d. (Gaylass) by Confidence (367), g.g.g.d. — by Charity (375).

(2797) SWEET BRIAR.

Red with white face, calved September 17, 1863; bred by Mr. Thomas Powell, Bage, Madley, Hereford, the property of Mr. W. Dyer, Moorhampton, Abbeydore, Hereford; got by Portly (2165), dam (Matron) by a Son of Madley (1301), g.d. (Matron) by Zouave (1833), g.g.d. (Governess) by Governor the Second (2018), g.g.g.d. (Fairmaid) by Royal (331).

(2798) SWEETMEAT.

Red with white face, calved July 13, 1864; bred by Mr. James Bourn, Mawley Town Farm, Cleobury Mortimer, the property of Mr. P. Phillpots, Moorend Farm, Cleobury Mortimer; got by Cardinal (1526), dam (Laura) by Wigmore (1800), g.d. (Lady Grey) by Regent (1705), g.g.d. (Damsel) by Bowlegs (1617).

(2799) SYMMETRY.

Red with white face, calved November 27, 1862; bred by Mr. S. W. Urwick, Leinthall, Ludlow, the property of Mr. Fenn, Burrington, Ludlow; got by Tatnor (1754), dam (Whitehorn) by Newton (1667), g.d. — by Sir David (349).

(2800) TADPOLE.

Red with white face, calved January 12, 1863; bred by the Rev. Archer Clive, Whitfield, Hereford, the property of Mr.

John Morris, Town House, Madley, Hereford; got by Bertram (1513), dam Noke, bred by Mr. Turner, Noke.

(2801)   TAMBARINE THE SECOND.

Red with white face, calved January 17, 1863, bred by and the property of Mr. William Taylor, Showle Court, Ledbury; got by Tambarine (2254), dam (Dewsall) by Hereford (968).

Tambarine the Second was a winner of the First Prize in his class at the Cirencester meeting of the Gloucestershire Agricultural Society, 1863, and Third at their Cheltenham meeting, 1864; First at the Tredegar meetings, 1863 and 1864; Second at the Bristol meeting of the Bath and West of England Society, and First at Ledbury, 1864.

(2802)   TATTLER.

Red with white face, calved September 11, 1864; bred by and the property of Mr. William Taylor, Showle Court, Ledbury; got by Tambarine (2254), dam (Dewsall the Third) by Holmer (2043), g.d. (Dewsall) by Hereford (968).

(2803)   TEMPLAR.

Red with white face, calved August 1, 1863; bred by Mr. William Taylor, Showle Court, Ledbury, the property of Mr. Smith, Lower Eaton, Hereford; got by Tambarine (2254), dam (Browny) by Triton (1106), g.d. (Prettymaid) by Tomboy (1097), g.g.d. (Miss Carpenter) by Commerce (354).

(2804)   TEMPLE BAR.

Red with white face, calved September 20, 1862; bred by and the property of Mr. J. M. Read, Elkstone, Cheltenham; got by Caliban (1163), dam (Clara) by Chancellor (929), g.d. (Cherry) by Big Ben (248), g.g.d. — by Severn (245).

BULLS.

(2805)   TEMPLETON.

Red with white face, calved August 16, 1863; bred by Mr. William Taylor, Showle Court, Ledbury, the property of Mr. Parry, Great Hardwick, Abergavenny; got by Tambarine (2254), dam (Fairmaid) by Holmer (2043), g.d. Fairmaid by Tomboy (1097).

(2806).   TENANT FARMER.

Red with white face, calved July 27, 1864; bred by and the property of Mr. William Taylor, Showle Court, Ledbury; got by Tambarine (2254), dam (Cowslip) by Prince of Wales (687), g.d. (Old Cowslip).

(2807)   TENBURY.

Red with white face, calved in the year 1861; bred by Mr. W. Lort, The Great Heath, Ludlow, the property of Mr. E. Farmer, Kyrewood, Tenbury; got by Gambler (1247), dam (Duchess) by Malcolm (1305), g.d. (Croft) by Andrew the Second (619).

(2808)   THE APPRENTICE.

Red with white face, calved February 23, 1864; bred by Mr. W. Beaumand, Vron End, Clun, Salop, the property of Mr. Richard Beaumand, Vron End, Clun, Salop; got by Mameluke the Second (2632), dam (Westonbury Lass) by Westonbury (1452), g.d. (Lydbury Lass) by Quicksilver (353), g g.d. (Broady) by Albert (330).

(2809)   THE ANCHORITE

Red with white face, calved December 5, 1862; bred by and the property of Mr. J. M. Read, Elkstone, Cheltenham; got by

Ariconium (1498), dam (Theora) by Sebastopol (1381), g.d. (Cherry the Seventh) by Hotspur (855), g.g.d. (Cherry the Fifth) by Cholstrey (868), g.g.g.d (Cherry the Fourth) by Green's Grey Bull (850A), g.g.g.g.d. (Cherry the Third) by Chancellor (156), — (Cherry the Second) by Thickset (1769), — (Cherry) bred by the late Mr. Knight, Downton Castle.

(2810) THE CHARACTER.

Red with white face, calved March 11, 1862; bred by Mr. Richard Shirley, Baucott, Munslow, Church Stretton, the property of Messrs. Instone and Milner, Callaughton, Much Wenlock; got by The Grove (1764), dam (Big Beauty) by Beefy Ben (1869), g.d. — by Cholstrey (217), g.g.d. — by Zest of Oxford (2352), g.g.g.d. by — Discount (339).

(2811) THE COUNT.

Red with white face, calved November 10, 1863; bred by and the property of Mr. P. W. Bowen, Shrawardine Castle, Salop; got by Claret (1177), dam (Countess of Shrewsbury the Second) by Gratitude (1261), g.d. (Countess of Shrewsbury) by Conrad (1183), g.g.d. (Countess) by Venison (1441), g.g.g.d. — by Dinedor (395), g.g.g.g.d. — by Trojan (542), — by Hector (535), — by Son of Waterloo (49).

(2812) THE DANE.

Red with white face, calved October 22, 1863; bred by and the property of Mr. Richard Hill, Orleton Court, Ludlow; got by Pilot (1036), dam (Villa) by Chanticleer (1173), g.d. (Jenny Lind) by Chance, g.g.d. (Patch) by Zealous (2348), g.g.g.d. (Patty) by Lottery (410).

BULLS.

(2813) THE DIGGER.

Red with white face, calved October 5, 1863; bred by Mr. Edward Tanner, Hopton Castle, Clun, Salop, the property Mr. Thomas Farmer, Stanton Lacy, Ludlow; got by The Doctor (1083); dam (Lark) by Buckton (1891), g.d. (Moorhen) by Northampton (600), g.g.d. Lottery.

(2814) THE DOCTOR.

Red with white face, calved December 18, 1863; bred by and the property of Messrs. R. and T. Lewis, Stapleton Castle, Presteigne; got by Prosperous (2177), dam (Alexandra) by The Count (1760), g.d. (Princess) by Stapleton, (1743), g.g.d. (Princess), bred by the late Mr. R. Lewis, Presteigne.

(2815) THE FARM.

Red with white face, calved in the month of November, 1853; bred by and the property of Mr. E. Jones, The Moat, Knighton; got by Nelson (1021).

(2816) THE GENERAL.

Red with white face, calved December 1, 1864; bred by and the property of Mr. F. W. Stone, Moreton Lodge, Guelph, Canada, West; got by Guelph (2023), dam (Gentle) by Carlisle (923), g.d. (Lady) by The Knight (185), g.g.d. — by Monarch (504), g.g g.d. —, bred by the late Mr. Turner, Noke, Leominster.

(2817) THE GENERAL.

Red with white face, calved May 1, 1857; bred by Mr. W. Magness, Bullingham, Hereford; the property of Mr. J. R. Paramore, Dinedor Court, Hereford; got by General (1251), dam (Tulip) by Sir David (349).

BULLS.

(2818) THE GUELPH BARON.

Red with white face, calved June 12, 1863; bred by Mr. F. W. Stone, Moreton Lodge, Guelph, Canada West, the property of Mr. Henry Haines, Canada West; got by Sailor (2200), dam (Baroness) by Carlisle (923), g.d. (Little Beauty) by Andrew the Second (619), g.g.d (Dainty) by Vulcan (1446), g.g.g.d. bred by the late Mr, Turner, Noke, Leominster.

The Guelph Baron was a winner of the Second Prize in his class at the Kingstone meeting of the Upper Canada Provincial Exhibition, 1863.

(1819) THE MARQUIS.

Red with white face, calved August 16, 1863; bred by his Grace the Duke of Bedford, Woburn Park, Beds., the property of Mr. Henry Alington, Little Barford, St. Neots, Hunts.; got by Victory (2298), dam (Blossom) by The Prince (1092), g.d. (Fanny) by Stripling (582), g.g.d. (Bright Eyes) by Bonaparte (1152), g.g.g.d. (Bright Eyes) by Napoleon (1334), g.g.g.g.d. (Old Bilberry) by Lupin.

(2820) THE PRINCE OF WALES.

Red with white face, calved July 6, 1863; bred by Mr. Thomas Morris, Therrow, Llyswen, Hay, the property of Colonel Feilden, Dulas Court, Hereford; got by Stratagem (1745), dam (Curly the Fourth) by Aberhonddu (903), g.d. (Curly the Second) by Young Byron (832), g.g.d. (Noble the First) by Noble (543), g.g.g.d. (Favourite) by Sovereign (404), g.g.g.g.d. (Damsel) by Young Wellington (505).

The Prince of Wales was a winner of the First Prize in his class at the meeting of the Breconshire and Herefordshire Agricultural Societies, 1864.

BULLS.

(2821) THE ROVER.

Red with white face, calved in the month of May, 1861; bred by Mr. William Vaughan, Cholstrey, the joint property of Mr. W. D. Turner, Lynch Court, Leominster, and Mr. Edward Stedman, Burghill, Hereford; got by Pirate by Plunder (1038), dam — bred by the late Mr. David Williams, Newton.

(2822) THE SHERIFF.

Red with white face, calved in the month of July, 1864; bred by Mr. J. B. Green, Marlow, Leintwardine, Salop, the property of Mr. Rogers, The Grove, Pembridge, Leominster; got by Zeal (2342), dam (Gem) by Beefy Ben (1869), g.d. (Graceful) by Cholstrey (217), g.g.d. (Gracious) by Zest of Oxford (2352), g.g.g.d. (Gentle) by Discount (339).

(2823) THE YOUNG KNIGHT.

Red with white face, calved July 25, 1862; bred by Mr. John Naylor, Leighton Hall, Montgomeryshire, the property of Mr. John Pryce, Kerry, Montgomeryshire; got by Volunteer (2299), dam (Emeline) by Tom of Lincoln (1099), g.d. (Ellen) by The Knight (185), g.g.d. (Lady Elinor) by Big Ben (248), g.g.g.d. (Butterfly) by Prince (251), g.g.g.g.d. (Nell) by a Son of Sir Andrew (183).

(2824) TOBY.

Red with white face, calved June 14, 1863; bred by Mr. Philip Turner, The Leen, Pembridge, Leominster; the property of Mr. J. P. Apperley, Fownhope, Hereford; got by Bolingbroke (1883), dam (Bessie) by Silurian (1064), g.d, (Comely) by (a son of Confidence (367), g.g.d. (Gaudy) by (Defiance) (1209), g.g.g.d. (Beauty) by Old Court (306).

BULLS.

(2825)   TOMBOY.

Red with white face, calved September 16, 1862; bred by Mr. Edward Tanner, Hopton Castle, Clun, Salop, the property of Mrs. Weyman, Purslow Hall, Salop; got by The Doctor (1083), dam (Mulberry) by Northampton (600), g.d. Oakley.

(2826)   TOMBOY.

Red with white face, calved in the month of July, 1864; bred by Mr. J. B. Green, Marlow, Leintwardine, Salop, the property of Mr. William Blakeway, Wooton, Onibury, Ludlow; got by Zeal (2342), dam (Fairmaid) by Beefy Ben (1869), g.d. (Fancy) by Cholstrey (217), g.g.d. (Faithful) by Zest of Oxford (2352), g.g.g.d. (Fanny) by Discount (339).

(2827)   TOM BRIGHT.

Red with white face, calved November 11, 1862; bred by and the property of Mr. William Taylor, Showle Court, Ledbury; got by Tambarine (2254), dam (Daisy) by Mercury (1317).

(2828)   TOM BROWN.

Red with white face, calved August 19, 1862; bred by and the property of Mr. William Taylor, Showle Court, Ledbury; got by Admiration (1140), dam (Cowslip) by Prince of Wales (687), g.d. (Old Cowslip).

Tom Brown was a winner of the Second Prize in his class at the Hereford Meeting of the Bath and West of England Society.

(2829)   TOM KING.

Red with white face, calved in the year 1863; bred by Mr. J. B. Green, Marlow, Leintwardine, the property of Capt. Crawshay, Danypark, Crickhowell; got by Sir Benjamin (1387), dam (Governess) bred by the late Mr. Jeffries, The Grove, Pembridge, Leominster.

BULLS.

(2830) TOM KING.

Red with white face, calved July 4, 1862; bred by and the property of Mr. Naylor, Leighton Hall, Montgomeryshire; got by Volunteer (2299), dam (Mary Ann) by Tom of Lincoln (1099), g.d. (Mary) by Silvester (797), g.g.d. (Lady Elinor) by Big Ben (248), g.g.g.d (Butterfly) by Prince (251), g.g.g.g.d. — by a Son of Sir Andrew (183).

(2831) TOM LONG.

Red with white face, calved August 5, 1862; bred by and the property of Mr. William Taylor, Showle Court, Ledbury; got by Tambarine (2254), dam (Beauty) by Glasbury (709), g.d. (Alice) by Quicksilver (353).

(2832) TOM THUMB.

Red with white face, calved November 28, 1863; bred by and the property of Mr. J. S. Elliott, Holm Lacey, Hereford; got by Champion (1904), dam (Beauty) by Old Governor.

(2833) TREBARRIED.

Red with white face, calved November 16, 1861; bred by and the property of Mr. John Ricketts, Trebarried, Bronllys, Hay; got by Rifleman (2191), dam (Prettymaid) by Aberhonddu (903), g.d. (Therrow) by Robin (1053), g.g.d. — bred by the late Mr. Morris, Therrow.

Trebarried was the winner of the First Prize in his class at the meeting of the Breconshire Agricultural Society in the years 1862, 1863, and 1864.

BULLS.

(2834) TRERIBBLE.

Red with white face, calved August 8, 1859; bred by Mr. Edward Williams, Court of Llowes, Hay, the property of Mr. Edward Drinkwater, Treribble, Llangarren, Ross; got by Chieftain (930), dam (Glasbury) by Quicksilver (353), g.d. — by Prince (333).

(2835) TRIBUTE.

Red with white face, calved September 10, 1863; bred by and the property of Mr. William Taylor, Showle Court, Ledbury; got by Tambarine (2254), dam (Cowslip) by Prince of Wales (687), g.d. (Old Cowslip).

(2836) TRIUMPH.

Red with white face, calved March 16, 1864; bred by the late Mr. Thomas Rea, Westonbury, Leominster, the property of Mr. William Taylor, Showle Court, Ledbury; got by Sir Benjamin (1387), dam (Lucy) by Chieftain (930), g.d. (Fairmaid) by Regent (891), g.g.d. (Fairmaid) by Barrister (658), g.g.g.d. (Fairmaid) by Old Court (306).

Triumph was a winner of the First Prize in his class at the Hereford meeting of the Bath and West of England Society.

(2837) TRIUMPH.

Red with white face, calved August 18, 1863; bred by Mr. Thomas Roberts, Ivington Bury, Leominster, the property of Mr. Bedford, Hatfield, Leominster; got by Sir Thomas (2228), dam (Trinket) by Master Butterfly (1313), g.d. (Triumph) by King James (978), g.g.d. (Triumph) by Fairboy (617), g.g.g.d. (Tractable) by Original (216).

BULLS.

(2838) TROOPER.

Red with white face, calved October 17, 1862; bred by Mr. T. Powell, Bage, Madley, Hereford, the property of Mr. W. Price, Bullingham, Hereford; got by Troubadour (1780), dam (Governess the Second) by Greengage (1266), g.d. Governess by Young Governor (1815), g.g.d. (Fairmaid) by Royal (331).

(2839) TRUEBOY.

Red with white face, calved September 27, 1864; bred by and the property of Mr. J. Paramore, Dinedor Court, Hereford; got by Portly (2165), dam (Cherry) by Hotspur (972), g.d. (Noke) bred by the late Mr. Davies, Tarrington.

Trueboy and his dam were winners of a prize in the Extra Stock class at the meeting of the Herefordshire Agricultural Society, 1864; Trueboy was also a winner of the Second Prize in his class at the Plymouth meeting of the Royal Agricultural Society of England.

(2840) TRUELOVE.

Red with white face, bred by Mr. W. Vaughan, Cholstrey, Leominster.

(2841) TRUEMAN.

Red with white face, calved in the month of July, 1863; bred by Mr. William Bennett, North Cerney, the property of Mr. John Haynes, Llanrothall, Monmouth; got by Macaroni, dam (Countess) by Warrior, g.d. (Cowslip) by Champion (1906), g d. — by Fitzfavourite (441).

(2842) TRUMP.

Red with white face, calved in the month of August, 1855; bred by the late Mr. Rogers, Stocken, the property of Miss Abley, Norton, Presteign; got by The Count (1760), dam (Miss Noble) by Royal (331), g.d. (Wanton) by Prince (251), g.g.d. (Wanton) by Charity (375).

(2843) TRUMPET.

Red with white face, calved in the month of September, 1857; bred by Mr. Roberts, Trippleton, Leintwardine, the property of Mr. R. Edwards, Stanton Lacy, Ludlow; got by Tatteridge (1755), dam (Lightfoot the Third) by a Son of Walford (871), g.d. (Lightfoot the Second) by Cromwell (889), g.g.d. (Lightfoot the First) by Bullingham, g.g.g.d. (Silver) by The Sheriff (356).

(2844) TRUMPETER.

Red with white face, calved October 31, 1863; bred by Mr. Thomas Roberts, Ivington Bury, Leominster, the property of Lord Hatherton, Teddesley Park, Penkridge; got by Sir Thomas (2228), dam (Trump the Second) by Master Butterfly (1313), g.d. (Trump) by Arthur Napoleon (910).

(2845) TRUSTFUL.

Red with white face, calved August 28, 1861; bred by Mr. Edward Price, Court House, Pembridge, the joint property of Mr. Taylor, Stretford Court, and Mr. Griffiths, King's Pyon, Hereford; got by Salisbury (2204), dam (Gipsy Queen) by Magnet (823), g.d. (Lily) by Sir David (349), g.g.d. (Lily the Second) by Prince Dangerous (362), g.g.g.d. (Modesty) by The Sheriff (356), g.g.g.g.d. (Snowdrop) by Forester (398).

(2846) TRUSTY.

Red with white face, calved in the month of July, 1860; bred by Mr. Price, Town House, Bredwardine, Hereford, the property of Mr. J. Jones, Llwyn-y-gaer, Raglan; got by Salisbury (2204), dam (Blossom) by Pembridge (721), g.d. (Countess) by Sir David (349), g.g.d. (Snowdrop) by Forester (398), g.g.g.d. (Red Rose) by Crabstock (303).

BULLS.

(2847) TRUSTY.

Red with white face; bred by Mr. Moor, Norton, Presteign, the property of Mr. Aaron Rogers, The Homme, Dilwyn; got by Moor's old Bull, dam — bred by the late Mr. Jeffries, The Grove.

(2848) TRUSTY BEN.

Red with white face, calved September 8, 1864; bred by and the property of Mr. B. Rogers, The Grove, Pembridge; got by Grove the Second (2556), dam (Daisy the Third) by The Grove (1764), g.d. (Young Prettymaid) by Severus (1062), g.g.d. (Prettymaid the Second) by Young Royal (1470), g.g.g.d. (Prettymaid) by Prince (251).

(2849) TUGFORD.

Red with mottle face, calved in the month of October, 1855; bred by Mr. John Blockey, Tugford, Munslow, Salop, the property of Mr. H. Instone, Leighton, Bromfield, Salop; got by a Son of Old Court (306), dam — by a Son of Sir George (405), g.d. — bred by Mr. Downes, Aston Hall.

(2850) TURNASTONE.

Red with white face, calved August 16, 1863; bred by and the property of Mr. W. H. Apperley, Withington, Hereford; got by Rover (2730), dam (Browny) bred by Mr. Probert, Turnastone, Hereford.

(2851) TURTLE.

Red with white face, calved November 27, 1862; bred by Mr. T. L. Meire, Cound Arbour, Salop, the property of Mr. Ward, Crick Heath, Salop; got by Franky (1243), dam (Beauty) by Cound (1193), g.d. (Old Beauty) by Layman (767), g.gd. (Old Lovely) by Dinedor (395), g.g.g.d. (Old Lawton) by Hero (458).

BULLS.

(2852) UPSTART.

Red with white face, calved January 5, 1850; bred by Mr. Edward Price, Pembridge, Leominster, the property of Mr. J. Walker, Holmer, Hereford; got by Contest (593), dam — by Hope (411), g.d. — by Sovereign (404).

(2853) VAINHOPE.

Red with white face, calved August 14, 1863; bred by Mr. John Monkhouse, The Stow, Hereford, the property of Mr. L. Watkins, Pipton, Brecon; got by Chieftain (930), dam (Vanity) by Madoc (899), g.d. (Lofty) by Phantom (1035), g.g.d. (Stately) by Sir Andrew (183), g.g.g.d. (Stately) by a Son of Sovereign (404).

(2854) VALIANT.

Red with white face, calved November 15, 1864; bred by and the property of the Hon. and Rev. H. Noel Hill, Cronkhill, Salop; got by Albert (2380), dam (Victorine) by Sir David (349), g.d. (Young Vic) by Wonder (420), g.g.d. (Victoria) by Hope (439), g.g.g.d. (Countess) by Young Chance (449).

(2855) VICTOR.

Red with white face, calved October 7, 1862; bred by Mr. Thomas Powell, Bage, Madley, Hereford, the property of Mr. J. Burlton, Luntley, Leominster; got by Zeno (1825), dam (Rosabelle) by Mameluke (1307), g.d (Old Rosabelle) by Pope (527), g.g.d. (Old Silver) by Old Wellington (507), g.g.g.d. (Beauty) by Sovereign (404), g.g.g.g.d. (Old Gentle) by Old Chance.

BULLS.

(2856) VICTOR.

Red with white face, calved September 23, 1863; bred by and the property of the Hon. and Rev. H. Noel Hill, Cronkhill, Salop; got by Van Tromp (2291), dam (Victorine) by Sir David (349), g.d. (Young Vic) by Wonder (420), g.g d. (Victoria) by Hope (439), g.g.g.d. (Countess) by Young Chance (449).

(2857) VICTOR.

Red with white face, calved June 18, 1864; bred by and the property of Mr. Naylor, Leighton Hall, Montgomeryshire; got by Gladstone (2547), dam (Apple Blossom) by Attingham (911), g.d. Grey (Oak Apple) by Tom Thumb (243), g.g.d. (Oak Apple) by Commerce (354), g.g.g.d. (Strawberry) bred by the late Mr. Jeffries, g.g.g.g.d. dam — by a Son of Guinea.

(2858) VINCENT.

Red with white face, calved August 15, 1861; bred by and the property of Mr. John Hewer, Vern House, Marden, Hereford; got by Van Tromp (1440), dam (Countess the Fourth) by Discount (530), g.d. (Miss Lottery) by Lottery (410), g.g.d. (Old Beauty) by Old Wellington (507).

(2859) VISCOUNT.

Red with white face, calved in the month of July, 1864; bred by and the property of Mr. J. B. Green, Marlow, Leintwardine; got by Zealous (2349), dam (Jenny) by Vanguard (1109), g.d. (Jessamine) by Beefy Ben (1869), g.g.d. (Juno) by Cholstrey (217), g.g.g d. (Jessie) by Zest of Oxford (2352).

(2860) VITELLIUS.

Red with white face, calved August 21, 1864; bred by and the property of Mr. W. H. Oatley, Wroxeter, Salop; got by

Nero (2650), dam (Agrippina) by Julius Cæsar (2054), g.d. (Long Horns) by Surprise (779), g.g.d. (Princess Royal) by Emigrant (1980), g.g.g.d. (Sovereign Cow) by Sovereign (404).

(2861)   VOLUNTEER.

Red with white face, calved September 29, 1859; bred by Mr. Tudge, Adforton, Ludlow, the joint property of Mr. Badham, Aylton Court, and Mr. Holmes, The Castle, Ledbury; got by The Doctor (1083), dam (Lily the Fourth) by Orleton (901), g.d. (Lily the Third) by Turpin (300), g.g.d. (Lily the Second) by a Tully Bull, g.g.g.d. (Lily) by Crabstock (303).

(2862)   VOLUNTEER.

Red with white face, calved in the month of March, 1863; bred by Mr. Burlton, Luntley Court, Leominster, the property of Mr. William Yapp, The Town, Middleton, Leominster; got by Rifleman (2189), dam (Cherry the Third), by Havelock (1609), g.d. (Cherry the Second) by Sampson (1061), g.g.d. (Cherry the First) by Red Ben (768), g.g.g.d. (Cherry) by The Count (351), g.g.g.g.d. (Old Cherry) by a Son of Goldfinder (383).

(2863)   VOLUNTEERIST.

Red with white face, calved February 24, 1864; bred by Mr. John Burlton, Luntley Court, Leominster, the property of Mr. Yeld, Twyford, Leominster; got by Rifleman (2189), dam (Cherry the Third) by Havelock (1609), g.d. (Cherry the Second) by Sampson (1061), g.g.d. (Cherry the First) by Red Ben (768), g.g.g.d. (Cherry) by The Count (351), g.g.g g.d. (Cherry) by a Son of Goldfinder (383).

BULLS.

(2864) WALLEND.

Red with white face, calved September 20, 1859; bred by Mr. William Perry, St. Oswald, Cholstrey, Leominster, the property of Mr. John Cave, Wallend, Monkland; got by Salisbury (2204), dam (Worcester Lass) by Goldfinder the Second (959), g.d. (Duchess) by Prince Royal (554), g g.d. (Duchess late Blossom) by Marden (564), g.g g.d. (Countess) by Albert (330).

(2865) WANDERER.

Red with white face, calved August 28, 1864; bred by Mrs. Gwillim, Breinton, Hereford, the property of Mr. Thomas Powell, Bage, Madley, Hereford; got by Battenhall (2406), dam (Sylph) by Murphy (1331).

(2866) WARLOCK.

Red with white face, calved January 10, 1864; bred by and the property of Mr. John Williams, St. Mary's, Kingsland, Leominster; got by Witchend the Second (2315), dam (Bud the Second) by Showman (2211), g.d. (Bud) by Fairboy (617), g.g.d. (Beauty) by Young Confidence (653), g.g.g.d. (Sidenhead) by Goldfinder (383).

(2867) WARRIOR.

Red with white face, calved November 1, 1863; bred by and the property of Mr. S. W. Urwick, Leinthall, Ludlow; got by Tatnor (1754), dam (Prettymaid) by Young Royal (1469).

(2868) WASHINGTON.

Red with white face, calved December 15, 1862; bred by the late Mr. James Rea, Monaughty, Knighton, the property of Mr.

Thomas Bennett, Clehonger, Hereford; got by Sir Benjamin (1387), dam (Forget-me-not) by Pilot (1037), g.d. (Lily of the Valley) by Chieftain (930), g.g.d. (Lily) by Confidence (367), g.g.g.d. (Lily) by Old Court (306).

(2869)　　　　　WAXY.

Red with white face, calved September 1, 1864; bred by and the property of Mr. Naylor, Leighton Hall, Welshpool, Montgomeryshire; got by Salisbury (2204), dam (Patience) by Silvester (797), g.d. (Blowdy) by Big Ben (248), g.g.d. (Mottle) by Prince (251), g.g.g.d. (Beauty) by Claret (253), g.g.g.g.d. (Spot) by Trump (490).

(2870)　　　　　WELLINGTON HERO.

Red with white face, calved October 12, 1864; bred by and the property of Mr. F. W. Stone, Moreton Lodge, Guelph, Canada West; got by Sailor (2200), dam (Cherry) by Albert Edward (859), g.d. (Cherry the Thirteenth) by Walford (871), g.g.d. (Red Cherry) by Tom Thumb (243), g.g.g.d. (Cherry the Fifth) by Cholstrey (868), g.g.g.g.d. (Cherry the Fourth) by Green's Grey Bull (850A), — (Cherry the Third) by Chancellor (156), — (Cherry the Second) by Thickset (1769).

(2871)　　　　　WENLOCK.

Red with white face, calved in the month of October, 1863; bred by the late Mr. James Rea, Monaughty, Knighton, the property of Mrs. Jeffries, The Downes, Much Wenlock; got by Sir Benjamin (1387), dam (Emerald) by Treasurer (1105), g.d. (Spangle) by Chieftain (930), g.g.d (Young Venus) by Cholstrey (217), g.g.g.d. (Venus) by Albert (330), g.g.g.g.d. (Countess) by Old Court (306).

BULLS.

(2872) WENTWOOD.
Red with white face, calved July 27, 1863; bred by and the property of Mr. J. J. Stone, Scyborwen, Llantrissent, Monmouth; got by Prince of Wales (2172), dam (Jenny) by Attingham (911), g.d. (Barbara) by Albert Edward (859), g.g.d. (Big Damsel) by The Count (351), g.g.g.d. (Prettymaid) by Cholstrey (868), g.g.g.g.d. (Old Damsel) by Coleman's Bull (1547), — (Old Daisy) by Chancellor (156), — (Cherry the Second) by Thickset (1769).

(2873) WEST OF ENGLAND.
Red with white face, calved January 9, 1864; bred by Mr. T. S. Bradstock, Cobrey Park, Ross, the property of Messrs. Wilmott, Llanthewy Farm, Monmouth; got by Young Rambler (2335), dam (Maria) by The Jew (2266), g.d. (Fairy) by Daniel (1201), g.g.d. (Freckle) by Deluge (1210), g.g.g.d. (Blossom) by Young Royal (1468).

(2874) WESTON.
Red with white face, calved October 22, 1864; bred by and the property of Mr. Thomas Woolley, Weston Court, Pembridge; got by Monkland the Third (1013), dam (Blossom) by Monkland (552), g.d. (Silver) by Young Goldfinder (1126).

(2875) WIGMORE.
Red with white face, calved in the month of July, 1855; bred by Mr. Charles Nott, Bury House, Wigmore, the property of Mr. Stedman, High Ercle Hall, Wellington, Salop; got by Mortimer (896), dam (Darling) by Little Ben (769), dam (Old Darling) by Son of Confidence (367), g.g.d. — by Old Court (306), g.g.g.d. — by Cotmore the Second (1191).

BULLS.

(2876) **WILD BOY.**
Red with white face, calved March 8, 1864; bred by and the property of Mr. J. Palmer, Hampton-on-the-Hill, Warwick; got by My Lord (2647), dam (Caroline) by Cardinal Wiseman (1168), g.d. (Hampton Lass) by Mark (424), g.g.d. (Miss Hampton) by Garrick (1248), g.g.g.d. (Lady Hampton) by Reform (508), g.g.g.g.d. (Moss Rose) by Hope (411).

(2877) **WOBURN DUKE.**
Red with white face, calved July 19, 1863; bred by his Grace the Duke of Bedford, Woburn Park, Beds., the property of Mr. Gilbert, Cautley, Norwich; got by Carbonel (1525), dam (Bertha) by Goldfinder the Second (959), g.d. (Blossom) by Pembridge (721), g.g.d. (Countess) by Sir David (349), g.g.g d. (Snowdrop) by Forester (398), g.g.g.g.d. (Red Rose) by Crabstock (303).

(2878) **WORCESTER.**
Red with white face, calved August 17, 1862; bred by and the property of Mr. R. Harcourt Capper, The Northgate, Ross; got by Lord Wellington (2094), dam (Ada) by Attingham (911), g.d. (Silver) by Emperor (221).

Worcester was a winner of the Second Prize in his class at the Worcester meeting of the Royal Agricultural Society of England, and First at the meeting of the Herefordshire Agricultural Society, 1863.

(2879) **WORKMAN.**
Red with white face, calved June 24, 1863; bred by Mr. E. Price, Pembridge, Leominster, the property of Mr. J. Barton, Colne, Fairford; got by Earl Derby the Second (2510), dam (Victoria the Second) by Goldfinder the Second (959), g.d. (Victoria) by Sir David (349), g.g.d. (Curly) by Prince Dangerous (362), g.g.g.d. (Countess) by The Sheriff (356), g.g.g.g.d. (Tidy) by Forester (398), — (Silk) by Crabstock (303).

BULLS.

(2880) **YEOMAN.**

Red with white face, calved May 2, 1863; bred by the late Mr. James Rea. Monaughty, Knighton, the property of Mr. E. V. Wheeler, Kyrewood House, Tenbury; got by Agriculturist (1842), dam (Spangle the Second) by Wellington (1112), g.d. (Spangle) by Chieftain (930), g.g.d. (Venus) by Cholstrey (217), g.g.g.d. (Venus) by Albert (330), g.g.g.g.d. (Countess) by Old Court (306).

(2881) **YOUNG BUCKTON.**

Red with white face, calved in the month of June, 1862; bred by and the property of Mr. Edward Tanner, Hopton Castle, Clun, Salop; got by Buckton (1891), dam (Prettymaid).

(2882) **YOUNG CARDINAL WISEMAN.**

Red with white face, calved March 7, 1859; bred by and the property of Mr. John Barton, Colne, Fairford; got by Cardinal Wiseman (1168), dam (Beauty) by Guinea (963), g.d. — bred by Mr. Peter Matthews.

(2883) **YOUNG CARLISLE.**

Red with white face, calved April 28, 1862; bred by and the property of Mr. P. R. Jackson, Blackbrook, Skenfrith, Monmouth; got by Carlisle (923), dam (Miss Gay) by Gaylad (400), g.d. — by Berrington (435)

(2884) **YOUNG CROMWELL.**

Red with white face; bred by Mr. B. Rogers, The Grove, Pembridge, Leominster, the property of Mr. Aaron Rogers, The Homme, Dilwyn; got by Cromwell (889), dam — by Stone Cote, bred by the late Mr. John Rogers.

(2885) YOUNG DAVID.
Red with white face; bred by Mr. Aaron Rogers, The Homme, Dilwyn, the property of Mr. Marstone, Willersley, Hay; got by David (1204).

(2886) YOUNG DRUMMER.
Red with white face, calved November 28, 1863; bred by and the property of Mr. William Perry, St. Oswald, Cholstrey, Leominster; got by Kettledrum (2057), dam (Duppa the Second late Gloucester Flower) by Noble Boy (1337), dam (Worcester Flower) by Goldfinder the Second (959), g.g.d. (Duchess the Second) by Prince Royal (554), g.g.g.d. (Duchess) by Marden (564).

(2887) YOUNG GAYLAD,
Red with white face; bred by and the property of Mr. Aaron Rogers, The Homme, Dilwyn; got by Gaylad the Second (1589), dam — by Heartsease, g.d. — by Portrait (372).

(2888) YOUNG GROVE.
Red with white face, calved August 16, 1864; bred by and the property Mr. Thomas Edwards, Wintercott, Leominster; got by Adforton (1839), dam (Lady the Second) by Croft (937), g.d. (Lady) by Paddock (773), g.g.d. (Lovely) by Coningsby the Second (1552).

(2889) YOUNG HERO.
Red with white face, calved October 16, 1862; bred by Mr. Thomas Edwards, Wintercott, Leominster, the property of Mr. Lane, Compton Casey, Andoversford; got by Hero (2040), dam (Flora) by Sir Newton (1731), g.d. (Clara) by Croft (937), g.g.d. (Prettymaid) by Coningsby the Second (1552), g.g.g.d. (Beauty) by Big Ben (248).

BULLS.

(2890) YOUNG MATCHLESS.

Red with white face, calved November 6, 1864; bred by and the property of Mr. John Rogers, Letchmoor, Presteign; got by Matchless (2110), dam (Prettymaid) by Priam (1039), g.d. — by Young Dewsall (1125), g.g.d. — bred by the late Sir Robert Price, Foxley.

(2891) YOUNG NOBLE.

Red with white face, calved February 24, 1863; bred by Mr. John Perry, Paunsford Court, Much Cowarne, the property of Mr. Wm. Perry, St. Oswald, Cholstrey, Leominster; got by Noble Boy (1337), dam (Fancy) by Chancellor (927), g.d. (Silver) by Mr. George Tomkins's Bull.

(2892) YOUNG ROYAL.

Red with white face, calved September 21, 1864; bred by the Rev. Archer Clive, Whitfield, Hereford, the property of Mr. R. H. Capper, The Northgate, Ross; got by Garibaldi (2008), dam (Peeress) by Wellington (1112), g.d. (Purity) by Regent (891), g.g.d. (Duchess) by Gallant (239), g.g.g.d. — by Old Court (306).

(2893) YOUNG SALISBURY.

Red with white face, calved November 5, 1860; bred by Mr. E. Price, Pembridge, Leominster, the property of Mr. Bemand, Bockleton, Leominster; got by Salisbury (2204), dam (Beauty) by Pembridge (721), g.d. (Blowdy) by Sir David (349), g.g.d. (Wagtail) by Prince Dangerous (362), g.g.g.d. (Morocco) by The Sheriff (365), g.g.g.g.d. (Partridge) by Crabstock (303).

(2894)   YOUNG SOVEREIGN.

Red with white face; bred by the late Mr. Jeffries, The Grove, Pembridge, Leominster, sold to Mr. Lewis, Newchurch, Kinnersley; got by Sovereign (404).

(2895)   YOUNG SOVEREIGN.

Red with white face, calved in the year 1839; bred by the late Mr. Turner, Noke Court, Leominster, sold to the late Mr. James, Mappowder, Blandford, Dorset; got by a Son of Sovereign (404), dam (Countess the Third) by Sir Charles (1388), g.d. (Countess the Second) by Sovereign (404), g.g.d. (Old Prettymaid) by Wellington (507), g.g.g.d. (Old Primrose) by Silver (540).

(2896)   YOUNG SQUIRE.

Red with white face, calved August 25, 1862; bred by Mr. Warren Evans, Llandowlas, Usk, the property of Mr. Alfred Gethen, Penrose, Caerleon, Monmouth; got by Monaughty (2117), dam (Snowdrop) by Symmetry (1073), g.d. (Duchess) by Gaylad (400), g.g.d (Fillpail) by Llanarth (2077), g.g.g.d. (Beauty) by Stumpfoot (2345).

(2897)   YOUNG WALFORD.

Red with white face; bred by the late Mr. Edwards, The Cwm; got by Walford (871).

(2898)   YOUNG WITCHEND.

Red with white face, calved May 1, 1861; bred by and the property of Mr. John Perry, Paunsford Court, Much Cowarne; got by Witchend the Second (2315), dam (Fancy) by Chancellor (927), g.d. (Silver) by Young Goldfinder (1126), g.g.d. (Old Silver) by Mr. George Tomkin's Bull.

BULLS.

(2899) YOUNGSTER *alias* YOUNG GROVE.

Red with white face, calved in the month of December, 1833; bred by the late Mr. Jeffries, The Grove, Leominster, the property of Mr. G. Lobb, Lawhitton, Launceston, Cornwall; got by Grove (370), dam — Alpha the Second (457).

(2900) ZEBULON.

Red with white face, calved in the month of July, 1864; bred by and the property of Mr. J. B. Green, Marlow, Leintwardine, Salop; got by Zealous (2349), dam (Heiress) by Zealot (2344), g.d. (Hopeful) by Beefy Ben (1869), g.g.d. (Hannah) by Cholstrey (217), g.g.d. (Handsome) by Zest of Oxford. (2352).

(2901) ZENO.

Red with white face, calved in the month of August, 1863; bred by and the property of Mr. John Thomas, Cholstrey, Leominster; got by Caractacus (2445), dam (Zenobia) by Sir David (349), g.d. — bred by Mr. C. W. Allen, The Moor, Kington.

(2902) ZENO THE SECOND.

Red with white face, calved June 29, 1863; bred by and the property of Mr. James Lewis, Milton, Pembridge, Leominster got by Zeno (1825), dam (Miss Governess) by Golding (1257), g.d. (Governess) by Gratitude (1261), g.g.d. (Governess) by Monarch (504).

(2903) ZERO THE SECOND.

Red with white face, calved April 5, 1864; bred by and the property of Mr. W. B. Peren, Compton, South Petherton; got by Zero (1827), dam (Duchess the Fourth) by Zimmerman (1830), g.d. (Duchess the Third) by Fair Brother (949), g.g.d.

BULLS.

(Duchess the Second) by Quicksilver (353), g.g.g.d. (Duchess) by Waverly (106) g.g g.g.d. (Venus) by a bull of Mr. Child's, Wigmore Grange, — (Spot) by Young Sovereign (379).

(2904) ZOUAVE.

Red with white face, calved July 7, 1864; bred by and the property of Mr. H. R. Evans, Jun., Swanstone Court, Leominster; got by Chatham (1914), dam (Gentle the Second) by Rambler (1046), g.d. (Gentle) by King James (978), g.g.d. (Silver) by Coningsby (718)), g.g.g.d. (Beauty) bred by the Rev. N. Penoyre, The Moor, Hay.

(2905) ZOUAVITE.

Red with white face, calved June 30, 1864; bred by and the property of Mr. Lewis Loyd, Monks Orchard, Addington, Surrey; got by Victor (2294), dam (Butterfly) by the Doctor (1083), g.d. (Redrose) by Orleton (901), g.g.d. (Redrose) by Nelson (1021), g.g.g.d. (Redrose) by Turpin (300).

END OF BULLS.

COWS,

AND THEIR PRODUCE.

ADA.

Red with white face, of 1857, vol. v., p. 122.

Bred by the late Lord Berwick, Cronkhill, Salop, the property of Mr. W. B. Peren, Compton, South Petherton; got by Attingham (911), dam (Princess), g.d. (Brecon) by Young Hope (343), g.g.d. (Duchess the Second) by Youug Byron (832), g.g.g.d. (Duchess) by Chance (348).

PRODUCE IN	NAME.	BY WHAT BULL.	BY WHOM BRED.
1862, Aug. 26, r.--w.f. H	Adelaide	Cardinal 1526	Mr. Bourn.
1863, July 14, r.—w.f. B	Ajax 2378	do.	do.
1864, Sept. 17, r.—w.f. H	Bella	do.	do.

ADA.

Red with white face, calved July 12, 1861.

Bred by and the property of Mr. T. Olver, Penhallow, Grampound, Cornwall; got by Volunteer (2300), dam (Annie) by Duke of Cornwall (1569), g.d. — bred by the Earl of St. Germans.

1864, Sept. 30, r.—w.f. B		Zippor 2354	Mr. T. Olver.

ADA.
Red with white face.

Bred by Capt. Power, Hill Court, Ross, the property of Mr. William Jones, Hill of Eaton, Ross; got by The Jew (2266), dam by Uncle Tom (1108), g.d. — bred by Mr. P. Turner, The Leen, Leominster.

PRODUCE IN	NAME.	BY WHAT BULL.	BY WHOM BRED.
1862, Dec. 5, r.—w.f. H	Ada	Geologist 2012	Mr. Wm. Jones.
1863, Sept. 16, r.—w.f. H	Woodlark	do.	do.
1864, Sept. 9, r.—w.f. B	Steer	Chieftain 2nd 1917	do.

ADA.
Red with white face, calved May 3, 1858.

Bred by and the property of Mr. J. R. Paramore, Dinedor Court, Hereford; got by Hotspur (972), dam (Laura) by Noke, g.d. — bred by Dr. Lamb, Henwood, Dilwyn.

1861, March 8, r.—w.f. H	Ada 2nd	General 2817	Mr. Paramore.

ADA THE SECOND.
Red with white face, calved March 8, 1861.

Bred by and the property of Mr. J. R. Paramore, Dinedor Court, Hereford; got by The General (2817), dam (Ada) — by Hotspur, (972), g.d. (Laura) by Noke, g.g.d. — bred by Dr. Lamb, Henwood, Dilwyn.

1864, Feb. 21, r.—w.f. B		Portly 2165	Mr. Paramore.

ADELA.
Red with white face, calved October 4, 1861.

Bred by Mr. Bourn, Mawley Town Farm, Cleobury Mortimer, the property of Mr. W. B. Peren, Compton, South Petherton; got by Cardinal (1526), dam (Ada) by Attingham (911), g.d.

(Princess), g.g.d. (Brecon) by Young Hope (343), g.g.g.d. (Duchess the Second) by Young Byron (832), g.g.g.g.d. (Duchess) by Chance (348).

PRODUCE IN	NAME.	BY WHAT BULL.	BY WHOM BRED.
1864, Aug. 7, r.—w.f. H	Etty	Captain 2443	Mr. Bourn.

ADELINA.

Red with white face, calved October 23, 1861.

Bred by and the property of Mr. J. Baldwin, Luddington, Stratford-on-Avon; got by Severn (1382), dam (Agnes) by Attingham (911), g d. (Silver) by Emperor (221).

1864, May 29, r.—w.f. B	Royal Oak 2734	Battersea 1865	Mr. Baldwin.

* Adelina was a winner of the second prize in her class, at the Exeter meeting of the Bath and West of England Society, and first at the meeting of the Vale of Evesham Agricultural Society 1863.

ADELAIDE.

Red with white face, calved August 29, 1858.

Bred by Lord Berwick, Cronkhill, Salop, the property of Mr. Richard S. Fetherstonhaugh, Rockview, Killucan, Ireland; got by Attingham (911), dam (Silver) by Emperor (221).

1861, Aug., r.—w.f. B	Silver Stream 2214	Severn 1382	Lord Berwick.
1862, Sept. 9, r.—w.f. B	Argentine 2390	Cropper 1559	Mr. Fetherstonhaugh
1863, r.—w.f. B	Steer	Leominster 2071	do.

AGATHA.

Red with white face, calved August 4, 1859.

Bred by Lord Berwick, Cronkhill, Salop, the property of Mr. John J. Stone, Scyborwen, Llantrissant, Monmouth; got by Attingham (911), dam (Silver) by Emperor (221).

1862, Aug. 10, r.—w.f. B	Mountain Chief 2645	Franky 1243	Mr. Stone.
1864, April 22, r.—w.f. H	Silver Queen	Monaughty 2117	do.

COWS.

AGATHA.
Red with white face, calved July 14, 1860.

Bred by and the property of Mr. Naylor, Leighton Hall, Montgomeryshire; got by Tom of Lincoln (1099), dam (Alice) by Nelson (1661), g.d. Miss Woodbatch.

PRODUCE IN	NAME.	BY WHAT BULL.	BY WHOM BRED.
1863, July 14, r.—w.f. H	Agatha 2nd	Salisbury 2204	Mr. Naylor.

AGRIPINA.
Red with white face, calved in the month of December 1857.

Bred by and the property of Mr. W. H. Oatley, Wroxeter, Salop; got by Julius Cæsar (2054), dam (Longhorns) by Surprise (779), g.d. (Princess Royal) by Emigrant (1980), g.g.d. (Sovereign Cow) by Sovereign (404).

PRODUCE IN	NAME.	BY WHAT BULL.	BY WHOM BRED.
1860, Nov. 22, r.—w.f. H	Agripina 2nd	Son of Walford	Mr. Oatley.
1862, Jan. 15, r.—w.f. B	Nero 2650	Tiberius Cæsar 2272	do.
1863, Feb. 27, r.—w.f. {B / H}	Otho 2663 / Martin	Claudius Cæsar 1922	do.
1864, Mar. 15, r.—w.f. B	Galba 2527	Franky 1243	do.

ALEXANDRIA.
Red with white face, calved in the year 1862.

Bred by and the property of Mr. Thomas Morris, Therrow, Llyswen, Hay; got by Druid (1220), dam (Curly the Second) by Newton (344), g.d. (Curly) by Young Byron (832), g.g.d. (Noble First) by Noble (543), g.g.g.d. (Favourite) by Sovereign (404).

PRODUCE IN	NAME.	BY WHAT BULL.	BY WHOM BRED.
1864, August, r.—w.f. B	Steer.	Prince Imperial 2171	Mr. Morris.

ALEXANDRIA.
Red with white face, calved in the month of July, 1859.

Bred by and the property of Messrs. R. and T. Lewis, Stapleton Castle, Presteign; got by The Count (1760), dam (Princess) by Stapleton (1743), g.d. (Princess), bred by the late Mr. R. Lewis, Presteign.

PRODUCE IN	NAME.	BY WHAT BULL.	BY WHOM BRED.
1863, Dec. 18, r.—w.f. B	The Doctor 2814	Prosperous 2177	Messrs. Lewis.

COWS.

ALMA.
Red with white face, calved July 10, 1859.

Bred by and the property of Mr. T. Olver, Penhallow, Grampound, Cornwall; got by Great Eastern (1598), dam (Annie) by Duke of Cornwall (1569).

PRODUCE IN	NAME.	BY WHAT BULL.	BY WHOM BRED.
1863, Sept. 16, r.—w.f. B	Achilles 2370	Conservative 1931	Mr. Olver.
1864, Aug. 19, r.—w.f. B	Rambler 2711	Zippor 2354	do.

Alma was a winner of the first prize in her class at the Truro meeting of the Royal Cornwall Agricultural Society.

AMETHYST (A TWIN).
Red with white face, calved July 14, 1858.

Bred by and the property of Mr. Philip Turner, The Leen, Pembridge; got by Felix (953), dam (Brilliant) by Andrew the Second (619), g.d. (Gun) by Sir Walter (352), g.g.d. (Jewel) by Commerce (354), g.g.g.d. (Mischief) by a Bull bred by Mr. Bowen, of Monkland.

1862, May 15, r.—w.f. B	Steer	Bolingbroke 1883	Mr. Turner.
1863, July 24, r.—w.f. H	Purity	do.	do.

ANNIE.
Red with white face, of 1855, vol. iv., p. 80.

Bred by and the property of Mr. Naylor, Leighton Hall, Welshpool; got by Silvester (797), dam (Elinor) by The Knight (185), g.d. (Lady Elinor) by Big Ben (248), g.g.d. (Butterfly) by Prince (251), g.g.g.d. (Nell) by a son of Sir Andrew (183).

1859, Aug. 2, r.—w.f. H	Dead	Tom of Lincoln 1099	Mr. Naylor.
1860, Aug. 11, r.—w.f. B	Steer	do.	do.
1861, July 26, r.—w.f. B	Steer	do.	do.
1862, Aug. 4, r.—w.f. H	Dead	Admiral 1481	do.
1863, June 24, r.—w.f. H	Annie 2nd	Blondin 1880	do.
1864, June 25, r.—w.f. B	Radical 2707	Salisbury 2204	do.

APPLE BLOSSOM.

Grey, with white face, calved October 21, 1859.

Bred by Lord Berwick, Cronkhill, Salop, the property of Mr. Naylor, Leighton Hall, Montgomeryshire; got by Attingham (911), dam (Grey Oak Apple) by Tom Thumb (243), g.d. (Oak Apple) by Commerce (354), g.g.d. (Strawberry) by a Son of Guinea, g.g.g.d. — bred by the late Mr. E. Jeffries.

PRODUCE IN	NAME.	BY WHAT BULL.	BY WHOM BRED.
1862, Aug. 20, r.—w.f. B	Lichfield 2597	Salisbury, 2204	Mr. Naylor.
1863, June 29, r.—w.f. B	Cronkhill 2486	Blondin 1880	do.
1864, June 18, r.—w.f. B	Victor 2857	Gladstone 2547	do.

ARIEL.

Red with white face, calved July 22, 1858.

Bred by and the property of Mr. Philip Turner, The Leen, Pembridge; got by Felix (953), dam (Brunette) by Duke of St. Alban's (945), g.d. (Belle) by Sir Walter (352), g.g.d. (Myrtle) by Commerce (354), g.g.g.d. (Sylph) by Old Court the Second (1341).

1863, April 26, r.—w.f. H	Gipsy Queen	Bolingbroke 1883	Mr. Turner.
1864, July 12, r.—w.f. H	Sunbeam	do.	do.

ARROGANCE.

Red with white face, calved July 28, 1860.

Bred by and the property of Mr. John Monkhouse, The Stow, Hereford; got by Madoc (899), dam (Lofty) by Phantom (1035), g.d. (Stately) by Sir Andrew (183).

1863, July 4, r.—w.f. B	Steer	Chieftain 930	Mr. Monkhouse.
1864, July 9, r.—w.f. H	Pride	do.	do.

Arrogance with Pantomine were winners of the First Prize in their class at the meetings of the Herefordshire Agricultural Society, 1861 and 1862.

AUGUSTA.

Red with white face, calved in the month of December, 1859.

Bred by and the property of Mr. W. H. Oatley, Wroxeter, Salop; got by Carausius (2446), dam (Long Horns) by Surprise (779), g.d. (Princess Royal) by Emigrant (1980), g.g.d. (Sovereign Cow) by Sovereign (404).

PRODUCE IN	NAME.	BY WHAT BULL.	BY WHOM BRED.
1862, July 16, r.—w.f. H		Claudius Cæsar 1922	Mr. Oatley.
1863, Aug. 12, r.—w.f. B	Steer	do.	do.
1864, Aug. 21, r.—w.f. B	Vitellius 2860	Nero 2650	do.

AUNT ESTHER.

Red with white face, calved March 27, 1857.

Bred by and the property of Mr. W. D. Turner, Lynch Court, Leominster; got by Newton (1667), dam (Rosa) by Andrew the Third (908), g.d. (Broady) by Waverly (1793), g.g.d. — bred by the late Mr. P. Turner, Aymstrey.

1860, Nov. 2, r.-w.f. H	Esther	Logic 2079	Mr. Turner.
1861, Sept. 10, r.—w.f. B	Steer	do.	do.
1862, Sept. 1, r.—w.f. H	Alice	do.	do.
1863, Nov. 14, r.—w.f. H	Laura	Earl Derby 2nd 2510	do.

BARONESS.

Red with white face, calved in the year 1857.

Bred by and the property of Mr. Richards, Cound, Shrewsbury; got by The Knight (185), dam — bred by Mr. Rammel.

1860, Nov. 6, r.—w.f. H	Primrose	Baronet 1860	Mr. Richards.
1861, Nov. 10, r.—w.f. H	Dead	do.	do.
1862, Nov. 16, r.—w.f. H	Daisy	Pirate 2158	do.
1863, Nov. 26, r.—w.f. H		do.	do.
1864, Oct. 6, r.—w.f. H		Cound 2484	do.

Baroness was a winner of the First Prize in her class at the meeting of the Bridgnorth Agricultural Society 1859 and 1862.

COWS.

BARONESS THE SECOND.
Red with white face, calved December 3rd, 1861.

Bred by and the property of Mr. F. W. Stone, Moreton Lodge, Guelph, Canada West; got by Patriot (2150), dam (Baroness) by Carlisle (923), g.d. (Little Beauty) by Andrew the Second (619) g.g.d. (Dainty) by Vulcan (1446), g.g.g.d. — bred by the late Mr. Turner, Noke Court, Leominster.

PRODUCE IN	NAME.	BY WHAT BULL	BY WHOM BRED.
1864, July 13, r.—w.f. B	Captain 2442	Sailor 2200	Mr. Stone.

Baroness the Second was a winner of the First Prize in her class at the Toronto, Kingston, and Hamilton meetings of the Canadian Agricultural Society.

BEATRICE.
Red with white face, of 1858, vol. v., p. 125.

Bred by the late Mr. J. Rea, Monaughty, Knighton; got by Pilot (1037), dam (Fanny) by Cholstrey (217), g.d. (Fanny) by Hope (439), g.g.d. — by The Sheriff (356).

1862, Dec. 25, r.—w.f. H	Bashful	Sir Benjamin 1387	Mr. Rea.

Bashful sold to Mr. T. Cadle, Longcroft, Westbury-on-Severn.

BEATRICE.
Red with white face, calved September 19, 1860.

Bred by and the property of Mr. T. Olver, Penhallow, Grampound, Cornwall; got by Earl Derby (1979), dam (Beauty) by Young Walford (1820).

1864, Jan. 3, r.—w.f. B	Burnside 2438	Zippor 2354	Mr. Olver.

BEAUTIFUL.
Red with white face, of 1858, vol. v., p. 125.

Bred by the late Mr. Thomas Rea, Westonbury, Leominster, the property of Mr. J. R. Paramore, Dinedor Court, Hereford; got by Sir Benjamin (1387), dam (Beauty the Fourth) by Young

COWS.

Conrad (2322), g.d. (Beauty the Third) by Nero (884), g.g.d. (Beauty the Second) by Albert (330), g.g.d. (Beauty) by Old Court (306).

PRODUCE IN	NAME.	BY WHAT BULL.	BY WHOM BRED.
1863, Feb. 11, r.—w.f. B	Sir William 2780	Lord Nelson 2088	Mr. Rea.
1864, Mar. 26, r.—w.f. H	Beauty of Westonbury	Artful 2391	do.

Beautiful was one of seven, winners of the First Prize in their class at the meetings of the Herefordshire and the Leominster Agricultural Societies, 1863. Beauty of Westonbury sold to Mr. J. R. Paramore.

BEAUTIFUL.
Red with white face, calved July 22, 1859.

Bred by and the property of Mr. Henry Evans, jun., Swanstone Court, Leominster; got by Rambler (1046), dam (Young Lovely) by Emperor (373), g.d. (Lovely) by Young Trueboy (1475), g.g.d. (Lovely) by Ashley Moor White Bull (870), g.g.g.d. (Old Damsel) by Coleman's Bull (1547), g.g.g.g.d. (Old Daisy) by Chancellor (156).

PRODUCE IN	NAME.	BY WHAT BULL.	BY WHOM BRED.
1862, Mar. 26, r.—w.f. B	Steer	Benicia Boy 1872	Mr. Evans.
1863, June 18, r.—w.f. B	Steer	Chatham 1914	do.
1864, July 8, r.—w.f. B	Steer	do.	do.

BEAUTY.
Red with white face, calved in the year 1858.

Bred by Mr. William Berrow, The Green, Allensmore, Hereford, the property of Mr. Samuel Gilliland, Brook Hall, Londonderry; got by Widgeon (1792), dam (Original) by Original the First (455).

PRODUCE IN	NAME.	BY WHAT BULL.	BY WHOM BRED.
1861, Jan. 28, r.—w.f. H	Beauty 2nd	Jolly Miller 4th 2582	Mr. Gilliland.
1862, Jan. 10, r.—w.f. B	Jolly Miller 13th 2586	do. [2508	do.
1863, Jan. 11, r.—w.f. H	Beauty 3rd	Duke of Wellington	do.
1864, May 20, r.—w.f. B	Jolly Miller 14th 2587	Jolly Miller 10th 2584	do.

Beauty was a winner of second prizes at the Londonderry and North-West Agricultural Societies Meetings 1861, 1862, and 1863; and first at the North-West and second at the Londonderry in 1864.

cows.

BEAUTY.

Red with white face, calved in the month of February, 1856.

Bred by and the property of Mr. Henry Haywood, Blakemere, Hereford; got by Preston (2688), dam — by Wilmaston (1455), g.d. — by Jupiter (511).

PRODUCE IN	NAME.	BY WHAT BULL.	BY WHOM BRED.
1862, Feb. 1, r.—w.f. H	Beatrice Princess	Cholstrey 1918	Mr. Haywood.
1863, Jan. 1, r.—w.f. H	Be-true	do.	do.
1864, Jan. 20 r.—w.f. H	Belle of the Village	do.	do.

BEAUTY.

Red with white face, calved in the month of March, 1856.

Bred by and the property of Mr. John Barton, Coln St. Aldwin's, Fairford; got by Guinea (963), dam — bred by Mr. Peter Matthews, Coomb End, Cirencester.

1859, Mar. 7, r.—w.f. B	Young Cardinal Wiseman 2882	Cardinal Wiseman 1168	Mr. Barton.
1860, Feb. 12, r.—w.f. B	Steer	Eastington	do.
1861, Mar. 18, r.—w.f. H	Topsy	Cardinal Wiseman 1168	do.
1862, Mar. 22, r.—w.f. B	Steer	Eastington	do.
1863, Jan. 8, r.—w.f. B	Steer	do.	do.

BEAUTY.

Red with white face, calved October 14, 1860.

Bred by Mr. H. E. Powell, Great Brampton, the property of Mr. Thomas Broad, The Castle, Madley, Hereford; got by Dutiful (1978), dam (Lofty) by Prince Royal (280), g.d. (Lady) by Young Sovereign (379), g g.d. (Miss Cotmore Second) by Lottery Second (408), g.g.g.d. (Miss Cotmore) by Cotmore (376).

1864, March, r.—w.f. H	Bella	Mentor 2112	Mr. Broad.

BEAUTY.

Red with white face, calved in the year 1861.

Bred by Mr. Sheriff, Coxall, Ludlow, the property of Mr. R. H. Capper, The Northgate, St. Weonards, Ross; got by Sir Colin (2216), dam (Tinsell) by Young Royal (1470).

PRODUCE IN	NAME.	BY WHAT BULL.	BY WHOM BRED.
1864, July 23, r.—w.f. H	Rosa	Garibaldi 2008	Mr. Capper.

BEAUTY.

Red with mottled face, calved in the year 1859.

Bred by Mr. T. Davies, Burlton Court, Hereford, the property of Mr. J. R. Paramore, Dinedor Court, Hereford; got by Garrick (1248), dam (Beauty) bred by the late Mr. Tunstall, Burlton Court.

1863, July 9, r.—w.f. H	Graceful	Courtier 1194	Mr. T. Davies.
1864, Aug. 17, r.—w.f. H	Beauty 2nd	The Jew 2266	Mr. Paramore.

Graceful sold to Mr. J. R. Paramore.

BEAUTY.

Red with white face, calved in the month of May, 1852.

Bred by the late Mr. Prosser, Meadow Farm, the property of Mr. J. Prosser, Honeybourne Grounds, Worcester; got by Berrington (435).

1859, June, r.—w.f. H	Beauty 2nd	Medallist 1009	Mr. Prosser.
1860,	Steer	do.	do.
1861,	Steer	do.	do.
1862, Jan. r.—w.f. H	Dell	Lacey 2063	do.
1863, Sept. r.—w.f. B	Berrington 2415	The Jew 2266	do.

BEAUTY.

Red with white face, of 1856, vol. iv., p. 81.

Bred by the late Mr. Josiah Davies, Ivington, Leominster, the property of Mrs. A. Davies, Ivington, Leominster; got by Corner Cop

(2481), dam (Young Hereford), by Truelove (2840), g.d. (Hereford), bred by Mr. Edwards, Brinsop.

PRODUCE IN		NAME.	BY WHAT BULL.	BY WHOM BRED.
1859, Aug.,	r.—w.f. H	Beauty 2nd	Claret 1542	Mr. Davies.
1860, July,	r.—w.f. H	Ellen	Yg. Cholstrey 1808	do.
1861, July,	r.—w.f. H	Lively	Duke of Marlboro' 1974	do.
1862, Sept. 8,	r.—w.f. B	Steer	Conqueror	do.
1863, Sept. 26,	r.—w.f. H	Miss Ivington	do.	do.
1864, Sept. 22,	r.—w.f. B	Steer	Chance 1908	do.

BEAUTY THE SECOND.

Red with white face, calved in the month of August, 1859.

Bred by the late Mr. J. Davies, the property of Mrs. Ann Davies, Chipp's House, Leominster; got by Claret (1542), dam (Beauty) by Corner Cop (2481), g.d. (Young Hereford) by Truelove (2840), g.g.d. (Hereford), bred by Mr. Edwards, Brinsop Court, Hereford.

1862, Oct. 1,	r.—w.f. B	Steer	Conqueror	Mr. Davies.
1863, Oct. 19,	r.—w.f. B	do.	do.	do.
1864, Sept. 8,	r.—w.f. B		do.	do.

BEAUTY THE SECOND.

Red with white face, calved June, 1859.

Bred by and the property of Mr. John Prosser, Honeybourne Grounds, Worcester; got by Medallist (1009), dam (Beauty) by Berrington (435).

1862, June,	r.—w.f. H	Gaylass	The Jew 2266	Mr. J. Prosser.
1863, August,	r.—w.f. H	Blossom	do.	do.
1864, Dec. 10,	r.—w.f. H	Beauty 3rd	do.	do.

BEAUTY THE SECOND.

Red with white face, calved June 29, 1860.

Bred by and the property of Mr. Henry R. Evans, Swanstone Court, Leominster; got by Sir Franklin; (1068), dam (Beauty) by Swanstone (1072), g.d. (Young Lovely) by Emperor (373), g.g.d. (Lovely) by Young Trueboy (1475), g.g.g.d. (Lovely) by Ashley Moor White Bull (870), g.g.g.g.d. (Old Damsel) by Coleman's bull (1547), — (Old Daisy) by Chancellor (156).

PRODUCE IN	NAME.	BY WHAT BULL.	BY WHOM BRED.
1863, May 31, r.—w.f. B	Steer	Chatham 1914	Mr. Evans.
1864, Aug. 23, r.—w.f. B	Steer	England's Glory 1983	do.

BEAUTY THE SECOND.
Red with white face, of 1855, vol. iv., p. 83.

Bred by and the property of Mr. William Tudge, Adforton, Ludlow; got by Young Walford (1820), dam (Beauty) by Nelson (1021), g.d. (Daisy) by Turpin (300).

1860, Jan. 12, r.—w.f. H	Young Beauty	The Doctor 1083	Mr. Tudge.
1860, Oct. 1, r.—w.f. H	Bonnie	Carbonel 1525	do.
1861, Oct. 2, r.—w.f. B	Pilot 2156	The Grove 1764	do.

Young Beauty was a winner of the First Prize in her class and the Silver Cup at Abingdon, 1863; the Second Prize at Birmingham; and the First Prize and Silver Medal at the meeting of the Smithfield Club, the same year.

BELLA.
Red with white face, of 1855, vol. v., p. 130.

Bred by the late Mr. Rea, Monaughty, Knighton, the property of Mr. J. Monkhouse, The Stow, Hereford; got by Grenadier (961), dam (Cherry) by Regent (891), g.d. (Cherry) by Commerce (354).

1862, Mar. 26, r.—w.f. H	Bella 3rd	Lord Nelson 2088	Mr. Rea.

Bella was one of seven winners of the First Prize in their class at the meetings of the Herefordshire and Leominster Agricultural Societies, 1860.

BELLA THE SECOND.
Red with white face, calved April 18, 1861.

Bred by Mr. Thomas Rea, Westonbury, Pembridge, the property of Mr. John Wigmore, Bickerton Court, Much Marcle, Dymock;

got by Silvius (1726), dam (Bella) by Grenadier (961), g.d. (Cherry) by Regent 891, g.g d. (Cherry), by Commerce (354).

PRODUCE IN	NAME.	BY WHAT BULL.	BY WHOM BRED.
1863, June 27, r.—w.f.H	Bella 4th	Artful 2391	Mr. Rea.
1864, March 4, r.—w.f. B	Steer	do.	do.

Bella the Second with her sire and dam were winners of the First Prize in their class at the meeting of the Herefordshire Agricultural Society, 1861, and the Challenge Sweepstake at Leominster the same year.

Bella the Fourth, sold to Mr. T. Cadle, Longcroft, Westbury-on-Severn.

BELLONA.

Red with white face, calved in the year 1857.

Bred by Mr. H. E. Powell, Great Brampton, Madley, Hereford, the property of Mr. Thomas Rogers, Coxall, Ludlow; got by Musselman (1333), dam (Baroness) by Edgar (946), g.d. (Miss Cotmore the Second) by Lottery the Second (408), g.g.d. (Miss Cotmore) by Cotmore (367), g.g.g.d. — by Conqueror (415).

1864, April, r.—w.f. B	Steer	Vincent 2858	Mr. Powell.

BERTHA.

Red with white face, calved August 29, 1859.

Bred by Mr. Edward Price, Pembridge, Leominster, the property of the Duke of Bedford, Woburn, Beds.; got by Goldfinder the Second (959), dam (Blossom) by Pembridge (721), g.d. (Countess) by Sir David (349), g.g.d. (Snowdrop) by Forester (398), g.g.g.d. (Redrose) by Crabstock (303),

1862, July 1, r.—w.f. H	Dead	Carbonel 1525	Duke of Bedford.
1863, July 17, r.—w.f. B	Woburn Duke 2877	do.	do.
1864, June 12, r.—w.f. H	Bonnett	Victory 2296	do.

BESSIE.

Red with white face, of 1857, vol. v., p. 131.

Bred by Lord Bateman, Shobdon Court, Leominster, the property of Mr. J. R. Paramore, Dinedor Court, Hereford; got by Carlisle (923), dam (Laurel) by The Duke (495).

PRODUCE IN	NAME.	BY WHAT BULL.	BY WHOM BRED.
1861, Nov. 30, r.—w.f. H	Beauty	Bolstone, 1884	Mr. Garrold.
1862, Oct. 10, r.—w.f. H	Bessie 2nd	Melon 2111	do.
1863, Sept. 30, r.—w.f. H	Jewess	The Jew 2266	Mr. Paramore
1864, Aug. 25, r.—w.f. H	Comely	Chelmsford 1915	do.

Bessie 2nd sold to Mr. Paramore, Dinedor Court.

BIRTHDAY.

Red with white face, calved August 29, 1860.

Bred by and the property of Mr. E. T. Goldingham, Grimley, Worcester; got by Gaylad (1586), dam (Young Bedstone) by Grateful (1260), g.d. (Bedstone Prettymaid) by Conrad (1183), g.g.d. (Prettymaid) by Dinedor (395).

1862, Nov. 17, r.—w.f. H	Banjo	Tasso 1753	Mr. Goldingham.
1864, May 17, r.—w.f. H	Vesta	Mars 2107	do.

BLANCHE.

Red with white face, calved October 28, 1858.

Bred by and the property of Mr. Thomas Olver, Penhallow, Grampound, Cornwall; got by Young Sir David (1818), dam (Beauty) by Young Walford (1820), g.g.d. —, bred by Mr. Thomas Longmore, Buckton, Salop.

1862, Jan. 3, r.—w.f. B	Dead	Earl Derby 1979	Mr. T. Olver.
1862, Dec. 19, r.—w.f. B	Steer	Conservative 1931	do.
1863, Oct. 20, r.—w.f. B	Steer	Zippor 2354	do.
1864, Sept. 3, r.—w.f. B	Reveller 2723	do.	do.

COWS.

BLOOMER.
Red with white face, calved March 28, 1858.

Bred by and the property of Mr. J. M. Read, Elkstone, Cheltenham; got by Sebastopol (1381), dam (Cockhorn) by Young Dangerous (1809), g.d. (Old Cockhorn) by Liston (414).

PRODUCE IN	NAME.	BY WHAT BULL.	BY WHOM BRED.
1861, Jan. 21, r.—w.f. B	Steer	Caliban 1163	Mr. Read.
1862, Feb. 7, r.—w.f. H	Redrose	do.	do.
1863, Jan. 25, r.—w.f. B	Rapsgate	do.	do.

BLOSSOM THE FOURTH.
Grey with white face, calved in the month of August, 1860.

Bred by and the property of Mr. Benjamin Rogers, The Grove, Pembridge, Leominster; got by the Grove (1764), dam (Blossom the Third) by Young Royal (1470), g.d. (Blossom the Second) by Sovereign the Second (1739), g.g.d. (Blossom the First) by Portrait (372).

1863, July, r.—w.f. H	Blossom 5th	Interest 2046	Mr. Rogers.
1864, June 5, grey H	Grey	North Star 2138	do.

BLOSSOM.
Red with white face, calved in the year 1861.

Bred by and the property of Mr. John Rogers, Letchmoor, Presteign; got by Young David (2885), dam — by Young Dewshall (1125).

1864, June r.—w.f B		Matchless 2110	Mr. Rogers.

BLOWDY.
Red with white face, calved May 11, 1860.

Bred by and the property of Mr. Thomas Edwards, Wintercott, Leominster; got by Sir Newton (1731), dam (Pink the Third) by Wellington (1113), g.d. (Pink) by Stretford (1749), g.g.d. (Lady) by Coningsby the Second (1552).

1862, Oct. 29, r.—w.f. B	Steer	Nelson 2129	Mr. Edwards.
1863, Aug. 27, r.—w.f. B	Adforton 4th 2373	Adforton, 1839	do.
1864, Aug. 22, r.—w.f. H	Young Blowdy	do.	do.

BLOWDY.
Red with white face, calved in the month of August, 1860.

Bred by and the property of Mr. John Rogers, Letchmoor, Presteign; got by Young David (2885), dam — by Young Cromwell (2884), g.d. — by Gay Lad the Second (1589).

PRODUCE IN		NAME.	BY WHAT BULL.	BY WHOM BRED.
1863, Aug.	r.—w.f. B		Matchless 2110	Mr. Rogers.
1864, July,	r.—w.f. B		do.	do.

BLOWDY THE SECOND.
Red with white face.

Bred by Mr. J. Burlton, Luntley, Leominster, the property of Mr. B. Rogers, The Grove, Pembridge; got by Sampson (1061), dam (Blowdy) by Red Ben (768).

1863, Dec. 19,	r.—w.f. B	Corporal 2483	Rifleman 2189	Mr. Rogers.
1864, Nov. 25,	r.—w.f. H	Blowdy 3rd	Grove 2nd 2556	do.

BLUE BELL.
Red with white face, calved December 12, 1856.

Bred by and the property of Mr. J. R. Paramore, Dinedor Court, Hereford; got by Brecon (918), dam Pretty, bred by the late Mr. Davies, Tarrington.

1859,	r.—w.f. B	Steer	The General 2817	Mr. Paramore.
1860,	r.—w.f. B	Steer	do.	do.
1861, Mar. 5,	r.—w.f. H	Blue Bell 2nd	do.	do.
1862, July 8,	r.—w.f. H	Blue Bell 3rd	do.	do.
1863, Aug. 29,	r.—w.f. B	Steer	The Jew 2266	do.
1864, Oct. 5,	r.—w.f. H	Pretty	Portly 2765	do.

Blue Bell the Third was awarded a prize in the Extra Stock Class at the meeting of the Herefordshire Agricultural Society, 1864.

BLUE BELL.
Red with white face, calved in the year 1861.

Bred by and the property of Mr. R. Green Price, M.P., Norton Manor, Presteign; got by Havelock (2563).

1864, Aug. 16,	r.—w.f. H	Fanny	Lord of the Manor 2622	Mr. Price.

COWS.

BLUE BELL.
Red with white face, calved September 21, 1858.

Bred by Mr. Rea, Monaughty, Knighton, the property of Mr. E. T. Goldingham, Grimley, Worcester; got by Revenge (1051), dam (Marchioness) by Chieftain (930), g.d. (Gaylass) by Canute (890).

PRODUCE IN	NAME.	BY WHAT BULL.	BY WHOM BRED.
1864, Aug. 26, r.—w.f. B	Knight Errant	Sir Richard, 1734	Mr. T. Rea

BLUE BELL THE SECOND.
Red with white face, calved March 5th, 1861.

Bred by and the property of Mr. J. R. Paramore, Dinedor Court, Hereford; got by The General (2817), dam (Blue Bell) by Brecon (918), g.d. (Pretty), bred by the late Mr. Davies, Tarrington.

1863, Sept. 8, r.—w.f. H	Daisy	The Jew 2266	Mr. Paramore.
1864, Nov. 19, r.—w.f. B	General 2539	Portly 2165	do.

BOUNTY.
Red with white face, calved August 14, 1859.

Bred by the late Lord Berwick, Cronkhill, Salop, the property of the Hon. and Rev. Noel Hill, Berrington, Salop; got by Sir David (349), dam (Bessie) by Walford (871), g.d. (Dorcas the Third) by Tom Thumb (243), g.g.d. (Dorcas the Second) by Wonder (420), g.g.g.d. (Dorcas) by Ashley Moor White Bull (870), g.g.g.g.d. (Old Damsel) by Coleman's Bull (1547), — (Old Daisy) by Chancellor (156), — (Cherry the Second) by Thickset (1769), — (Cherry the First), bred by Mr. Knight, Downton Castle.

1863, March, r.—w.f. H	Bountiful	Van Tromp 2291	Hon. and Rev. Noel Hill.
1864, Jan. 26, r.—w.f. H	Beauty	do.	do.
1864, Dec. 15, r.—w.f. { H / H	Beauty 2nd / Beauty 3rd }	Albert 2380	do.

BRANDY.

Red with white face, calved November 3, 1860.

Bred by and the property of Mr. James P. Apperley, Fownhope, Hereford; got by Coroner (1555), dam (Bolt) by Partner (1031), g.d. (Bateman) by Thingehill, g.g.d. (Blossom) by Yorick (1461), g.g.g.d. (Blossom) by Temple (1406).

PRODUCE IN	NAME.	BY WHAT BULL.	BY WHOM BRED.
1863, Sept. 4, r.—w.f. H	Bounty	Cornet 1834	Mr. Apperley.

BRANDY.

Red with white face, calved in the year 1858.

Bred by Mr. Edward Price, Court House, Pembridge, Herefordshire, the property of Mr. J. O. G. Pollock, Mountainstown, Navan, Ireland; got by Goldfinder the Second (959), dam (Blossom) by Pembridge (721), g.d. (Countess) by Sir David (349), g.g.d. (Snowdrop) by Forester (398), g.g.g.d. (Redrose) by Crabstock (303).

1861, July 5, r.—w.f. B	Steer	Salisbury 2204	Mr. Price.
1862, Aug. 7, r.—w.f. H	Brandy 2nd	do.	do.
1863, Aug. 24, r.—w.f. B	Steer	Earl Derby 2nd 2510	do.
1864, Sept. 17, r.—w.f. H	Brandy 3rd	do.	Mr. Pollock.

BRONITH.

Red with white face, calved January 11, 1856.

Bred by and the property of Mr. John Monkhouse, The Stow, Hereford; got by Madoc (899), dam (Jenny Clyro) by Cantab (717), g.d. (Clyro) by Guy Fawkes (581), g.g.d. (Clyro) by Johnny Turpin, g.g.g.d. (Clyro) by Sir Andrew (183), bred by the late Mr. Thomas Tully, Clyro.

1858, Sept. 1, r.—w.f. B	Steer	Formidable 1240	Mr. Monkhouse.
1860, July 9, r.—w.f. B	Steer	Chieftain 930	do.
1861, Oct. 8, r.—w.f. B	Steer	do.	do.
1862, Oct. 17, r.—w.f. H	Mary Ann	do.	do.
1863, Nov. 8, r.—w.f. H	Bronllys	Llowess 2608	do.
1864, Nov. 1, r.—w.f. B	Steer	Chieftain 930	do.

COWS.

BROWN BESS.
Red with white face, calved June 30, 1858.

Bred by and the property of Mr. T. Olver, Penhallow, Grampound, Cornwall; got by Great Eastern (1598), dam (Beauty) by Young Walford (1820), g.d. — bred by the late Mr. Longmore, Bucton, Ludlow.

PRODUCE IN	NAME.	BY WHAT BULL.	BY WHOM BRED.
1862, July 12, r.—w.f. H	Buttercup	Conservative 1931	Mr. T. Olver.
1863, June 24, r.—w.f. H	Butterfly	do.	do.
1864, April 21, r.—w.f. H	Brunette	do.	do.

BROWN DUCHESS.
Red with white face, calved February 4, 1860.

Bred by and the property of Mr. W. D. Turner, Lynch Court, Leominster; got by The Doctor (1083), dam (Blossom) by Burton (1159), g d. (Pigeon) by Andrew the Third (908), g.g.d. (Woodpigeon) by Cassio (1528), g.g.g.d. — bred by the late Mr. Turner, Aymstrey Court.

1862, April 20, r.—w.f. H	Browny	Logic 2079	Mr. Turner.
1863, Sept. 8, r.—w.f. H	Dead	The Rover 2821	do.
1864, Aug. 14, r.—w.f. H	Lively	do.	do.

BROWNY.
Red with white face, calved in the year 1858.

Bred by and the property of Mr. John Rogers, Letchmoor, Presteign; got by Young Gaylad (2886).

1861, r.—w.f. B	Steer	Young David 2884	Mr. Rogers.
1862, June r.—w.f. H	Young Browny	Matchless	do.

BUD.
Red with mottle face, calved September 16, 1859.

Bred by and the property of Captain Peploe, Garnstone, Weobley; got by Musician (725), dam (Blossom) by Tyrant (737), g.d. (Blossom) by Tyro (692), g.g.d. (Lily) by Victory (2297), g.g.g.d. (Venus) by Victor (73).

1862, Sept. 25, r.—m.f. H	Bloom	The Twin 1420	Captain Peploe.
1863, Nov. 7, r.—m.f. H	Blossom	Leo 2070	do.
1864, Nov. 19, r.—m.f. H	Dead	do.	do.

BUTTERCUP.

Red with white face, of 1856, vol. v., p. 140.

Bred by Mr. W. H. Apperley, Withington, Hereford, the property of Mr. W. R. Grose, Penpont, Wadebridge, Cornwall; got by Partner (1031), dam (Cherry) by Corporal (1190), g.d. (Bell) by Noke.

PRODUCE IN		NAME.	BY WHAT BULL.	BY WHOM BRED.
1862, Oct. 24,	r.—w.f. H	Dead	Conservative 1931	Mr. Grose.
1863, Nov. 20,	r.—w.f. B	Steer	Sir Hugh 2223	do.
1864, Aug. 31,	r.—w.f. H	Betsy	Dick the Dustman 2495	do.

BUTTERCUP.

Red with white face, calved in the month of January, 1857.

Bred by Mr. H. E. Powell, Great Brampton, Hereford, the property of Mr. Colebatch, Rosemaund, Felton, Bromyard; got by Mameluke (1307), dam (Baroness) by Edgar (946), g.d. (Miss Cotmore the Second) by Lottery the Second (408), g.g.d. (Miss Cotmore) by Cotmore (376), g.g.g.d. — by Conqueror (412).

1859, Dec.,	r.—w.f. H	Bryony	Dutiful 1978	Mr. Powell.
1861,	r.—w.f. B	Steer	do.	do.
1862,	r.—w.f. B	Steer	do.	do.
1863, April,	r.—w.f. H	Bertha	Mentor 2112	do.

Bertha, sold to Mr. G. Pye, Cublington, Madley, Hereford.

BUTTERFLY.

Red with white face, calved September 9, 1859.

Bred by Mr. W. Tudge, Adforton, Ludlow, the property of Mr. Lewis Loyd, Monks' Orchard, Addington, Surrey; got by the Doctor (1083), dam (Redrose) by Orleton (901), g.d. (Redrose) by Nelson (1021), g.g.d. (Redrose) by Turpin (300).

1863, July 26,	r.—w.f. H	Moth	Victor 2294	Mr. Loyd.
1864, June 30,	r.—w.f. B	Zouavite 2905	do.	do.

COWS.

CAMILLA.

Red with white face, calved August 12, 1860.

Bred by and the property of Mr. William Tudge, Adforton, Ludlow; got by Kyrewood, dam (Comely) by Orleton (901), g.d. (Young Spot) by Nelson (1021), g.g.d. (Spot) by Turpin (300), g.g.g.d. (Cherry) by a Tully Bull.

PRODUCE IN	NAME.	BY WHAT BULL.	BY WHOM BRED.
1862, Aug. 14, r.—w.f. B	Steer	Sir Colin 2216	Mr. Tudge.
1863, Sept. 1, r.—w.f. B	Dominie Sampson 2501	Pilot 2156	do.
1864, Aug. 22, r.—w.f. H	Catharina	do.	do.

CARLISLE.

Red with white face, of 1854, vols. iv. and v., p.p. 93, 141.

Bred by the late Lord Berwick, Cronkhill, Salop, the property of Mr. T. Duckham, Baysham Court, Ross; got by Albert Edward (859), dam (Silver) by Emperor (221).

1862, Aug. 8, r.—w.f. B	Commodore 2472	Castor 1900	Mr. Duckham.
1863, Aug. 27, r.—w.f. B	Lord Clyde 2614	Victory 2296	do.
1864, July 24, r.—w.f. B	Rob Roy 2726	Cato 1902	do.

CAROLINE.

Red with white face, of 1857, vol. v., p. 141.

Bred by Mr. J. Hewer, Vern House, Marden, Hereford, the property of Mr. John Palmer, Hampton-on-the-Hill, Warwick; got by Cardinal Wiseman (1168), dam (Hampton Lass) by Mark (424), g.d. (Miss Hampton) by Garrick (1248), g.g.d. (Lady Hampton) by Reform (508), g.g.g.d. (Mossrose) by Hope (411).

1863, Mar. 24, r.—w.f. B	Kenilworth Duke 2592	Sir Edmund Lyons 2219	Mr. Palmer.
1864, Mar. 8, r.—w.f. B	Wild Boy 2876	My Lord 2647	do.

COWS.

CAROON.
Red with mottled face, calved November 13, 1859.

Bred by and the property of Captain Savery, Hardwick Lodge, Chepstow; got by Pedlar (1681), dam (Mottle) bred by Mr. Proctor, Wellington, Hereford.

PRODUCE IN	NAME.	BY WHAT BULL.	BY WHOM BRED.
1862, July 4, r.—w.f. H	Rosalie	Rob Roy 1708	Capt. Savery.
1863, Sep. 24, r.—w.f. B	Steer	Didley 1566	do.

CARROON.
Red with white face, of 1857, vol. 5, p. 141.

Bred by and the property of Mr. J. P. Apperley, Fownhope, Hereford; got by Partner (1031), dam (Bennett) by Yorick (1461), g.d. (Blossom) by Temple (1406), g.g.d. (Beauty) by Burton, g.g.g.d. (China) by Ploughman.

1862, Nov. 6, r.—w.f. H	Blowdy	Cornet 1934	Mr. Apperley.
1863, Oct. 13, r.—w.f. H	Dead	do.	do.
1864, Sep. 16, r.—w.f. H	Belle	Capt. Perry 2444	do.

CATHERINE.
Red with white face, calved September 21, 1859.

Bred by and the property of Mr. J. Monkhouse, The Stow, Hereford; got by Madoc (899), dam (Lofty) by Phantom (1035), g.d. (Stately) by Sir Andrew (183), g.g.d. (Stately) by a son of Sovereign (404).

1862, July 2, r.—w.f. H	Kitty	Monkland 3rd 1013	Mr. Monkhouse.
1863, June 8, r.—w.f. B	Peter the Great 2677	Chieftain 930	do.
1864, July 14, r.—w.f. H	Theresa	do.	do.

CELIA.
Red with white face, calved November, 6, 1859.

Bred by and the property of Mr. P. Turner, The Leen, Pembridge; got by Veracity (1443), dam (Princess) by Andrew the Second (619), g.d. (Brenda) by Viscount (816), g.g.d. (Rarity) by Cupid (1950), g.g.g.d. (Mayflower) bred by the late Mr. Turner, Aymestry.

PRODUCE IN		NAME.	BY WHAT BULL.	BY WHOM BRED.
1863, July 16,	B H	Dead	Bolingbroke 1883	Mr. Turner.
1864, June 17, r.—w.f.	B	Cervantes 2450	do.	do.

CERES.
Red with white face, of 1857, vol. 5, p. 142.

Bred by Lord Berwick, Cronkhill, Salop, the property of Gen. the Hon. A. N. Hood, Cumberland Lodge, Windsor; got by Albert Edward (859), dam (Cherry the Thirteenth) by Walford (871), g.d. (Red Cherry) by Tom Thumb (243), g.g.d. (Cherry the Fifth) by Cholstrey (868).

1863, Feb. 5, r.—w.f. H	Georgina	Ajax 1843	Hon. A N. Hood

CHERRY.
Red with white face, calved in the year 1861.

Bred by the late Mr. Williams, Chapel Clun, Salop, the property of Mr. John Rogers, Letchmoor, Presteign; got by Jerry (976).

1864, June, r.—w.f. B		Plato 2161	Mr. Rogers.

CHANCE.
Red with white face, calved November 12, 1859.

Bred by and the property of Capt. Savery, Hardwick Lodge, Chepstow; got by Desford (1213), dam (Dainty) by Coporal (1190), g.d. (Dainty) by Withington (1457), g.g.d. (Violet) by Yorick (1461),

1862, May 3, r.—w.f. B	Steer	Rob Roy 1708	Capt. Savery.
1863, Aug. 30, r.—w.f. B	Steer	Didley 1566	do.

COWS.

CHARITY.

Red with white face, of 1858, vol. v., p. 143.

Bred by Mr. John Smith, Sevenhampton, the property of Mr. John Barton, Coln, Fairford; got by Vanquish (1439), dam (Fatima) by Prince of Wales (1041), g.d. (Young Faith) by The Baby (1079), g.g.d. (Faith) by Hope (439).

PRODUCE IN	NAME.	BY WHAT BULL	BY WHOM BRED.
1862, April 10, r.—w.f. B	Sevenhampton 2748	Cotswold 1938	Mr. Smith.
1863, Feb. 7, r.—w.f. B	Steer	Cardinal Wiseman 2nd 1898	Mr. Barton.
1863, Dec. 24, r.—w.f. H	Countess	Coleshill 1923	do.

CHARMER.

Red with white face, calved in the year 1859.

Bred by Mr. Sobey, Tencreek, Liskeard, the property of Mr. W. R. Grose, Penpont, Wadebridge, Cornwall; got by Big Ben (1875), dam (Felton) by Coningsby (718), g.d. — bred by Mr. Good, Felton, Bromyard.

1862, Jan. 20, r.—w.f. H	Miss Rose	Penhallow 2154	Mr. Grose.

CHEERFUL.

Red with white face, calved March 3rd, 1859.

Bred by and the property of Mr. Thomas Olver, Penhallow, Grampouud, Cornwall; got by Great Eastern (1598), dam (Patience) by Colossus (591), g.d. (Cheerful) by Invincible (592), g.g.d. (Cherry) by Reform (508), g.g.g.d. (Rosebud) by Byron (440).

1862, Jan. 3, r.—w.f. B	Steer	Volunteer 2300	Mr. T. Olver.
1862, Nov. 4, r.—w.f. B	Carlton 2449	Sir Hugh 2223	do.

COWS.

CHERRY.
Red with white face, calved in the year 1855.

Bred by the late Mr. Thomas Longmore, Buckton, Salop, the property of Mr. R. H. Capper, The Northgate, Ross; got by Young Walford (1820).

PRODUCE IN	NAME.	BY WHAT BULL.	BY WHOM BRED.
1863, Sept. 8, r.—w.f. B	Sir Richard 2776	Carlisle 923	Mr. Jackson.
1864, Oct. 1, r.—w.f. B	Steer	Worcester 2878	Mr. Capper.

CHERRY.
Red with white face, calved February 18, 1856.

Bred by the late Mr. Keene, Pencraig, the property of Mr. Rees Keene, Pencraig, Caerleon, Monmouth; got by Young Chance, dam (Jenny) by Young David (2325), g.d. (Rose) by Foxhall (2520), g.g.d. Primrose.

1859, Mar. r.—w.f. B	Steer	Gen. Wyndham 1590	Mr. Keene.
1860, Mar. r.—w.f. B	Steer	do.	do.
1861, Feb. r.—w.f. H	Fancy	Odd Trick 1674	do.
1862, Feb. r.—w.f. H	Jenny	do.	do.
1863, Jan. r.—w.f. H	Willey	Pencraig 2671	do.
1864, Feb. r.—w.f. B	Bold Boy 2427	Cholstrey 2nd 1919	do.

CHERRY.
Red with white face, calved February 17, 1859.

Bred by and the property of Mr. J. R. Paramore, Dinedor Court, Hereford; got by Hotspur (972), dam (Noke) by a bull bred by the late Mr. Turner, Noke, g.d. bred by the late Mr. Davies, Tarrington.

1861, Aug. 15, r.—w.f. H	Cherry 2nd	The General 2817	Mr. Paramore.
1862, Aug. 23, r.—w.f. B	Dead	do.	do.
1863, Aug. 26, r.—w.f. H	Morella	The Jew 2266	do.
1864, Sep. 27, r.—w.f. B	Trueboy 2839	Portly 2165	do.

Cherry and Trueboy were awarded a prize in the Extra Stock Class at the meeting of the Herefordshire Agricultural Society, 1864.

CHERRY THE SECOND.
Red with white face, calved August 15, 1861.

Bred by and the property of Mr. J. R. Paramore, Dinedor Court, Hereford; got by The General (2817), dam (Cherry) by Hotspur (972), g.d. (Noke) by a bull bred by the late Mr. Turner, Noke) g.g.d. bred by the late Mr. Davies, Tarrington.

PRODUCE IN	NAME.	BY WHAT BULL.	BY WHOM BRED.
1864, Feb. 22, r.—w.f. H	Bigarrean	Portly 2165	Mr. Paramore.

CHERRY.
Red with white face, calved November 15, 1861.

Bred by and the property of Mr. J. P. Apperley, Fownhope, Hereford; got by Cornet (1934), dam (Bolt) by Partner (1031), g.d. (Bateman) by Thingehill, g.g.d. (Blossom) by Yorick (1461), g.g.g.d. (Blossom) by Temple (1406).

1864, May 21, r.—w.f. H	Bonny	Capt. Perry 2444	Mr. Apperley.

CHERRY.
Red with white face, calved in the month of September, 1858.

Bred by Mr. J. Stephens, Sheep House, Hay, the property of Mr. J. Jones, Llwyn-y-gaer, Raglan; got by Alma (1144), dam (Old Cherry) by Conqueror (684), g.d. (Gally) by Young Sovereign (683), g.g.d. (Cowslip) by Wonder (682).

1862, June 1, r.—w.f. B	Dead	King David 2060	Mr. J. Jones,
1863, April 10, r.—w.f. B	Steer	Bold David 1881	do.
1864, Feb. 7, r.—w.f. H	Curly	Red Man 2715	do.
1864, Dec. 20, r.—w.f. B	Banting	Trusty 2846	do.

CHERRY.
Red with white face, calved in the month of March, 1859.

Bred by Mr. J. Burlton, Luntley, Leominster, the property of Mr. B. Rogers, the Grove, Pembridge, Leominster; got by Samson (1061), dam (Cherry) by Red Ben (768), g.d. (Cherry) by a bull bred by Mr. Perry, Cholstrey.

1862, Jan., r.—w.f. B	Steer	Rifleman 2189	Mr. Burlton.
1863, Dec. r.—w.f. B	Corporal 2483	do.	Mr. Rogers.

COWS.

CHERRY.

Red with white face, calved May 4, 1860.

Bred by and the property of Mr. John Richards, Cound, Shrewsbury; got by Baronet (1860), dam (Duchess) by The Knight (185).

PRODUCE IN	NAME.	BY WHAT BULL.	BY WHOM BRED.
1863, Dec. 7, r.—w.f. H	Cherry 2nd	Pirate 2158	Mr. Richards.
1864, Oct. 15, r.—w.f. H	Cherry 3rd	do.	do.

CHERRY.

Red with white face, of 1852, vol. v., p. 146.

Bred by Mr. Spencer, Yarpole, Leominster, the property of Mr. James Taylor, Stretford Court, Leominster; got by Andrew (1495), dam — by Cotmore (376), g.d. — by Eyton (557).

1862, Aug. 10, r.—w.f. B	Steer	Croft 937	Mr. Taylor.

Cherry was one of five winners of a First Prize in their class at the meeting of the Leominster Agricultural Society, 1862; also as one of seven at the Hereford Meeting, the same year.

CHERRY THE SECOND.

Red with white face, of 1855, vol. v., p. 146.

Bred by Mr. John Taylor, Stretford Court, Leominster, the property of Mr. James Taylor, Stretford Court, Leominster; got by Sir William (1736), dam (Cherry) by Andrew (1495), g.d. — by Cotmore (376), g.g.d. — by Eyton (557).

1862, July 25, r.—w.f. {B/B}	Steers	Croft 937	Mr. Taylor.
1863, Sep. 18, r.—w.f. H	Steer	Unity 2287	do.
1864, Sep. 20, r.—w.f. H	Cherry 7th	Trustful 2845	do.

One of the twins of 1862, was one of four winners of the First Prize in their class at the meeting of the Leominster Agricultural Society, 1863; and a Second Prize at Ludlow, 1864.

CHERRY THE THIRD.
Red with white face, calved August 24, 1860.

Bred by Mr. J. Taylor, Stretford Court, Leominster, the property of Mr. John Baldwin, Luddington, Stratford-on-Avon; got by Croft (937), dam (Cherry) by Andrew (1495), g.d. — by Cotmore (376), g.g.d. — by Eyton (557).

PRODUCE IN	NAME.	BY WHAT BULL.	BY WHOM BRED.
1863, Aug. 19, r.—w.f. H	Cherry 4th	Sir George	Mr. Baldwin.

CHERRY THE FOURTH.
Red with white face, of 1858, vol. v., p. 147.

Bred by and the property of Mr. James Taylor, Stretford Court, Leominster; got by St. Oswall (1378), dam (Cherry the Second) by Sir William (1736), g.d. (Cherry) by Andrew (1495), g.g.d. — by Cotmore (376), g.g.g.d. — by Eyton (557).

1862, July 23, r.—w.f. B	Steer	Croft 937	Mr. Taylor.
1863, Aug. 23, r.—w.f. B	do.	Trustful 2845	do.
1864, Aug. 4, r.—w.f. B	do.	do.	do.

The Steer of 1862, was one of four winners of the First Prize in their class at the meeting of the Leominster Agricultural Society, 1863; and a Second Prize at Ludlow, 1864.

CHERRY THE FIFTH.
Red with white face, calved August 7, 1859.

Bred by and the property of Mr. James Taylor, Stretford Court, Leominster; got by St. Oswall (1378), dam (Cherry the Second) by Sir William (1736), g.d. (Cherry) by Andrew (1495), g.g.d. — by Cotmore (376), g.g.g.d. — by Eyton (557).

1862, Feb. 13, r.—w.f. B	Steer	Croft 937	Mr. Taylor.
1863, July 24, r.—w.f. B	do.	Garibaldi 2004	do.
1864, Aug. 26, r.—w.f. B	do.	Trustful 2845	do.

The Steer of 1863, was one of four winners of the First Prize in their class at the meeting of the Ludlow Agricultural Society, 1864; also as one of a pair at the Leominster and Hereford Agricultural Meetings, the same year.

CHERRY THE THIRD.
Red with white face, calved in the month of March, 1860.

Bred by and the property of Mr. John Burlton, Luntley Court, Leominster; got by Havelock (1609), dam (Cherry the Second) by Sampson (1061), g.d. (Cherry the First) by Red Ben (768), g.g.d. (Cherry) by the Count (351), g.g.g.d. (Cherry) by a son of Goldfinder (383).

PRODUCE IN	NAME.	BY WHAT BULL.	BY WHOM BRED.
1863, March, r.—w.f. B	Volunteer 2862	Rifleman 2189	Mr. Burlton.
1864, Feb. 24, r.—w.f. B	Volunteerist 2863	do.	do.

CHERRY THE FIFTH.
Red with white face, calved in the year 1858.

Bred by and the property of Mr. Evan Davies, Patton, Wenlock, Salop; got by Tyrant (1784), dam (Old Cherry) by a bull bred by the late Mr. Tarte, of the Bache, Corvedale.

1861, July 6, r.—w. f. H	Cherry Cheeks	Cound 1193	Mr. Evan Davies.
1862, July 10, r.—w. f. H	Dead	Lincoln 2076	do.
1863, July 15, r.—w. f. B	Steer	do.	do.
1864, July 29, r.—w. f. H	Columbine	do.	do.

CHERRY THE SIXTH.
Red with white face, calved in the month of October, 1865.

Bred by and the property of Mr. Hollings, Hillend, Hereford; got by Noke (1338), dam (Cherry the Fifth) by Voltigeur (1445), g.d. (Cherry the Fourth) by Reveller (2722), g.g.d. (Cherry the Third) by Cornet (1933), g.g.g.d. (Cherry the Second) by Young Waterloo (2341), g.g.g.g.d. (Cherry the First) by Foxley, bred by the late Sir R. Price.

1859, August, r.—w. f. B	Dead	Woodman 1460	Mr. Hollings.
1860, Sept., r.—w. f. B	Steer	do.	do.
1861, Oct. 9, r.—w. f. B	Steer	St. Clement 2201	do.
1862, Oct. 23, r.—w. f. H	Cherry 7th	do.	do.
1863, Nov. 4, r.—w. f. H	Cherry 8th	do.	do.

COWS.

CHERRY BLOSSOM.
Red with white face, calved in the year 1861.

Bred by and the property of Mr. R. Green Price, M.P., Norton Manor, Presteign; got by Stanage (1742).

PRODUCE IN	NAME.	BY WHAT BULL.	BY WHOM BRED.
1864, Sept. 10, r.—w.f. B		Lord of the Manor 2622	Mr. Price.

CHERRY BLOSSOM.
Red with white face, calved August 29, 1858.

Bred by the late Mr. Rea, Monaughty, Knighton, the property of Mr. P. J. Kearney, Miltown House, Clonmellon, Ireland; got by Borderer (1153), dam (Cherry) by Chieftain (930), g.d. (Litton) by Barrister (658), g.g.d. (Litton) by Old Court (306).

1863, Oct. 29, r.—w.f. H	Cherry Fruit	Sir Benjamin 1387	Mr. Kearney.
1864, Oct. 14, r.—w.f. H	Corah	Silverstream 2214	do.

CHLOE.
Red with white face, calved January 11, 1862.

Bred by and the property of Mr. Naylor, Leighton Hall, Montgomeryshire; got by Admiral (1481), dam (Clara) by Silvester (797), g.d. (Kate) by Big Ben (248), g.g.d. (Kitty) by Prince (251), g.g.g.d. (Kate) by Eclipse (252), g.g.g g.d. (Venus) by Tobias (487).

1864, July 13, r.—w.f. H	Chloe 2nd	Lichfield 2597	Mr. Naylor.

CHUB.
Red with white face, calved December 20, 1860.

Bred by and the property of Mr. Naylor, Leighton Hall, Montgomeryshire; got by Admiral (1481), dam (Clara) by Silvester (797), g.d. (Kate) by Big Ben (248), g.g.d. (Kitty) by Prince (251), g.g.g.d. (Kate) by Eclipse (252), g.g.g.g.d. (Venus) by Tobias (487).

1863, June 28, r.—w.f. H	Chub 2nd	Salisbury 2204	Mr. Naylor.
1864, July 11, r.—w.f. H	Chub 3rd	do.	do.

cows.

CHURCH HOUSE.
Red with white face, of 1853, vol. v., p. 149.

Bred by Mr. John Turner, Noke, Leominster, the property of Mr. C. H. Hinckesman, The Poles, Ludlow; got by Andrew the Second (619), dam (Church House) by Cotmore (376).

PRODUCE IN	NAME.	BY WHAT BULL.	BY WHOM BRED.
1862, July 14, r.—w.f. H	Church House 3d	The Friar 1085	Mr. Hinckesman.
1863, Oct. 31, r.—w.f. H	Church House 4th	Berwick 1874	do.
1864, Oct. 12, r.—w.f. B	Berwick 2nd 2418	do.	do.

CINDERELLA.
Red with white face, calved December 19, 1858.

Bred by and the property of Mr. John Monkhouse, The Stow, Hereford; got by Tipperary Boy (1424), dam (Celendine the Second) by Madoc (899), g.d. (Celendine) by Phantom (1035), g.g.d. (Pansy) by Sir Andrew (183).

1862, July 4, r.—w.f. H	Slipper	Chieftain 930	Mr. Monkhouse.
1864, Jan. 13, r.—w.f. H	Princess	do.	do.

CLARA.
Red with white face, of 1855, vol. v., p. 150.

Bred by Mr. Castree, Gloucester, the property of Mr. John Barton, Coln, Fairford; got by Venison the Second (1442), dam (Clara) by The Duke (493), g.d. — bred by Earl Radnor.

1863, Nov. 12, r.—w.f. H	Carnation	St. Michael 1718	Mr. Barton.

CLARA.
Red with white face, calved October 12, 1860.

Bred by the Rev. Archer Clive, Whitfield, Hereford, the property of Mr. J. R. Paramore, Dinedor Court, Hereford; got by General (1251), dam (Countess) by a Son of Governor (464), g.d. (Mottle), bred by the late Mr. Bowen, Monkland.

1863, Sep. 24, r.—w.f. B		The Jew 2266	Mr. Paramore.

COBWEB.

Red with mottled face, calved in the month of August, 1860.

Bred by and the property of Mr. John Monkhouse, The Stow, Hereford; got by Chieftain (930), dam (Sylph) by Madoc (899) g.d. (Sylph) by Guy Fawkes (581), g.g.d. (Sprite) by Sir Andrew (183).

PRODUCE IN	NAME.	BY WHAT BULL.	BY WHOM BRED.
1863, Aug. 9, r.—w.f. H	Spider	Artful 2391	Mr. Monkhouse.
1864, July 6, r.—m.f. B	Steer	Llowes 2608	do.

COLUMBINE.

Red with white face, of 1857, vol. v., p. 151.

Bred by and the property of Mr. John Monkhouse, The Stow, Hereford; got by Formidable (1240), dam (Celendine the Second) by Madoc (899), g.d. (Celendine) by Phantom (1035), g.g.d. (Pansy) by Sir Andrew (183).

1862, July22, r.—w.f. B	Steer	Chieftain 930	Mr. Monkhouse.
1864, Sep. 29, r.—w.f. H	Geraldine	do.	do.

COMELY.

Red with white face, calved in the month of December 1858.

Bred by and the property of Mr. T. S. Bradstock, Cobrey Park, Ross; got by Daniel (1201), dam (Cocky) by Longitude, g.d. (Prettymaid) by Sovereign the Third, g.g.d. (Eaton) by Lottery the Second (408).

1861, June18, r.—w.f. B	Steer	Young Rambler 2335	Mr. Bradstock.
1862, Aug. 1, r.—w.f. B	Bathurst 2405	do.	do.
1863, Aug. 6, r.—w.f. B	Steer	do.	do.
1864, Sep.24, r.—w.f. B	Barling 2398	do.	do.

COMELY.

Red with white face, calved July 13, 1860.

Bred by and the property of Mr. John Monkhouse, The Stow, Hereford; got by Madoc (899), dam (Young Lofty) by Cantab

COWS.

(717), g.d. (Lofty) by Phantom (1035), g.g.d. (Stately) by Sir Andrew (183).

PRODUCE IN	NAME.	BY WHAT BULL.	BY WHOM BRED.
1863, June 7, r.—w.f. B	Steer	Chieftain 930	Mr. Monkhouse.
1864, Sep.16, r.—w.f. H	Lassie	do.	do.

COMELY THE SECOND.
Red with white face, calved October 26, 1860.

Bred by and the property of Mr. J. P. Apperley, Fownhope, Hereford; got by Coroner (1555), dam (Comely) by Partner (1031), g.d. (Bran) by Favourite (952), g.g.d. (Brandy) by Yorick (1461), g.g.g.d. (Brandy) by Noke.

1863, Dec.15, r.—w.f. B	Steer	Cornet 1834	Mr. Apperley.
1864, Dec 31, r.—w.f. H	Cockney	Abbot 2367	do.

COMELY.
Red with white face, of 1859, vol. v., p. 152.

Bred by Mr. Mason, Priors Court, Ledbury, the property of Capt. Savery, Hardwick Lodge, Chepstow; got by Goldsmith (1258), dam (Blossom) by Garrick (1248).

1864, Aug.22, r.—w.f. H	Cricket	Didley 1566	Capt. Savery.

Cricket, sold to Mr. T. Cadle, Westbury-on-Severn.

COMELY THE SECOND.
Red with white face, calved August 29, 1859.

Bred by and the property of Mr. Wm. Tudge, Adforton, Leintwardine, Ludlow; got by The Doctor (1083), dam (Comely) by Orleton (901), g.d. (Young Spot) by Nelson (1021), g.g.d. (Spot) by Turpin (300), g.g.g.d. (Cherry) by a Tully bull.

1862, Aug.31, r.—w.f. H	Curly	Sir Colin 2216	Mr. Tudge.
1863, Aug.18, r.—w.f. H	Carlotta	Pilot 2156	do.

CONSTANCE.

Red with white face, calved in the month of December, 1860.

Bred by the late Mr. Powell, Great Brampton, Hereford, the property of Mr. J. Prosser, Honeybourne Grounds, Worcester; got by Dutiful (1978), dam (Countess) by Mameluke (1307), g.d. (Miss Cotmore the Second) by Lottery the Second (408), g.g.d. (Miss Cotmore) by Cotmore (376), g.g.g.d. by Conqueror (412).

PRODUCE IN	NAME.	BY WHAT BULL.	BY WHOM BRED.
1863, Nov. 7, r.—w.f. B	Steer	Mentor 2112	Mr Prosser.

COQUETTE.

Red with white face, calved July 17, 1859.

Bred by Mr. John Hewer, Vern House, Marden, Hereford, the property of Mr. Wm. Powell, Eglwysnunydd, Taibach, Glamorgan; got by Van Tromp (1440), dam (Lofty the Second) by Governor (464), g.d. (Lovely) by Hope (411), g.g.d. (Old Lofty) by Original (455), g.g.g.d. (Blossom) by Old Wellington (507), g.g.g.g.d. (Old Lofty) by Silver (540), — (Old Blossom) bred by Mr. W. Hewer, The Hardwicke.

1861, Sep. 10, r.—w.f. H	Flirt	Newcastle 2651	Mr. Hewer.
1863, Feb. 9, r.—w.f. B	Avon 2393	General 1251	Mr. Powell.

CORA.

Red with white face, calved in the year 1859.

Bred by Mr. John Sheriff, Burrington, Ludlow, the property of Mr. William Stallard, Brockhampton, Ross; got by Coxall (1196), dam by Confidence (367), g.d. by Byron (440).

1861, June 9, r.—w.f. B	Steer	Sir Colin 2216	Mr. Stallard.
1862, June 4, r.—w.f. H	Clementine	St. Clement 2201	do.
1863, June 3, r.—w.f. H	Clara	Chieftain 2nd 1917	do.
1864, May 30, r.—w.f. B	Steer	do.	do.

COWS.

CORA.

Red with mottled face, calved September 5, 1857.

Bred by Mr. Edward Price, Pembridge, Leominster, the property of the Duke of Bedford, Woburn Park, Beds.; got by Goldfinder the Second (959), dam (Lovely) by Pembridge (721), g.d. (Lucksall) by Prince Dangerous (362), g.g.d. (Mottle) by The Sheriff (356), g.g.g.d. (Lady) by Forester (398), g.g.g.g.d. (Pink) by Crabstock (303).

PRODUCE IN	NAME.	BY WHAT BULL	BY WHOM BRED.
1861, Aug. 1, r.—m.f. B	Steer	Salisbury 2204	Duke of Bedford.
1862, Dec. 18, r.—m.f. B	Steer	Carbonel 1525	do.
1863, Nov. 17, r.—w.f. H		do.	do.

CORA THE SECOND.

Red with white face, calved in the month of November, 1858.

Bred by the late Mr. Josiah Davies, Ivington, Leominster, the property of Mrs. Ann Davies, Ivington. Leominster; got by Giant (1411), dam (Cora) by Sutton (1752), g.d. bred by Mr. Boughton, Clifton-on-Teme.

1861, Nov. r.—w.f. H	Cora 3rd	Master Butterfly 1313	Mr. Davies.
1863, Feb. 14, r.—w.f. B	Steer	Conqueror	do.
1864, Jan. 1, r.—w.f. H	Milkmaid	do.	do.
1864, Nov. 24, r.—w.f. B	Steer	Chance 1908	do.

CORA THE THIRD.

Red with white face, calved in the month of November, 1861.

Bred by the late Mr. Josiah Davies, Ivington, Leominster, the property of Mrs. Ann Davies, Ivington, Leominster; got by Master Butterfly (1313), dam (Cora the Second) by The Giant (1411), g.d. (Cora) by Sutton (1752), g.g.d. — bred by Mr. Boughton, Clifton-on Teme.

1863, Nov. 12, r.—w.f. B	Steer	Chance 1908	Mr. Davies.
1864, Oct. 26, r.—w.f. H	Pansy	do.	do.

COWS.

COUNTESS.

Red with white face.

Bred by Mr. Edward Price, Court House, Pembridge, the property of Mr. J. O. G. Pollock, Mountainstown, Navan, Ireland; got by Salisbury (2204), dam (Countess) by Sir David (349), g.d. (Snowdrop) by Forester (398), g.g.d. (Redrose) by Crabstock (303).

PRODUCE IN	NAME.	BY WHAT BULL.	BY WHOM BRED.
1863, Aug. 12, r.—w.f. B		Earl Derby 2nd 2510	Mr. Price.

COUNTESS.

Red with white face, calved in the month of May, 1861.

Bred by Earl Somers, Eastnor Castle, Ledbury, the property of Mr. Thomas Smith, Bodenham, Much Marcle, Dymock; got by Abdel Kader (1837), dam — bred by Mr. Mason, Tarrington, g.d. — bred by Mr. Turner, Noke, Leominster.

1863, Aug. 13, r.—w.f. B	King Charles 2nd 2594	King Charles 2593	Mr. Deakin.
1864, Sept. 2, r.—w.f. H	Countess 2nd	do.	Mr. Smith.

COUNTESS.

Red with white face, calved September 21, 1857.

Bred by and the property of Mr. William Tudge, Adforton, Ludlow; got by The Doctor (1083), dam (Duchess) by Stanage (1741), g.d. (Laurel) by Nelson (1021), g.g d. (Lily the Third) by Turpin (300). g.g.d. (Lily the Second) by a Tully Bull.

1860, Sep. 10, r.—w.f. H	Clementina	Carbonel 1525	Mr. Tudge.
1861, Sep. 1, r.—w.f. B	Anglo-Saxon 1851	The Grove 1764	do.
1862, Sep. 5, r.—w.f. B	Blucher 2424	Sir Colin 2216	do.
1863, Aug 15, r.—w.f. H	Cloribel	Pilot 2156	do.

cows.

COUNTESS.
Red with white face, calved May 25, 1859.

Bred by and the property of Mr. James Gregg, Fencote Abbey, Leominster; got by Goldsmith (1258), dam (Miss Thingehill the Fourth) by Young Sir David (1137), g.d. (Miss Thingehill) by Thingehill (546), g.g.d. (Thingehill Pigeon) by Reform (508), g.g.d. (Hampton Pigeon) by Young Sovereign (506).

PRODUCE IN	NAME.	BY WHAT BULL.	BY WHOM BRED.
1862, June 18, r.—w.f. H	Prettymaid	Fencote 1989	Mr. Gregg.

COUNTESS.
Red with white face, calved in the year 1854.

Bred by the late Mr. J. Powell, Great Brampton, Hereford, the property of Mr. J. H. Whitehouse, Ipsley Court, Redditch; got by Mameluke (1307), dam (Miss Cotmore the Second) by Lottery the Second (408), g.d. (Miss Cotmore) by Cotmore (376), g.g.d. — by Conqueror (412)

1860, Dec., r.—w.f. H	Constance	Dutiful 1978	Mr. H. E. Powell.
1861, Dec., r.—w.f. B	Steer	do.	do.
1862, Feb., r.—w.f. H	Castanet	Mentor 2112	do.
1863, Feb., r.—w.f. B	Steer	do.	do.
1864, r.—w f. B		Vincent 2858	

Castanet, sold to Mr. R. S. Fetherstonhaugh, Rockview, Killucan, Ireland.

COUNTESS.
Red with white face, calved October 5, 1856.

Bred by and the property of Mr. John Richards, Cound, Shrewsbury; got by The Knight (185).

1859, Nov. 2, r.—w.f. H	Jessamine	Baronet 1860	Mr. Richards.
1860, Nov. 8, r.—w.f. H	Lizzie	do.	do.
1861, Nov. 17, r.—w.f. H	Salvia	do.	do.
1862, Dec. 4, r.—w.f. B	Cound 2484	Pirate 2158	do.
1863, Nov. 20, r.—w.f. B	Steer	do.	do.

COWS.

COUNTESS THE SECOND.
Red with white face, calved in the month of January, 1856.

Bred by Mr. J. Hollings, the property of Mr. J. A. Hollings, Hillend, Hereford; got by a bull bred by Mr. Vaughan, Cholstrey, dam (Countess) by Albert Edward (754), dam (Countess) by Monarch (504).

PRODUCE IN	NAME.	BY WHAT BULL.	BY WHOM BRED.
1860, Jan. 3, r.—w.f. B	Steer	Woodman 1460	Mr. Hollings.
1861, Jan. 20, r.—w.f. B	Steer	do.	do.
1861, Dec. 28, r.—w.f. B	Steer	St. Clement 2201	do.
1862, Dec. 19, r.—w.f. H	Lady	do.	do.
1863, Nov. 30, r.—w.f. B	Steer	do.	do.
1864, Dec. 7, r.—w.f. H	Countess 3rd	do.	do.

COUNTESS OF SHREWSBURY THE THIRD.
Red with white face, calved September 10, 1859.

Bred by Mr. Stedman, Bucknall House, Shrewsbury, the property of Mr. John Baldwin, Luddington, Stratford-on-Avon; got by Grateful (1260), dam (Countess of Shrewsbury) by Conrad (1183), g.d. (Countess) by Venison (1441), g.g.d. — by Dinedor (395), g.g.g.d. — by Trojan (542).

1863, June 6, r.—w.f. H	Dead	Battersea 1865	Mr. Baldwin.
1864, June 1, r.—w.f. H	Maid of Shrewsbury	do.	do.

Countess of Shrewsbury the Third was awarded a Prize as Extra Stock at the meeting of the Vale of Evesham Agricultural Society, 1863

COUNTESS THE FOURTH.
Red with white face, calved February 1, 1860.

Bred by and the property of Mr. James Taylor, Stretford Court, Leominster; got by St. Oswall (1378), dam Fancy (Leominster) by King John (830), g.d. (Countess) bred by the late Mr. James Bowen, Monkland.

1862, Aug. 17, r.—w.f. H	Countess 6th	Croft 937	Mr. Taylor.
1864, April 15, r.—w.f. B	Steer	Trustful 2845	do.

cows.

COUNTESS THE FIFTH.

Red with white face, calved May 10, 1861.

Bred by and the property of Mr. J. Taylor, Stretford Court, Leominster; got by Croft (937), dam (Fancy Leominster) by King John (830).

PRODUCE IN	NAME.	BY WHAT BULL.	BY WHOM BRED.
1864, May 19, r.—w.f. H	Countess 8th	Trustful 2845	Mr. Taylor.

Countess the Fifth was one of six winners of the First Prize in their class at the meeting of the Herefordshire Agricultural Society, 1862; and one of seven at their meeting in 1864; also as one of four at Leominster, the same year.

COWSLIP.

Red with white face, calved in the year 1856.

Bred by Mr. Sheriff, Coxall, Ludlow, the property of Mr. R. G. Price, M.P., Norton Manor, Presteign; got by Brilliant (1518), dam — by Young Royal (1470), g.d. — by Young Emperor (1811), g.g.d. — by Byron (440).

1863, July 15, r.—w.f. B		Sir Colin 2216	Mr. Sheriff.
1864, Dec. 3, r.—w.f. H	Rose of the Valley	Lord of the Manor 2622	Mr. Price.

COWSLIP.

Red with white face, calved in the year 1859.

Bred by Messrs. F. and C. Bodenham, Hereford, the property of Mr. John Rogers, Letchmoor, Presteign; got by Priam (1039), dam — by Young Dewshall (1125), g.d. — bred by the late Sir Robert Price, Foxley, Hereford.

1862, July r.—w.t. B	Steer	Matchless 2110	Mr. Rogers.
1863,	Dead	do.	do.
1864, June r.—w.f. H	Cowslip 2nd	do.	do.

COWSLIP.

Red with white face, calved in the year 1858.

Bred by Mr. J. B. Green, Marlow, Leintwardine, Salop, the property of Capt. Crawshay, Danypark, Crickhowell; got by Beefy Ben (1869), dam — by Cholstrey (217), g.d. — (Zest of Oxford) (2352), g g.d. — by Discount (339).

PRODUCE IN		NAME.	BY WHAT BULL.	BY WHOM BRED.
1861,	r,—w.f. H	Vanity	Vanguard 1109	Mr. Green.
1862,	r,—w.f. B		Zealot 2344	do.
1863,	r,—w.f. H	Violet	do.	do.
1864,	r,—w.f. H	Primrose	Zealous 2349	Capt. Crawshay.

COWSLIP.

Red with white face, calved September, 24, 1861.

Bred by and the property of Mr. H. Gibbons, Hampton Bishop, Hereford; got by Shamrock the Second (2210), dam (Silk) by The Admiral (1078), g.d. (Young Beauty) by Young Gaylad (1463).

1864. June 15, r,—w.f. B		Trumpeter 2282	Mr. Gibbons.

COWSLIP THE SECOND.

Red with white face, of 1853, vol. v., p. 157.

Bred by Mr. John Taylor, Stretford Court, the property of Mr. James Taylor, Stretford Court, Leominster; got by King John (830), dam (Cowslip), bred by the late Mr. Mason, Middleton, Tenbury.

1862, Aug. 27, r.—w.f. B	Steer	Unity 2287	Mr. Taylor.
1863, July 12, r.—w.f. H	Cowslip 4th	Garibaldi 2004	do.
1864, Aug. 4, r.—w.f. H	Cowslip 5th	Trustful 2845	do.

Cowslip the Second was one of six, winners of the First Prize in their class at the meeting of the Leominster Agricultural Society, 1862; and one of seven at Hereford the same year.

COWSLIP THE THIRD.

Red with white face, calved in the year 1859.

Bred by and the property of Mr. T. Morris, Therrow, Llyswen, Hay; got by Telegraph (1404), dam (Cowslip) by Young Hope (343), g.d. (Queen) by Young Byron (832), g g.d. (Lily) by Prince Llewellyn (713), g.g.g.d. (Princess) by Charity the Second (516).

PRODUCE IN		NAME.	BY WHAT BULL.	BY WHOM BRED.
1863,	r,—w.f. B	Steer	Prince Imperial 2171	Mr. Morris.
1864,	r,—w.f. B		do.	do.

CRAFTY.

Red with white face, calved August 11, 1855.

Bred by and the property of Mr. George Lobb, Lawhitton, Launceston; got by Protector the Second (1362), dam (Roundhorns) by Youngster (2899), g.d. — bred by Earl St. Germans, from the stock of Mr. Hewer, Vern House, Hereford.

1858,	r,—w.f. B	Lucky	Woodbine 1120	Mr. Lobb.
1859,	r,—w.f. H	Dead	do.	do.
1860,	r,—w.f. B	Router 2729	Young Orleton 1476	do.
1861,	r,—w.f. H	Dead	do.	do.
1863,	r,—w.f. H	Crafty 2nd	do.	do.
1864,	r,—w.f. H	Crafty 3rd	Brown Willie 2433	do.

Crafty was a winner of the First Prize in her class at the meetings of the Lifton, Tavistock, and Launceston Agricultural Societies, 1863.

CRINOLINE.

Red with white face, calved April 15, 1859.

Bred by Mr. J. Hewer, Vern House, Marden, Hereford, the property of Messrs. T. and W. Vaughan, Lawton, Leominster; got by Purifier (1364), dam (Curly) by Garrick (1248), g.d.

Moss Rose) by Hope (411), g.g.d. (Old Moss Rose) by Old Wellington (507).

PRODUCE IN	NAME.	BY WHAT BULL.	BY WHOM BRED.
1862, Feb., r.—w.f. B	Steer	Mameluke 1307	Mr. Vaughan.
1863, Jan. 2, r.—w.f. B	Alfred the Great 2385	Chelmsford 1915	Messrs. Vaughan.
1863, Oct. 15, r.—w.f. B	Mackney 2628	Cholstrey 2nd 1919	do.
1864, Sept. 22, r.—w.f. H	Crinoline 2nd	The Sheriff	do.

Crinoline with her sire and dam were winners of a prize at the Cirencester Meeting of the Gloucestershire Agricultural Society.

CROCUS.

Red with white face, calved September 15, 1859.

Bred by Lord Berwick, Cronkhill, Salop, the property of Mr. Naylor, Leighton Hall, Montgomeryshire; got by Sir David (349), dam (Verbena) by Attingham (911), g.d. (Young Rebecca) by Young Hope (343), g.g.d. (Rebecca) by Governor (464), g.g.g.d. Old Prettymaid) by Young Sovereign (1472), g.g.g.g.d. — by Whitenob (345), — by Young Wellington (505).

1862, July 27, r.—w.f. B	Dead	Volunteer 2299	Mr. Naylor.
1863, Aug. 1, r.—w.f. B	Mars 2634	Blondin 1880	do.
1864, June 22, r.—w.f. H	Crocus 2nd	Gladstone 2547	do.

CURLY.

Red with white face, calved in the month of July, 1861.

Bred by and the property of Mr. Rogers, The Grove, Pembridge; got by The Grove (1764), dam (Curly) by Severus (1062), g.d. (Curly) by Young Royal (1470), g.g.d. (Curly) by Prince (251), g.g.g.d. (Curly) by Charity 2nd (1535).

1864, May 10, r.—w.f. B	Admiral 2374	North Star 2138	Mr. Rogers.

CURLY.
Red with white face, calved in the year 1859.

Bred by Messrs. F. and C. Bodenham, Hereford, the property of Mr. John Rogers, Letchmoor, Presteign; got by Curly (1561), dam — by Young Dewshall (1125), g.d. — bred by the late Sir R. Price, Foxley, Hereford.

PRODUCE IN		NAME.	BY WHAT BULL.	BY WHOM BRED.
1862, July,	r.—w.f. B	Steer	Matchless 2110	Mr. Rogers.
1863, July,	r.—w.f. B	do.	do.	do.
1864, June,	r.—w.f. H	Miss Curly	do.	do.

CURLY.
Red with white face, calved October 29, 1859.

Bred by and the property of Mr. John Wigmore, Bickerton Court, Dymock; got by Forester (1238), dam (Primrose) by Dolphin (2500), g.d. — bred by Mr. Jones, The Hollow, Dinedor.

1863, April 17, r.—w.f. B	Steer	Melon 2111	Mr. Wigmore.
1864, Mar. 25, r.—w.f. B		Speculator 2240	do.

Curly was a winner of the First Prize in her class at the meeting of the Gloucestershire Agricultural Society, 1862; and one of eight winners of a Second Prize at the Hereford Meeting, 1863.

CURLY.
Red with white face, calved December 25, 1860.

Bred by and the property of Mr. W. Lane, Compton Casey, Cheltenham; got by Casey (1899), dam (Lily) by Hospodar (1621), g.d. (Broad) by Planet (1690).

1864, Jan. 25, r.—w.f. H	Curly 2nd	Hardy 2027	Mr. Lane.

CURLY.
Red with white face, of 1855, vol. v., p. 160.

Bred by Mr. Hewer, Vern House, Marden, the property of Mr. John Palmer, Hampton-on-the-Hill, Warwick; got by Garrick

(1248), dam (Moss Rose) by Hope (411), g.d. (Old Moss Rose) by Old Wellington (507), g.g.d. (Fanny) by Fitz Favourite (441).

PRODUCE IN	NAME.	BY WHAT BULL.	BY WHOM BRED.
1862, Mar. 4, r.—w.f. B	My Lord 2647	Abdel Kader 1837	Mr. Palmer.
1863, Aug. 22, r.—w f. H	Miss Lyons	SirEdmundLyons2219	do.

Curly was a winner of the First Prize in her class at the Leamington Meeting of the Warwickshire Agricultural Society, 1863.

CURLY THE SECOND.
Red with white face, calved January 6, 1861.

Bred by and the property of Mr. Thomas Thomas, St. Hilary, Cowbridge; got by Goldfinder the Second (959), dam (Curly) by Young Royal (1469).

1864, June 10, r.—w.f. H		Goldfinder 2nd 959	Mr. Thomas.

Curly the Second was a winner of the Second Prize in her class at the meeting of the Cowbridge Agricultural Society, 1862; and First at the meeting in 1864; also First at Newport, the same year.

CURLY THE THIRD.
Red with white face, of 1856, vol. iv., p. 104.

Bred by and the property of Mr. T. Morris, Therrow, Llyswen, Hay; got by Telegraph (1404), dam (Curly) by Young Byron (832), g.d. (Miss Noble) by Noble (543), g g.d. (Favourite) by Sovereign (404), g.g.g.d. (Damsel) by Young Wellington (505).

1862, r.—w.f. H	Gaylass	Guardsman	Mr. Morris.

CURLY THE FIFTH.
Red with white face, calved in the month of July, 1860.

Bred by and the property of Mr. Thomas Morris, Therrow, Llyswen, Hay; got by Telegraph (1404), dam (Curly the Second) by Newton (344), g.d. (Curly) by Young Byron (832). g.g.d. (Noble the First) by Noble (543), g.g.g.d. (Favourite) by Sovereign (404).

1863, r.—w.f. B	Steer	Prince Imperial 2171	Mr. Morris.
1864, Aug. r.—w.f. H	Cleopatra	do.	do.

CUSTARD.
Red with white face, calved in the year 1858.

Bred by and the property of Mr. Evan Davies, Patton, Wenlock, Salop; got by Tyrant (1784), dam (Cowslip) by Patton (1679).

PRODUCE IN	NAME.	BY WHAT BULL.	BY WHOM BRED.
1861, July 15, r.—w.f. H	Cornfit	Cound 1193	Mr. E. Davies.
1862, July 10, r.—w.f. H	Cream	Lincoln 2076	do.
1863, Oct. 11, r.—w.f. B	Steer	do.	do.
1864, Nov. 26, r.—w.f. B	Steer	do.	do.

DAHLIA.
Red with white face, of 1854, vols. iii. and v., pp. 146, 162.

Bred by and the property of Mr. John Monkhouse, the Stow, Hereford; got by Madoc (899), dam (Dahlia) by Guy Fawkes (581), g d. (Daffodil) by Charity (375), g.g.d. (Tulip) by Sir Andrew (183), g.g.g.d. (Tulip) by Young Chance.

1862, Aug. 25, r.—w.f. H	Peony	Chieftain 930	Mr. Monkhouse.
1863, Aug. 17, r.—w.f. B	Steer	do.	do.
1864, Aug. 19, r.—w.f. B	do.	do.	do.

DAINTY.
Red with white face, of 1851, vol. v., p. 164.

Bred by and the property of the Rev. Archer Clive, Whitfield, Hereford; got by son of Governor (464), dam — by Robinhood (664).

1863, Oct. 28, r.—w.f. H	Dolly	Bertram 1513	Rev. A. Clive.
1864, Oct. 22, r.—w.f. B		Plato 2160	do.

DAINTY.
Red with white face, calved in the year 1856.

Bred by and the property of Mr. Henry Gibbons, Hampton Bishop, Hereford; got by the Admiral (1078), dam (Lofty) by Young Gaylad (1463), g.d. (Staggy) by Zephyr (1826).

1859, Oct. 9, r.—w.f. B	Steer	Woodman 2nd 1459	Mr. Gibbons.
1860, Sept. 2, r.—w.f. H	Dead	do.	do.
1861, Aug. 18, r.—w.f. B	Frugality 1997	Shamrock 2nd 2210	do.
1862, Sept. 20, r.—w.f. H	Dainty Lass	do.	do.
1863, Aug. 2, r.—w.f. H	Dewdrop	do.	do.
1864, July 22, r.—w.f. H	Diana	do.	do.

COWS.

DAINTY THE SECOND.
Red with white face, calved in the month of March, 1861.

Bred by and the property of Mr. John Jones, Llwyn-y-Gaer, Raglan; got by Patron the Second (1678), dam (Dainty) by The Noke (2268), g.d. (Dewshall) by Dewshall (358).

PRODUCE IN	NAME.	BY WHAT BULL.	BY WHOM BRED.
1864, Feb. 10, r.—w.f. B	Steer	Red Man 2715	Mr. Jones.
1864, Dec. 15, r.—w.f. H	Little Dainty	do.	do.

Dainty was one of a pair, winners of the First Prize in their class at the meeting of the Abergavenny Agricultural Society, 1862; Second at the Monmouth and Tredegar Meetings, 1862; also a First Prize at Monmouth, 1863.

DAINTY THE SECOND.
Red with white face, calved November 16, 1859.

Bred by and the property of Mr. Naylor, Leighton Hall, Montgomeryshire; got by Admiral (1481), dam (Dainty) by Attraction (892), g.d (Lucy Long) by Big Ben (248), g.g.d. (Little Cherry) by Prince (251), g.g.g.d. (Spot) by Trump (490), g.g.g.g.d. (Cherry) by Red Robin (263)

1862, June 15, r.—w.f. B		Salisbury 2204	Mr. Naylor.
1863, July 10, r.—w.f. H	Dainty 3rd	do.	do.
1864, June 5, r.—w.f. H	Dainty 4th	do.	do.

DAISY.
Red with white face, calved January 1, 1856.

Bred by and the property of Mr. Henry Gibbons, Hampton Bishop, Hereford; got by The Admiral (1078), dam (Beauty) by Young Gaylad (1463). g.d. (Beauty) by Zephyr (1826).

1859, Sept. 25, r.—w.f. H	Silver	Woodman 2nd 1459	Mr. Gibbons.
1860, Aug. 18, r.—w.f. B	Lacey 2063	do.	do.
1861, Oct. 22, r.—w.f. H	Young Spot	Shamrock 2nd 2210	do.
1862, Aug. 19, r.—w.f. B	Dead	do.	do.
1863, Aug. 28, r.—w.f. B	Grateful 2555	do.	do.
1864, Aug. 9, r.—w.f. B	Gerald 2542	do.	do.

COWS.

DAISY.

Red with white face, calved September 8, 1857.

Bred by and the property of Mr. William Tudge, Adforton, Leintwardine, Ludlow; got by The Doctor (1083), dam (Dainty) by Orleton (901), g.d. (Prettymaid) by Nelson (1021), g.g.d. (Prettymaid) by Turpin (300).

PRODUCE IN		NAME.	BY WHAT BULL.	BY WHOM BRED.
1860, Sept. 10, r.—w.f.	H	Darling	Carbonel 1525	Mr. Tudge.
1861, Sep. 9, r.—w.f.	B B	Steers	The Grove 1764	do.
1862, Aug. 31, r.—w.f.	H	Dido	Sir Colin 2216	do.
1863, Nov. 5, r.—w.f.	B	Steer	Pilot 2156	do.
1864, Oct. 10, r.—w.f.	B	do.	do.	do.

DAISY.

Red with white face, calved in the year 1861.

Bred by and the property of Mr. John Rogers, Letchmoor, Presteign; got by Young David (2885), dam — by Young Dewshall (1125).

1864, June, r.—w.f. H	Young Daisy	Matchless 2110	Mr. Rogers.

DAISY.

Red with white face, of 1850, vol. v., p. 166.

Bred by Mr. J. Hollings, Hillend. Hereford, the property of Mr. J. A. Hollings, Hillend, Hereford; got by Byron (380), dam (Cherry the Fourth) by Reveller (2722), g.d. (Cherry the Third) by Cornet (1933), g.g.d. (Cherry the Second) by Young Waterloo (2341), g.g.g.d. (Cherry the First) by Foxley, bred by the late Sir R. Price.

1862, Nov. 6, r.—w.f. B	Steer	St. Clement 2201	Mr. Hollings.
1863, Nov. 29, r.—w.f. H	Clement's Daisy	do.	do.

COWS.

DAISY.

Red with white face, calved December 25, 1859.

Bred by and the property of Mr. W. D. Turner, Lynch Court, Leominster; got by The Doctor (1083), dam (Miss Bluefoot) by Cassio (1528), g.d. (Lady Bluefoot) by Wigmore, g.g.d. — bred by Mr. Child, Grange, Wigmore.

PRODUCE IN	NAME.	BY WHAT BULL.	BY WHOM BRED.
1862, Mar. 11, r.—w.f. B	Steer	Logic 2079	Mr. Turner.
1863, Feb. 19, r.—w.f. H	Cassio	The Rover 2821	do.
1864, Jan. 9, r.—w.f. B	Steer	do.	do.
1864, Dec. 13, r.—w.f. H	Sylph	Stockwell 2793	do.

DAISY.

Red with white face.

Bred by Mr. Vaughan, Cholstrey, Leominster, the property of Messrs. T. and W. Vaughan, Lawton, Leominster; got by Emperor (373), dam (Letton) by Charity (375)

1859, Oct. 6, r.—w.f. H	Fairmaid	Plunder 1038	Mr. Vaughan.
1860, Aug. 4, r.—w.f. B	Lawton	Gayboy 1584	do.
1861, Sept. 4, r.—w.f. B	Cholstry 2nd 1919	Sir Oliver 2nd 1733	do.
1862, Aug., r.—w.f. H	Ruby	Sambo 1720	do.
1863, Nov. 23, r.—w.f. B	Steer	Caractacus 2445	Messrs. Vaughan.
1864, Sept. 16, r.—w.f. H	Daisy 2nd	Sir Oliver 2nd 1733	do.

DAISY.

Red with white face, calved in the month of March, 1859.

Bred by Mr. B. Rogers, The Grove, Presteign, the property of Mr. Downes, Brynich, Brecon; got by Severus (1062), dam (Prettymaid the Second) by Young Royal (1470), g.d. (Prettymaid) by Prince (251), g.g.d. (Curly the Fourth) by Charity the Second (1535), g.g.g.d. (Curly the Third) by Portrait (372).

1862, July, r.—w.f. B	Liberal 2596	Bolingbroke 1883	Mr. Rogers.
1863, July, r.—w.f. B	Major 2629	Interest 2046	do.
1864, June 15, r.—w.f. B	Brecon 2430	North Star 2138	do.

COWS.

DAISY THE SECOND.
Red with white face, calved in the month of November, 1861.

Bred by and the property of Mr. B. Rogers, The Grove, Pembridge; got by The Grove (1764), dam (Prettymaid the Second) by Young Royal (1470), g.d. (Prettymaid) by Prince (251), g.g.d. (Curly) by Charity the Second (1535), g.g.g.d. (Curly) by Portrait (372).

PRODUCE IN	NAME.	BY WHAT BULL.	BY WHOM BRED.
1864, Dec. 25, r.—w.f. B	Hercules 2565	Grove 2nd 2556	Mr. Rogers.

DAINTY.
Red with white face, calved in the year 1859.

Bred by Mr. Sheriff, Coxall, Ludlow; the property of Mr. R. H. Capper, The Northgate, Ross; got by Young Sir David (1818), dam — by Brilliant (1518), g.d. — by Young Emperor (1811), g.g.d. — by Confidence (367).

1863, July 1, r.—w.f. B		Jerry 976	Mr. Sheriff.
1864, July 21, r.—w.f. H	Dainty 2nd	Garibaldi 2008	Mr. Capper.

DAINTY THE FOURTH (*alias* DELIGHT).
Red with white face, of 1855, vols. iv. and v., pp. 108, 169.

Bred by the late Mr. Rea, Monaughty, Knighton, the property of Mr. T. Duckham, Baysham Court, Ross; got by Granadier (961), dam (Dainty) by Regent (891), g.d. (Dainty) by Confidence (367), g.g.d. (Dainty) by Old Court (306), g.g.g.d. — by Duke (304), g.g.g.g.d. — by Regulator (360).

1862, July 15, r.—w.f. B	Steer	Garibaldi 2003	Mr. Duckham.
1863, July 5, r.—w.f. H	Dainty 5th	Victory 2296	do.
1864, July 29, r.—w.f. H	Dainty 6th	Cato 1902	do.

Dainty the Fifth, sold to Mr. John Malcolm, Knockalva Ramble, Jamaica.

DAINTY THE FIFTH.
Red with mottled face, calved July 3, 1858.

Bred by and the property of Mr. James Taylor, Stretford Court,

Leominster; got by St. Oswall (1378), dam (Dainty the Second) by King John (830), g.d. (Dainty).

PRODUCE IN	NAME.	BY WHAT BULL.	BY WHOM BRED.
1861, July 13, r.—m.f. H	Dainty 9th	Croft 937	Mr. Taylor.

Dainty the Ninth was one of four winners of the First Prize in their class at the meeting of the Ludlow Agricultural Society 1863; and one of six at Hereford, the same year; she was also a winner of the Third Prize in her class at the Bristol Meeting of the Bath and West of England Society; and Second at the Newcastle Meeting of the Royal Agricultural Society of England.

DAINTY THE SIXTH.

Red with white face, calved August 9, 1859.

Bred by and the property of Mr. James Taylor, Stretford Court, Leominster; got by St. Oswall (1378), dam (Dainty the Second) by King John (830), g.d. (Dainty).

1862, Aug. 7, r.—w.f. H		Croft 937	Mr. Taylor.
1863, Oct. 7, r.—w.f. H	Dainty 10th	Trustful 2845	do.
1864, Aug. 3, r.—w.f. B	Steer	do.	do.

DAMSON.

Red with white face, calved in the year 1856.

Bred by and the property of Mr. Evan Davies, Patton, Wenlock, Salop; got by Patton (1679), dam (Old Daisy) by a Bull bred by the late Mr. Tarte, of The Bache, Culmington, Corvedale.

1860, Sep. 29, r.—w.f. H	Dainty	Franky 1243	Mr. Davies.
1861, Oct. 3, r.—w.f. H	Delicate	Cound 1193	do.
1862, Sep, 20, r.—w.f. H	Darling	Lincoln 2076	do.
1864, Jan. 14, r.—w.f. B	Steer	do.	do.

DAME'S VIOLET THE SECOND.

Red with white face, calved August 13, 1860.

Bred by the late Mr. J. Rea, Monaughty, Knighton, the property of Mr. Warren Evans, Llandowlas, Usk; got by Wellington (1112), dam (Dame's Violet) by Young Conrad (2322), g.d. (Dahlia) by Caractacus (659), g.g.d. (Dahlia) by Old Court (306).

1864, April 1, r.—w.f. H	Jesse	Monaughty 2118	Mr. Evans.

COWS.

DAMSEL.

Red with white face, calved October 5, 1860.

Bred by Mr. John Smith, Sevenhampton, Cheltenham, the property of Mr. John Barton, Coln, Fairford; got by Chieftain (1537), dam (Blossom), bred by Mr. Bennett, North Cerney.

PRODUCE IN	NAME.	BY WHAT BULL.	BY WHOM BRED.
1863, Mar. 10, r.—w.f. B		Cardinal Wiseman 2nd 1898	Mr. Barton. do.

DARKIE.

Red with white face, calved in the year 1858.

Bred by and the property of Mr. Evan Davies, Patton, Wenlock, Salop; got by Tyrant (1784), dam (Old Daisy) by a Bull bred by the late Mr. Tarte, The Bache, Culmington, Corvedale.

1861, July 5, r.—w.f. H	Double X	Cound 1193	Mr. Davies.
1862, Sep. 12, r.—w.f. H	Daylight	Lincoln 2076	do.
1863, Aug. 15, r.—w.f. B	Steer	do.	do.
1864, June 25, r.—w.f. H	Daisycutter	do	do.

DART.

Red with white face, calved in the year 1860.

Bred by the Rev. Archer Clive, Whitfield, Hereford; the property of Lieut-Col. Feilden, Dulas Court, Hereford; got by Wellington (1112), dam (Daylight) by Treasurer (1105), g.d. (Skylark) by Cound (1193), g.g.d. (Frolic) by Cholstrey (217), g.g.g.d. (Lively) by Albert (330).

1863, r.—w.f. H	Deception		Col. Feilden.
1863, Nov. 10, r.—w.f. H	Young Dart	Gift 1254	do.
1864, Nov. 15, r.—w.f. B	Steer	do.	do.

DARLING.

Red with white face, calved in the year, 1855.

Bred by Mr. H. Gibbons, Hampton Bishop, Hereford, the property of Mr. J. R. Paramore, Dinedor Court, Hereford; got by The Admiral (1078), dam (Lily) by Young Gaylad (1463), g.d. (Lily) by Zephyr (1826), g.g.d. (Silk) by a son of Dewshall (358).

PRODUCE IN	NAME.	BY WHAT BULL.	BY WHOM BRED.
1861, Aug. 29, r.—w.f. B	Steer	Shamrock 2nd 2210	Mr. Gibbons.
1862, Sep. r.—w.f. H		do.	do.
1863, Sep. 27, r.—w.f. B		The Priest 1416	Mr. Paramore.
1864, Aug. 20, r.—w.f. B		Grateful 1596	do.

DARLING.

Red with white face, calved September 10, 1860.

Bred by and the property of Mr. Tudge, Adforton, Ludlow; got by Carbonel (1525), dam (Daisy) by The Doctor (1083), g.d. (Dainty) by Orleton (901), g.g.d. (Prettymaid) by Nelson (1021), g.g.g.d. (Prettymaid) by Turpin (300).

1863, Aug. 3, r.—w.f. H	Deborah	Pilot 2156	Mr. Tudge.
1864, Sep. 11, r.—w.f. B	Stanway 2790	do.	do.

DELIGHT.

Red with white face, of 1857, vol. v., p. 170.

Bred by and the property of Mr. Naylor, Leighton Hall, Welshpool; got by Silvester (797), dam (Daisy Queen) by the Knight (185), g.d. (Toby Pigeon) by Big Ben (248), g.g.d. (Duchess) by Tobias (487), g.g.g.d. (Duchess) by Sovereign (404), g.g.g.g.d. — bred by the late Mr. Turner, Noke.

1862, Oct. 3, r.—w.f. B	Major 2630	Salisbury 2204	Mr. Naylor.
1863, Aug. 15, r.—w.f. H	Delight 2nd	do.	do.
1864, Aug. 12, r.—w.f. H	Delight 3rd	Gladstone 2547	do.

DELIGHT.

Red with white face, of 1855, vols. iii. and v., pp. 151, 170.

Bred by and the property of Mr. T. Duckham, Baysham Court, Ross; got by Pope (527), dam (Eywood) by Cotmore the Second 1191), g.d. — bred by the late Earl of Oxford.

PRODUCE IN	NAME.	BY WHAT BULL.	BY WHOM BRED.
1862, Nov. 24, r.—w.f. B	Prince Arthur 2695	Castor 1900	Mr. Duckham.
1863, Nov. 14, r.—w.f. B	Steer	Victory 2296	do.
1864, Nov. 10, r.—w.f. H	Darling	Commodore 2472	do.

Delight was a winner of the third prize in her class at the Worcester Meeting of the Royal Agricultural Society of England, and a third prize at the Cirencester Meeting of the Gloucestershire Agricultural Society, 1863.

DELL.

Red with white face, calved in the month of January, 1862.

Bred by and the property of Mr. J. Prosser, Honeybourne Grounds; got by Lacey (2063), dam (Beauty) by Berrington (435),

1864, Nov. 30, r.—w.f. H	Dell 2nd	The Jew 2266	Mr. Prosser.

DEWBERRY.

Red with white face, calved in the year 1861.

Bred by and the property of Mr. R. Green Price, M.P., Norton Manor, Presteign; got by Stanage (1742)

1864, April 20, r.—w.f. H	Violet	Lord of the Manor 2622	Mr. Price.

DEWDROP.

Red with white face, calved March 7, 1860.

Bred by and the property of Mr. P. Turner, The Leen, Pembridge, Leominster; got by Bertram (1513), dam (Daphne) by Andrew the Second (619), g.d. (Sprite) by Sir Walter (352), g.g.d. (Sylph) by Viscount (816), g.g.g.d. (Sylph) by Old Court the Second (1341).

1862, Sep. 29, r.—w.f. B	Steer	Bolingbroke 1883	Mr. Turner.
1863, Sep. 8, r.—w.f. H	Cornflower	do.	do.

COWS.

DIANA.

Red with white face, calved in the month of October, 1859.

Bred by and the property of Mr. G. T. Forester, High Ercall, Wellington, Salop; got by Discord (1217), dam (Luna) by Darling (1202), g.d. (Hottentot) by Wonder (420), g.g.d (Venus) by Hope (439), g.g.g.d. (Miss Fitzfavourite) by Young Fitzfavourite, g.g.g.g.d. — by Fitzfavourite (441).

PRODUCE IN	NAME.	BY WHAT BULL.	BY WHOM BRED.
1862, Oct. 20, r.—w.f. B		Severn 1382	Mr. Forester.
1863, Aug. 25, r.—w.f. H	Stella	do.	do.
1864, July 6, r.—w.f. H	Sagitta	do.	do.

DIANA THE SECOND.

Red with white face, calved August 13, 1859.

Bred by Mr. Rea, Monaughty, Knighton, the property of Mr. Farr, Pontrilas, Hereford; got by Wellington (1112), dam (Diana) by Grenadier (961), g.d. (Spot the Second) by Cholstrey (217) g.g.d. (Spot) by Hope (439), g.g.g.d. (Spot) by Primate (204), g.g.g.g.d. — by Forester (112).

1864, May 18, r.—w.f. H	Diana 3rd	Sir Richard 1734	Mr. Farr.

Diana the Second was one of two winners of the first prize in their class at the Worcester Meeting of the Royal Agricultural Society of England. She was also a winner of the Second Prize at the Hereford Meeting of the Bath and West of England Society.

DINA.

Red with white face, calved April 10, 1859.

Bred by and the property of the Duke of Bedford, Woburn Park, Beds., got by Graftonbury (1259), dam (Damsel the Fifth) by Bruno (861), g.d. (Damsel the Fourth) by Bonaparte (1152), g.g.d. (Damsel the Third) by Napoleon (1334), g.g.g.d.

(Damsel the Second) by Trusty (15), g.g.g.g d. (Damsel) by Lupin.

PRODUCE IN	NAME.	BY WHAT BULL.	BY WHOM BRED.
1862, June 21, r.—w.f. B	Steer	Carbonel 1525	Duke of Bedford.
1863, July 12, r.—w.f. H	Dorothy	Victory 2298	do.
1864, July 15, r.—w.f. H	Dorothy 2nd	Victory 2296	do.

DOROTHY.

Red with white face, calved in the month of September, 1859.

Bred by and the property of Mr. T. S. Bradstock, Cobrey Park, Ross; got by Daniel (1201), dam (Darrel) by Deluge (1210), g.d. (Sister) by Benbow, g.g.d. (Prettymaid) by Sovereign the Third, g.g.g.d. (Eaton) by Lottery the Second (408).

1862, Nov. 30, r.—w.f. B		Young Rambler 2355	Mr. Bradstock.
1863, Dec. 20, r.—w.f. B		do.	do.

DOVE.

Red with white face, calved August 9, 1859.

Bred by the late Lord Berwick, Cronkhill, Salop, the property of the Hon. and Rev. Noel Hill, Berrington, Salop; got by Attingham (911), dam (Grey Dove) by Wonder (420), g.d. (Dove) by Ashley Moor (791).

1862, Oct. 14, r.—w.f. B	Steer	Conqueror 1929	Hon. and Rev. Noel Hill
1863, Oct. 7, r.—w.f. H	Ringdove	Van Tromp 2291	do.

DOVE.

Red with mottled face, calved August 30, 1860.

Bred by and the property of Capt. Peploe, Garnstone, Weobley; got by The Twin (1420), dam (Myrtle) by Musician (725), g.d. (Myrtle) by Victory (2297), g.g d. (Murphy) by Murphy Delany (36), g.g.g.d. (Blossom) by Young Waterloo (2340), g.g.g.g.d. (Blossom) by Semplon (58).

1863, July 24, r.—m.f. H		Leo 2070	Capt. Peploe.
1864, Oct. 26, r.—m.f. B		do.	do.

COWS.

DUCHESS.

Red with white face, calved in the month of February, 1860.

Bred by the late Mr. John Davies, Wormbridge, Hereford, the property of Mr. Thomas Wheeler, Wormhill, Eaton Bishop, Hereford; got by Muley (1330), dam (Prettymaid) by Young Royal (1136), g.d. (Prettymaid) by Reform (508).

PRODUCE IN	NAME.	BY WHAT BULL.	BY WHOM BRED.
1862, Aug. 15, r.—w.f. B	Charity 2454	Troubadour 1780	Mr. Wheeler.
1863, Sep. 11, r.—w.f. H	Duchess 2nd	Mentor 2112	do.
1864, Aug. 26, r.—w.f. H	Duchess 3rd	Washington 2868	do.

DUCHESS.

Red with white face, calved March 24, 1861.

Bred by and the property of Mr. J. R. Paramore, Dinedor Court, Hereford; got by General (1251), dam (Peggy) by a Bull bred by the late Mr. J. Turner, Noke Court, g.d. — bred by the late Mr. Davies, Tarrington, Hereford.

1863, Sep. 14, r.—w.f. B	Steer	The Jew 2266	Mr. Paramore.
1864, Aug. 22, r.—w.f. H	Duchess 2nd	Portly 2165	do.

DUCHESS.

Red with white face, calved January 5, 1856.

Bred by and the property of Mr. John Richards, Cound, Shrewsbury; got by the Knight (185), dam — bred by Mr. Rammell.

1858, Jan. 5, r.—w.f. H	Young Duchess	Baronet 1860	Mr. Richards.
1860, May 4, r.—w.f. H	Cherry	do.	do.
1862, May 12, r.—w.f. B	Steer	do.	do.
1863, June 12, r.—w.f. B	Duke 2506	Pirate 2158	do.

Duchess was a winner of the First Prize in her class at the meetings of the Bridgenorth Agricultural Society, 1860 and 1861.

DUCHESS.

Red with white face.

Bred by Mr. Price, Court House, Pembridge, Leominster; the property of Mr. J. O. G. Pollock, Mountainstown, Navan, Ireland; got by Salisbury, (2204, dam (Docksey) by Pembridge (721), g.d. (Victoria) by Sir David (349), g.g.d. (Curly) by (Prince Dangerous (362), g.g.g.d. (Princess) by the Sheriff (356), g.g.g.g.d. (Lofty) by Forester (398).

PRODUCE IN	NAME.	BY WHAT BULL.	BY WHOM BRED.
1863, July 25, r.—w.f. B		Earl Derby 2nd 2510	Mr. Price.

DUCHESS THE SECOND.

Red with white face, calved May 30, 1860.

Bred by and the property of Mr. P. R. Jackson, Blackbrook, Skenfrith, Monmouth; got by Young Sir David (1818), dam (Duchess) by Pope (527), g.d. (Duchess) by Darling (1202), g.g.d. (Miss Gay) by Gaylad (400), g.g.g.d. — by Berrington (435).

1863, Oct. 25, r.—w.f. H	Dainty	Carlisle 923	Mr. Jackson.

DUCHESS THE THIRD.

Red with white face, calved November 3, 1857.

Bred by and the property of Mr. W. Burchall Peren, Compton, South Petherton; got by Farebrother (949), dam (Duchess the Second) by Quicksilver (353) g.d. (Duchess) by Waverly (106), g.g.d. (Venus) by a bull of Mr. Child's, Wigmore Grange, g.g.g.d. (Spot) by Young Sovereign (379).

1860, Nov. 20, r.—w.f. H	Duchess 4th	Zimmerman 1830	Mr. Peren.
1862, Mar. 19, r.—w.f. B	Dead	Zero 1827	do.
1863, Feb. 4. r.—w.f. H	Duchess 5th	do.	do.
1864, Mar. 7, r.—w.f. H	Duchess 6th	do.	do.

COWS.

DUCHESS THE FOURTH.
Red with white face, calved November 20, 1860.

Bred by and the property of Mr. W. B. Peren, Compton, South Petherton; got by Zimmerman (1830), dam (Duchess the Third) by Farebrother (949), g.d. (Duchess the Second) by Quicksilver (353), g.g.d. (Duchess) by Waverley (106), g.g.g.d. (Venus) by a bull of Mr. Child's, Wigmore Grange, g.g.g.g.d. (Spot) by Young Sovereign (379).

PRODUCE IN	NAME.	BY WHAT BULL.	BY WHOM BRED.
1864, April 5, r.—w.f. B	Zero 2nd 2903	Zero 1827	Mr. Peren.

DUCHESS OF BEDFORD.
Red with white face of vol. v., p. 175.

Bred by and the property of Mr. T. Roberts, Ivington Bury, Leominster; got by Arthur Napoleon (910), dam — bred by Mr. Vaughan, Cholstrey.

1862, Sep. 27, r.—w.f. B	Duke of Bedford 2507	Sir Thomas 2228	Mr. Roberts.
1863, Aug. 25, r.—w.f. H	Duchess of Bedford 3rd	do.	do.
1864, July 22, r.—w.f. H	Duchess of Bedford 4th	do.	do.

DUCHESS OF BEDFORD THE SECOND.
Red with white face, calved September 24, 1861.

Bred by Mr. T. Roberts, Ivington Bury, Leominster, the property of Mr. John Baldwin, Luddington, Stratford-on-Avon; got by Sir Thomas (2228), dam (Duchess of Bedford) by Arthur Napoleon (910), g.d. — bred by Mr. Vaughan, Cholstrey.

1864, Sep. 10, r.—w.f. H	Rose of Bedford	Battersea 1865	Mr. Baldwin.

Duchess of Bedford the Second was a winner of the Second Prize in her class at the Worcester meeting of the Royal Agricultural Society of England; First at their Newcastle and Plymouth Meetings; Second at the Bristol and Hereford Meetings of the Bath and West of England Society; First at the Tewkesbury Meeting of the Gloucestershire Agricultural Society; and Second as one of a pair at the meeting of the Herefordshire Agricultural Society, 1863; besides other local prizes.

COWS.

DUCHESS OF HOLM.

Red with white face, calved in the year 1855.

Bred by and the property of Mr. J. Prosser, Honeybourne Grounds, Worcester; got by Fairfax (950).

PRODUCE IN		NAME.	BY WHAT BULL.	BY WHOM BRED.
1859, July,	r.—w.f. H	Duchess of Holm 2nd	Medallist 1009	Mr. J. Prosser.
1860, July,	r.—w.f. B	Steer		do.
1861, July,	r.—w.f. B	Steer		do.
1862, Aug.,	r.—w.f. H	Gipsy Queen	The Jew 2266	do.
1863, Sept.,	r.—w.f. H	Spot	do.	do.
1864, Aug. 10,	r.—w.f. H	Briony	do.	do.

DUCHESS OF HOLM THE SECOND.

Red with white face, calved in the month of July, 1859.

Bred by and the property of Mr. J. Prosser, Honeybourne Grounds, Worcester; got by Medallist (1009), dam (Duchess of Holm) by Fairfax (950).

1863, Feb.,	r.—w.f. H	Laura	The Jew 2266	Mr. J. Prosser.
1864, Jan.,	r.—w.f. H	Lily of the Vale	do.	do.
1864, Dec.,	r.—w.f. H	Primrose	do.	do.

DUCHESS OF YORK.

Red with white face, calved October 24, 1858.

Bred by and the property of Mr. R. H. Capper, The Northgate, Ross; got by Balaclava (1505), dam (Duchess) by Regent (891), g.d. (Duchess) by Barrister (658), g.g.d. (Duchess) by Gallant (239), g.g.g.d. (Old Duchess) by Oldcourt (306), g.g.g.g.d. — by Regulator (360), — by Crabstock (303).

1863, Nov. 10,	r.—w.f. B	Sir Benjamin 2nd 2759	Sir Benjamin 1387	Mr. Capper.

COWS.

ELEGANCE.

Red with white face, calved July 4, 1861.

Bred by and the property of Mr. T. Morris, Therrow, Llyswen, Hay; got by Druid (1220), dam (Lucy) by Newton (344), g.d. (Lucy) by Young Hope (343), g.g.d. (Lady) by Prince Llewellyn (713), g.g.g.d. (Lily) by Charity the Second (516).

PRODUCE IN			NAME.	BY WHAT BULL.	BY WHOM BRED.
1683, Sep.,	r.—w.f.	H	Zenobia	Don Salisbury 1969	Mr. Morris.
1864, July,	r.—w.f.	H	Beauty	Prince Imperial 2171	do.

ELINOR.

Red with white face, of 1852, vol. iv., p. 112.

Bred by and the property of Mr. Naylor, Leighton Hall, Montgomeryshire; got by The Knight (185), dam (Lady Elinor) by Big Ben (248), g.d. (Butterfly) by Prince (251), g.g.d. (Nell) by a son of Sir Andrew (183).

1859, July 20,	r.—w.f.	B	Duplicate 1976	Admiral 1481	Mr. Naylor.
1860, July 14,	r.—w.f.	B	Steer	do.	do.
1861, Aug. 6,	r.—w.f.	B	Nobleman 2134	do.	do.
1862, July 20,	r.—w.f.	B	Dead	Salisbury 2204	do.
1863, July 26,	r.—w.f.	B	Steer	Blondin 1880	do.

ELLEN.

Red with white face, of 1853, vol. v., p. 176.

Bred by Mr. Price, Pembridge, Leominster, the property of Mr. John Burlton, Luntley Court, Pembridge; got by Pembridge (721), dam — by Prince Dangerous (362), g.d. — by The Sheriff (356), g.g.d. — by Crabstock (303).

1862, Oct.,	r.—w.f.	H	Ellen 4th	Rifleman 2189	Mr. Burlton.
1863, Nov. 15,	r.—w.f.	B	Steer	do.	do.

COWS.

EMELINE.
Red with white face, calved February 2, 1858.

Bred by and the property of Mr. Naylor, Leighton Hall, Montgomeryshire; got by Tom of Lincoln (1099), dam (Ellen) by The Knight (185), g.d. (Lady Elinor) by Big Ben (248), g.g.d. (Butterfly) by Prince (251), g.g.g.d. (Nell) by a son of Sir Andrew (183).

PRODUCE IN	NAME.	BY WHAT BULL.	BY WHOM BRED.
1862, July 25, r.—w.f. B	The Young Knight 2823	Volunteer 2299	Mr. Naylor.

EMPRESS THE SECOND.
Red with white face, calved September 9, 1859.

Bred by and the property of Mr W. Tudge, Adforton, Ludlow; got by Cousin John (1195), dam (Empress) by The Doctor (1083), g.d. (Lady) by Orleton (901), g.g.d. (Stout) by Nelson (1021), g.g.g d. (Prettymaid) by Turpin (300).

1862, Dec. 15, r.—w.f. H	Eugenie	Sir Colin 2216	Mr. Tudge.
1863, Oct. 10, r.—w.f. H	Eva	Pilot 2156	do.
1864, Aug. 30, r.—w.f. H	Ella	do.	do.

EMPRESS THE FIFTH.
Red with white face, calved in the year 1858.

Bred by Mr. Stedman, Bedstone Hall, Salop, the property of Mr. John Barton, Coln, Fairford; got by Kinlet (1293), dam (Empress the Second) by Bedstone (2411), g.d. (Empress) by Dinedor (395), g.g.d. — by Trojan (542), g.g.g.d. — by Hector (535), g.g.g.g.d. — by a son of Waterloo (49).

1861, July 12, r.—w.f. B	Warrior	Little John	Mr. Stedman.
1862, Aug. 1, r.—w.f. H	Miss Stedman	Young Perfection	do.
1863, Aug. 7, r.—w.f. B	Steer	Young Cardinal Wiseman 2882	Mr. Barton.
1864, Aug. 2, r.—w.f. B	Steer	Sevenhampton 2748	do.

COWS.

EMPRESS THE FIFTH.

Red with white face, calved September 2, 1860.

Bred by and the property of Mr. James Taylor, Stretford Court, Leominster; got by Croft (937), dam (Empress) bred by the late Mr. James Bowen, Monkland.

PRODUCE IN	NAME.	BY WHAT BULL.	BY WHOM BRED.
1864, April 28, r.—w.f. B		Trustful 2845	Mr. Taylor.

Empress the Fifth was one of four winners of the First Prize in their class at the meeting of the Ludlow Agricultural Society, 1862; and as one of six at Hereford the same year; also as one of four at Leominster, and one of seven at Hereford, 1864.

EMPRESS THE EIGHTH.

Red with white face, calved September 8, 1859.

Bred by and the property of Mr. James Taylor, Stretford Court, Leominster; got by St. Oswall (1378), dam (Empress the Second) by Orleton (2144), g.d. (Empress) bred by the late Mr. James Bowen, Monkland.

1862, Aug. 16, r.—w.f. H		Croft 937	Mr. Taylor.
1863, Aug. 3, r.—w.f. B	Hansa 2559	Trustful 2845	do.
1864, Aug. 6, r.—w.f. H	Empress 10th	do.	do.

EMPRESS OF TEME.

Red with white face, calved in the month of August, 1860.

Bred by Mr Stedman, Bedstone Hall, Salop, the property of Mr. John Barton, Coln, Fairford; got by Young Perfection Dam (Empress the Second) by Bedstone (2411), g.d. (Empress) by Dinedor (395), g.g.d. — by Trojan (542), g.g.g.d. — by Hector (535), g.g.g.g.d. — by a son of Waterloo (49).

1862, Mar. 20, r.—w.f. B	Steer	Kinlet 1293	Mr. Barton.
1864, April 7, r.—w.f. H	Gaylass	Young Cardinal Wiseman 2882	do.

COWS.

ENGLAND'S BEAUTY.
Red with white face, calved September 1, 1858.

Bred by Mr. J. Rea, Monaughty, Knighton, the property of Mr. R. H. Ridler, Gattertop, Leominster; got by Pilot (1037), dam (Prudence) by Nelson (1021), g.d. (Lady) by Brampton (917), g.g.d. (Lady) by Monarch (219), g.g.g.d. — by Regulator (360), g.g.g.g.d. — by Crabstock (303).

PRODUCE IN	NAME.	BY WHAT BULL.	BY WHOM BRED.
1861, Jan. 1, r.—w.f. H	England's Beauty 2nd	Zealous 2348	Mr. Rea.
1864, Mar. 20, r.—w.f. H	Diana	Sir Richard 1734	Mr. Ridler.

EVA.
Red with white face, calved July 10, 1857.

Bred by and the property of Mr. T. S. Bradstock, Cobrey Park, Ross; got by Uncle Tom (1108), dam (Poston) by Foxwhelp (2522), g.d. (Daisy) by a son of Sovereign (404), bred by Mr. John Turner, Noke.

1860, July 29, r.—w.f. B	Steer	Daniel 1201	Mr. Bradstock.
1861, June 30, r.—w.f. B	Banter 2395	Melon 2111	do.
1862, July 20, r.—w.f. B	Discovery 2499	Geologist 2012	do.
1863, July 24, r.—w.f. B		Young Rambler 2335	do.
1864, July 21, r.—w.f. B	Buckman 2435	do.	do.

EVA.
Red with white face, calved October 5, 1861.

Bred by Mr. James Bourn, Mawley Town Farm, Cleobury Mortimer, the property of Mr. W. B. Peren, Compton, South Petherton; got by Cardinal (1526), dam (Eugenie) by Napoleon the Third (1019), g.d. (Fanny) by Walford (871), g.g.d. — bred by Mr. Bailey.

1864, July 22, r.—w.f. H	Eveline	Captain 2443	Mr. Bourn.

Eveline, sold to Mr. V. Gosport, Tanylan, Holywell.

EVA.

Red with white face, of 1852, vols. iii., iv., and v., pp. 156, 115, 179.

Bred by the late Mr. James Rea, Monaughty, Knighton, the property of Mrs. Rea, Westonbury, Leominster; got by Regent (891), dam (Spot) by Caractacus (659), g.d. (Spot) by Hope (439).

PRODUCE IN	NAME.	BY WHAT BULL.	BY WHOM BRED.
1863, July 28, r.—w.f. B	Sir Cornewall	Sir Benjamin 1387	Mr. Rea.

EVA.

Red with white face, calved March 17, 1860.

Bred by H.R.H. the Prince Consort, Windsor Castle, the property of Major-General the Hon. A. N. Hood, Cumberland Lodge, Windsor; got by Brecon (918), dam (Virginia) by Vanguard (1109), g.d. (Huntington) by Phantom (1035).

1862, June 14, r.—w.f. H	Florence	Windsor 1456	Hon. A. N. Hood.

FAIRLASS.

Red with white face, calved November 12, 1862.

Bred by and the property of Mr. T. Olver, Penhallow, Grampound, Cornwall; got by Earl Derby (1979), dam (Fairmaid) by Woodbine (1120), g.d. (Beauty) by Rory O'More (1711).

1864, Sep. 3, r.—w.f. B		Zippor 2354	Mr. Olver.

FAIRMAID.

Red with white face, calved in the year 1858.

Bred by Messrs. F. and C. Bodenham, Hereford, the property of Mr. J. Rogers, Letchmoor, Presteign; got by Young Dewshall (1125), dam — bred by the late Sir Robert Price, Foxley, Hereford.

1861, June, r.—w.f. B	Steer	Curly 1561	Mr. Rogers.
1862, July, r.—w.f. B	do.	Matchless 2110	do.
1863, July, r.—w.f. {B H}	Twins	do.	do.
1864, June, r.—w.f. {H H}	Fairmaid 2nd Beauty	do.	do.

COWS.

FAIRMAID.
Red with white face, calved October 6, 1859.

Bred by Mr. Vaughan Cholstrey, the property of Messrs. T. and W. Vaughan, Lawton, Leominster; got by Plunder (1038), dam (Daisy) by Emperor (373), g.d. (Letton) by Charity (375).

PRODUCE IN	NAME.	BY WHAT BULL.	BY WHOM BRED.
1862, July 20, r.—w.f. B	Lord Clifden 2613	Lord Wellington 2094	Messrs. Vaughan.
1863, Aug. 28, r.—w.f. H	Alexandra	Valentine 2288	do.
1864, Aug. 16, r.—w.f. H	Miss Grove	Baron Grove 2402	do.

FAIRMAID THE SECOND.
Red with white face, calved November 14, 1860.

Bred by and the property of Mr. Thomas Powell, The Bage, Madley, Hereford; got by Greengage (1266), dam (Fairmaid) by Royal (331), g.d. (Old Fairmaid).

1863, Sep. 14, r.—w.f. B	Pilot 2678	Portly 2165	Mr. Powell.
1864, Aug. 1, r.—w.f. B	Prince Royal 2698	do.	do.

FAIRMAID THE SECOND.
Red with white face, of 1857, vol. v., p. 182.

Bred by Mr. B. Rogers, The Grove, Pembridge, the property of Mr. Wigmore, Bickerton Court, Dymock; got by Mowley by Madley (1301), dam (Fairmaid) by Gaylad the Second (1589), g.d. (Fairmaid) by Prince (251),

1862, July, r.—w.f. H	Fairmaid 3rd	Bolingbroke 1883	Mr. Rogers.
1863, Aug., r.—w.f. H	Fairmaid 4th	Interest 2046	do.
1864, June 15, r.—w.f. B	Hereford 2567	Bolingbroke 1883	do.

FAIRMAID THE FIFTH.
Red with white face, calved July 10, 1860.

Bred by the late Mr. James Rea, Monaughty, Knighton, the property of Mr. James Gregg, Fencote Abbey, Leominster; got by Wellington (1112), dam (Fairmaid the Fourth) by Chieftain

COWS.

(930), g.d. (Fairmaid the Third) by Cholstrey (217), g.g.d. (Fairmaid the Second) by Gallant (239), g.g.g.d. (Fairmaid the First) by Portrait (372).

PRODUCE IN	NAME.	BY WHAT BULL.	BY WHOM BRED.
1863, May 20, r.—w.f. H	Mountain Maid	Fencote 1989	Mr. Gregg.

FAIRY.

Red with white face, calved in the month of January, 1859.

Bred by and the property of Mr. T. S. Bradstock, Cobrey Park, Ross; got by Daniel (1201), dam (Freckle) by Deluge (1210), g.d. (Blossom) by Young Royal (1468).

PRODUCE IN	NAME.	BY WHAT BULL.	BY WHOM BRED.
1860, Dec. 29, r.—w.f. H	Maria	The Jew 2266	Mr. Bradstock.
1861, Sep. 9, r.—w.f. B	Steer	Young Rambler 2335	do.
1862, Sep. 11, r.—w.f. H	Florence	do.	do.
1863, Dec. 20, r.—w.f. B		do.	do.
1864, Oct. 29, r.—w.f. B	Barter 2404	do.	do.

FAIRY THE THIRD.

Red with white face, calved August 22, 1861.

Bred by the late Mr. T. Rea, Westonbury, Leominster, the property of Mr. J. D. Allen, Pyt House, Tisbury, Wilts.; got by England's Glory (1983), dam (Fairy) by Grateful (1260), g.d. (Spot) by Gratitude (1261). g.g.d. (Spot) by Conrad (1183).

PRODUCE IN	NAME.	BY WHAT BULL.	BY WHOM BRED.
1864, Mar. 31, r.—w.f. B		Iris 2047	Mr. Allen.

FANCIFUL.

Red with white face, of 1853, vol. v., p. 185.

Bred by the late Viscount Hereford, Tregoyd, Hay, the property of Major General the Hon. A. N. Hood, Cumberland Lodge, Windsor; got by Phantom (1035), dam (Fancy the First).

PRODUCE IN	NAME.	BY WHAT BULL.	BY WHOM BRED.
1863, Sep. 26, r.—w.f. H	Constance	Ajax 1843	Hon. A. N. Hood

FANCY.
Red with white face, calved April 9, 1860.

Bred by Mr. William Perry, Cholstrey, Leominster, the property of Mr. J. H. Arkwright, Hampton Court, Leominster; got by Salisbury (2204), dam (Prettymaid) by Noble Boy (751), g.d. (Change) by Monkland (552), g.g.d. — by Goldfinder (383).

PRODUCE IN	NAME.	BY WHAT BULL.	BY WHOM BRED.
1864, Jan 20, r.—w.f. B		Witchend 2nd 2315	Mr. Perry.

FANCY.
Red with white face, calved in the month of January, 1859.

Bred by and the property of Mr. T. S. Bradstock, Cobrey Park, Ross; got by Daniel (1201), dam (Freckle) by Deluge (1210), g.d. (Blossom) by Young Royal (1468).

PRODUCE IN	NAME.	BY WHAT BULL.	BY WHOM BRED.
1862, Jan. 30, r.—w.f. B	Belamite 2412	Young Rambler 2335	Mr. Bradstock.
1863, Jan. 27, r.—w.f. H	Froudy	do.	do.
1864, Jan. 3, r.—w.f. H	Fanny	do.	6o.

FANCY.
Red with white face, calved in the month of January, 1860

Bred by Mr. Thomas, The Lodge, Salop, the property of Mr. J. Farr, Pontrilas, Hereford; got by Westonbury (1452).

PRODUCE IN	NAME.	BY WHAT BULL.	BY WHOM BRED.
1863, Nov. 23, r.—w.f. H	Fancy Fair	Sir Benjamin 1387	Mr. Farr.
1864, Sep. 25, r.—w.f. H	Clara	Salford 2738	do.

FANCY LEOMINSTER.
Red with white face, of 1853, vol. v., p. 186.

Bred by Mr. John Taylor, Stretford Court, the property of Mr. James Taylor, Stretford Court, Leominster; got by King John (830), dam (Countess) bred by the late Mr. James Bowen, Monkland.

PRODUCE IN	NAME.	BY WHAT BULL.	BY WHOM BRED.
1862, Aug. 29, r.—w.f. B	Steer	Unity 2287	Mr. Taylor.
1863, Nov. 24, r.—w.f. B	Reckless 2713	Trustful 2845	do.
1864, Nov. 20, r.—w.f. H	Countess 7th	do.	do.

Fancy Leominster was one of seven winners of the First Prize in their class at the meetings of the Leominster and the Herefordshire Agricultural Societies, 1862; as also Prizes in the Extra Stock classes at the meetings of those Societies, 1864.

FANNY THE SECOND.
Red with white face, calved July 9, 1860.

Bred by and the property of Mr. H. R. Evans, Jun., Swanstone Court, Leominster; got by Sir Franklin (1068), dam (Fanny) by Rambler (1046), g.d. (Venus) by Emperor (373), g.g.d (Lovely) by Young Trueboy (1475), g.g.g.d. (Lovely) by Ashley Moor White Bull, (870), g.g.g.g.d. (Old Damsel) by Coleman's Bull (1547), — (Old Daisy) by Chancellor (156).

PRODUCE IN	NAME.	BY WHAT BULL.	BY WHOM BRED.
1863, May 8, r.—w.f. H	Fanny 3rd	Chatham 1914	Mr. Evans.
1864, May 11, r.—w.f. H	Fanny 4th	do.	do.

Fanny the Third was one of four winners of the First Prize in their class at the meeting of the Leominster Agricultural Society, 1863.

FATTY.
Red with white face, calved November 29, 1859.

Bred by the late Lord Berwick, Cronkhill, Salop; the property of the Hon. and Rev. Noel Hill, Berrington, Salop; got by Attingham (911), dam (Fat Rumps) by Petchfield (649), g.d. — by Dawe's Grey Bull (1954A).

1862, Oct. 18, r.—w.f. B	Fat Boy	Van Tromp 2291	Hon. and Rev. N. Hill.
1863, Oct. 19, r.—w.f. H	Fat Lady	do.	do.
1864, Dec. 23, r.—w.f. H	Fat Girl	Conqueror 1929	do.

FAUSTINA.
Red with white face, calved in the year 1854.

Bred by and the property of Mr. W. H. Oatley, Wroxeter, Salop; got by Uriconium (598), dam (Fausta) by Bryony (599), g.d. (Princess Royal) by Emigrant (1980), g.g.d. (Sovereign Cow) by Sovereign (404).

1857,		r.—w.f. B	Steer	Julius Cæsar	Mr. Oatley.
1858,		r.—w.f. B	do.	do.	do.
1859,	Dec.,	r.—w.f. B	Tiberius Cæsar 2272	do.	do.
1860,		r.—w.f. B	Nameless 2649	Augustus Cæsar 1854	do.
1861,		r.—w.f. B	Steer	do.	do.
1862,		r.—w.f. B	do.	do.	do.

FENNY.

Red with white face, calved February 2, 1860.

Bred by the late Mr. Rea, Monaughty, Knighton, the property of the Duke of Bedford, Woburn Park, Beds.; got by Wellington (1112), dam (Faithful) by Cambria (887), g.d. (Lively) by Gallant (239), g.g.d. (Lively) by Old Court (306), g.g.g.d. — by Regulator (360), g.g.g.g.d. — by Noble (238).

PRODUCE IN		NAME.	BY WHAT BULL.	BY WHOM BRED.
1862, July 24, r.—w.f.	B	Steer	Carbonal 1525	Duke of Bedford.
1863, Oct. 31, r.—w.f.	H	Fauchette	do.	do.
1864, Oct. 6, r.—w.f.	B	Steer	Victory 2296	do.

FLAX.

Red with white face, calved in the month of December, 1861.

Bred by Mr. Rogers, Hereford, the property of Mr. Gilliland, Brook Hall, Londonderry, Ireland.

1864, May 21, r.—w.f.	H	Flax 2nd	Jolly Miller 10th 2584	Mr. Gilliland.

FLIRT.

Red with white face, calved July 8, 1860.

Bred by the late Mr. J. Rea, Monaughty, Knighton, the property of Mr. John Wigmore, Bickerton Court, Dymock; got by Sir Benjamin (1387), dam (Kate) by Chieftain (930), g.d. (Venus) by Albert (330), g.g.d. (Winifred) by Monaughty (220), g.g.g.d. (Venus the Fourth) by Duke (304), g.g.g.g.d. (Venus the Third) by Regulator (360).

1863, May 8, r.—w.f.	B	Steer	Melon 2111	Mr. Wigmore.
1864, July 20, r.—w.f.	B		Speculator 2240	do.

Flirt was a winner of the First Prize in her class at the meeting of the Herefordshire Agricultural Society, 1862,; and First at the Gloucester and Cirencester meetings of the Gloucestershire Society, 1862 and 1863; and as one of a pair at Ross and Monmouth, 1863; she was also one of eight, winners of a Second Prize at Hereford, 1863.

cows.

FLORA.

Red with white face, calved in the year 1857.

Bred by and the property of the Rev. Archer Clive, Whitfield, Hereford; got by Grateful (1596), dam (Fat Rumps) by Robin Hood (664).

PRODUCE IN	NAME.	BY WHAT BULL.	BY WHOM BRED.
1863, June 18, r.—w.f. H	Florence	Bertram 1513	Rev. A. Clive.
1864, July 3, r.—w.f. H	Faithful	Original 2660	do.

FLORA.

Red with white face, calved August 2, 1860.

Bred by and the property of Mr. T. Olver, Penhallow, Grampound, Cornwall; got by Earl Derby (1979) dam (Fairy) by Duke of Cornwall (1569).

1863, Jan. 1, r.—w.f. H		Zippor 2354	Mr. T. Olver.
1864, Nov. 22, r.—w.f. H	Fairy	do.	do.

FLORA.

Red with white face, calved January 19, 1860.

Bred by and the property of Mr. J. R. Paramore, Dinedor Court, Hereford; got by The General (2817), dam (Countess) by Noke, g.d. (Countess) bred by the late Mr. J. Bowen, Monkland.

1862, Aug. 27, r.—w.f. H		Shamrock 2nd 2210	Mr. Paramore.
1863, July 28, r.—w.f. H	Rosebud	do.	do.
1864, Aug. 15, r.—w.f. B	Steer	Grateful 1596	do.

FLORA.

Red with white face, calved July 6, 1860.

Bred by and the property of Mr. Thomas Edwards, Wintercott, Leominster; got by Sir Newton (1731), dam (Clara) by Croft (937), g d. (Prettymaid) by Coningsby the Second (1552), g.g.d. (Beauty) by Big Ben (248).

1862, Oct. 16, r.—w.f. B	Young Hero 2889	Hero 2040	Mr. Edwards.
1863, Oct. 11, r.—w.f. H	Florist	Adforton 1839	do.
1864, Sep. 4, r.—w.f. B	Steer	do.	do.

cows.

FORGET-ME-NOT.

Red with white face, calved in the year 1861.

Bred by Mr. Sheriff, Coxall, Ludlow, the property of Mr. R. H. Capper, The Northgate, Ross; got by Sir Colin (2216), dam (Comely) by Coxall (1196).

PRODUCE IN	NAME.	BY WHAT BULL.	BY WHOM BRED.
1864, Nov. 4, r.—w.f. B		Sir Benjamin 2nd 2757	Mr. Capper.

FORGET-ME-NOT.

Red with white face, calved August 21, 1858.

Bred by the late Mr. James Rea, Monaughty, Knighton, the property of Mr. P. J. Kearney, Miltown House, Clonmellon, Ireland; got by Pilot (1037) dam (Lily of the Valley) by Chieftain (930), g.d. (Lily) by Confidence (367), g.g.d. — by Old Court (306).

1864, Feb. 3, r.—w.f. B	Steer	Sir Benjamin 1387	Mr. Kearney.
1864, Dec. 31, r.—w.f. B		Silverstream 2214	do.

FRIDAY THE SEVENTH.

Red with white face, calved October 26, 1860.

Bred by and the property of Mr. J. M. Read, Elkstone, Cheltenham; got by Sebastopol (1381), dam (Friday the Fourth) by Albert Edward (859), g.d. (Friday the Third) by Walford (871), g.g.d. (Friday the Second) by Wonder (420), g.g.g.d. (Friday) by Commerce (354), g.g.g.g.d. (Prettymaid) by The Sheriff (356), — by Sovereign (404).

1863, Dec. 7, r.—w.f. B		Colesborne 2466	Mr. Read.

Friday the Seventh was a winner of a Prize at the Leamington Meeting of the Warwickshire Agricultural Society.

COWS.

FURY.
Red with white face, calved September 25, 1859.

Bred by Lord Berwick, Cronkhill, Salop, the property of Mr. Naylor, Leighton Hall, Montgomeryshire; got by Sir David (349), dam (Finchback) by Walford (871), g.d. (Young Damsel) by Tom Thumb (243), g.g.d. (Damsel) by Young Trueboy (1475), g.g.g.d. (Prettymaid) by Cholstrey (868), g.g.g.g.d (Old Damsel) by Coleman's bull (1547), — (Old Daisy) by Chancellor (156), — (Cherry the Second) by Thickset (1769).

PRODUCE IN	NAME.	BY WHAT BULL.	BY WHOM BRED.
1862, July 25, r.—w.f. B		Volunteer 2299	Mr. Naylor.
1863, June 25, r.—w.f. H	Fury 2nd	Blondin 1880	do.
1864, July 7, r.—w.f. B	Steer	Salisbury 2204	do.

GAGER THE SECOND. (A TWIN.)
Red with white face.

Bred by and the property of Mr. John Jones, Llwyn-y-Gaer, Raglan; got by Dolphin (2500), dam (Gager) by a Stockton Bull.

1861,	r.—w.f. H	Gager 3rd	Chancellor 1172	Mr. Jones.
1862,	r.—w.f. H	Gager 4th	do.	do.
1863, Dec. 1,	r.—w.f. B	Bold David 2nd 2428	Bold David 1881	do.
1864,	r.—w.f. H	Gager 5th	Vice Chancellor 2292	do.

Gager was one of a pair winners of a Prize at the meeting of the Monmouthshire Agricultural Society, 1862.

GAIETY.
Red with white face, calved in the month of November 1859.

Bred by Mr. Stedman, Bedstone Hall Salop; the property of Mr. R. S. Fetherstonhaugh, Rockview, Killucan, Ireland; got by Grateful (1260), dam (Gaylass) by Young Emperor (1811), g.d. — by Venison (1441), — by Lottery the Second (408), — by Rector (535), — by a son of Waterloo (49).

1862,		Steer	Kinlet 1293	Mr. Stedman.
1863,	r.—w.f. H	Gaiety 2nd	Silverstream 2214	Mr. Fetherstonhaugh.

GAIETY THE SECOND.
Red with white face, calved October 23, 1858.

Bred by Mr. James Rea, Monaughty, Knighton, the property of Mr. John Wigmore, Bickerton Court, Much Marcle; got by Sambo (1720), dam (Gaiety) by Regent (891), g.d. (Lively the Second) by Barrister (658), g.g.d. (Lively) by Gallant (239), g.g.g.d. (Pert) by Old Court (306), g.g.g.g.d. by Duke (304), — by Crabstock (303).

PRODUCE IN	NAME.	BY WHAT BULL.	BY WHOM BRED.
1863, Dec. 15, r.—w.f. H	Gaiety 3rd	Artful 2391	Mr. T. Rea.
1864, Nov. 24, r.—w.f. B	Steer	Zeno 1825	Mr. Wigmore.

Gaiety the Third, sold to Mr. Cadle, Longcroft, Westbury-on-Severn.
Gaiety the Second was one of seven, winners of the First Prize in their class at the meeting of the Herefordshire Agricultural Society, 1863.

GAIETY THE SECOND.
Red with white face, calved October 23, 1858.

Bred by the late Mr. T. Rea, Westonbury, Leominster, the property of Mr. Wigmore, Bickerton Court, Dymock; got by Sambo (1720), dam (Gaiety) by Regent (891), g.d. (Lively the Second) by Barrister (658), g.g.d. (Lively) by Gallant (239), g.g.g.d. (Pert) by Old Court (306), g.g.g.g.d. — by Duke (304), — by Crabstock (303).

1862, Feb. 13, r.—w.f. B	Steer	Lord Nelson 2088	Mr. Rea.
1863, Dec. 15, r.—w.f. H	Gaiety 3rd	Artful 2391	do.

Gaiety the Third, sold to Mr. Cadle, Longcroft, Westbury-on-Severn.

GARLIC.
Red with white face, calved June 6, 1858.

Bred by and the property of Mr. E. T. Goldingham, Grimley, Worcester,; got by General (1251), dam (Abercundrig) by Young Byron, g.d. — by Byron (440), g.g.d. — by Whitenob (345).

1860, Nov. 12, r.—w.f. B	Steer	Tasso 1753	Mr. Goldingham.
1861, Sep., r.—w.f. B	Tomboy	do.	do.
1862, Sep., r.—w.f. B	Steer	do.	do.
1863, July 31, r.—w.f. B	do.	Anonymous	do.
1864, June 14, r.—w.f. H	Amazon	Mars 2107	do.

COWS.

GAYLASS.

Red with white face, calved in the month of August, 1858.

Bred by and the property of Mr. J. H. Arkwright, Hampton Court, Leominster; got by Riff Raff (1052), dam (Gaily) by Quicksilver the Second, g.d. — by Jupiter (1289), g.g.d. — by Reliance (278).

PRODUCE IN		NAME.	BY WHAT BULL.	BY WHOM BRED.
1861, Sep.,	r.—w.f. H	Gay	Sheriff 2752	Mr. Arkwright.
1862, Sep.,	r.—w.f. B	Steer	do.	do.
1863, Aug. 28,	r.—w.f. H	Hampton Olive	Sir Oliver 2nd 1733	do.
1864, Aug. 20,	r.—w.f. B		Dan O'Connell 1952	do.

Gay was one of a pair winners of the Second Prize in their class at the Meeting of the Herefordshire Agricultural Society, 1862; and First at the Meeting of 1863.

GAYLASS.

Red with white face, calved September 18, 1861.

Bred by and the property of Mr. W. D. Turner, Lynch Court, Leominster; got by Bertram (1513), dam (Garland) by Waverley (1793), g.d. (Lady) by Old Court the Second (1341).

1863, Dec. 16, r.—w.f. B	Steer	The Rover 2821	Mr. Turner.
1864, Dec. 15, r.—w.f. B		do.	do.

GAYLASS.

Red with white face, calved January 12, 1860.

Bred by and the property of Mr. P. Turner, The Leen, Pembridge; got by Sorcerer (1737), dam (Gaudy) by Defiance (1209), g.d. (Beauty) by Old Court (306), g.g.d. bred by Mr. Child, Wigmore Grange.

1862, July 17, r.—w.f. H	Leah	Bolingbroke 1883	Mr. Turner.
1863, Nov. 3, r.—w.f. B	Steer	Shylock 2212	do.
1864, Dec. 15, r.—w.f. H	Lily	Grove 2nd 2556	do.

COWS.

GAYLASS.
Red with white face, calved in the year 1861.

Bred by the Misses Abley, Norton, Presteign, the property of Mr. R. Green Price, M.P., Norton Manor, Presteign; got by Havelock (2563).

PRODUCE IN	NAME.	BY WHAT BULL.	BY WHOM BRED.
1864, Aug. 17, r.—w.f. H	White Rose	Lord of the Manor 2622	Mr. Price.

GAYLASS.
Red with white face, calved February 2, 1860.

Bred by and the property of Mr. T. Edwards, Llanarth, Raglan; got by Kars (1291), dam (Gaylass) by Alma (1144), g.d. (Lady) by Royalty (1374).

1863, Dec. 29, r.—w.f. H	Gaily	Sunbeam 2249	Mr. Edwards.
1864, Dec. 30, r.—w.f. H	Beauty	Chancellor 1172	do.

GAYLASS THE THIRD.
Red with white face, calved in the year 1857.

Bred by Mr. Stedman, Bedstone Hall, Salop, the property of Mr. W. Powell, Eglwysnunydd, Taibach, Glamorgan; got by Grateful (1260), dam (Gaylass) by Young Emperor (1811), g.d. — by Venison (1441), g.g.d. — by Lottery The Second (408), g.g.g.d. — by Hector (535), g.g.g.g.d. — by a son of Waterloo (49).

1861, Feb.	r.—w.f. H	Sprightly	Young Perfection	Mr. Stedman.
1862, Nov.	r.—w.f. H	Clara	Kinlet 1293	Mr. Powell.
1863, Dec. 18,	r.—w.f. B	Cardiff 2447	General 1251	do.

GAYLASS THE THIRD.
Red with white face, calved December 19, 1857.

Bred by and the property of Mr. Tudge, Adforton, Leintwardine, Ludlow; got by The Doctor (1083), dam (Gaylass the

Second), by Stanage (1741), g.d. (Gaylass) by Turpin (300), g.g.d. (Comely) by a Tully Bull.

PRODUCE IN	NAME.	BY WHAT BULL.	BY WHOM BRED.
1860, Sep. 1, r.—w.f. B	Genuine	Kyrewood	Mr. Tudge.
1861, July 30, r.—w.f. B	Harold 2029	The Grove 1764	do.
1862, July 27, r.—w.f. B	Steer	Sir Colin 2216	do.
1863, Aug. 20, r.—w.f. B	President 2687	Pilot 2156	do.

GENTLE.

Red with white face, calved in the year, 1859.

Bred by Mr. John Turner, Burghill, Hereford, the property of Mr. T. Rogers, Coxall, Brampton Bryan; got by Half-brother to Sir David (349), dam — by Young Dewshall (1125).

1864, July 21, r.—w.f. H	Gentle Anne	Grove 2nd 2556	Mr. Rogers.

GENTLE.

Red with white face, calved in the month of October, 1857.

Bred by and the property of Mr. Hollings, Hillend; got by Noke (1338), dam (Daisy) by Byron (380), g.d. (Cherry the Fourth) by Reveller (2722), g.g.d. (Cherry the Third) by Cornet (1933), g.g.g.d. (Cherry the Second) by Young Waterloo (2341), g.g.g.g.d. (Cherry the First) by Foxley, bred by the late Sir Robert Price.

1860, Sep., r.—w.f. H	Lovely	Woodman 1460	Mr. Hollings.
1861, Sep. 28, r.—w.f. B	Steer	St. Clement 2201	do.
1862, Nov. 1, r.—w.f. H	Gentle	do.	do.

GENTLE THE SECOND.

Red with white face, calved July 28, 1860.

Bred by and the property of Mr. H. R. Evans, Swanstone Court, Leominster; got by Rambler (1046), dam (Gentle) by King

COWS.

James (978), g.d. (Silver) by Coningsby (718), g.g.d. (Beauty) by a bull bred by the Rev. N. Penoyre, Hay, g.g.g.d. (Spot).

PRODUCE IN	NAME.	BY WHAT BULL.	BY WHOM BRED.
1863, May 10, r.—w.f. H	Lady Bateman	Chatham 1914	Mr. Evans.
1864, July 7, r.—w.f. B	Zouave 2905	do.	do.

Lady Bateman was one of four winners of the First Prize in their class at the meeting of the Leominster Agricultural Society, 1863.

GENTLE THE SECOND.

Red with white face, calved February 27, 1861.

Bred by and the property of Mr. F. W. Stone, Moreton Lodge, Guelph, Canada West; got by Golden Horn (2015), dam (Gentle) by Carlisle (923) g.d. (Lady) by The Knight (185), g.g.d. — by Monarch (504), g.g.g.d. — bred by the late Mr. Turner, Noke, Leominster.

PRODUCE IN	NAME.	BY WHAT BULL.	BY WHOM BRED.
1864, July 5, r.—w.f. H	Gentle 5th	Sailor 2200	Mr. Stone.

Gentle the Second was a winner of the Second Prize in her class at the Toronto, Kingston and Hamilton Meetings of the Canadian Agricultural Society.

GENTLE THE THIRD.

Red with white face, calved February 26, 1862.

Bred by and the property of Mr. F. W. Stone, Moreton Lodge, Guelph, Canada West; got by Patriot (2150), dam (Gentle) by Carlisle (923), g.d. (Lady) by the Knight (185), g.g.d. — by Monarch (504), g.g.g.d. — bred by the late Mr. Turner, Noke, Leominster.

PRODUCE IN	NAME.	BY WHAT BULL.	BY WHOM BRED.
1864, Oct. 22, r.—w.f. H	Gentle 6th	Sailor 2200	Mr. Stone.

Gentle the Third was a winner of the Second Prize in her class at the Toronto, Kingston and Hamilton Meetings of the Canadian Agricultural Society.

GEORGIANA.
Light grey, calved June 10, 1861.

Bred by Mr. Sobey, Tencreek, Liskeard, the property of Mr. R. Davey, M.P., Polsne House, Grampound, Cornwall; got by Conservative (1931), dam (Lady Grey) bred by the late Mr. Knight, Downton Castle.

PRODUCE IN	NAME.	BY WHAT BULL.	BY WHOM BRED.
1864, Mar. 3, g.—w.f. B		Castor 1900	Mr. Davey.

GERANIUM THE SECOND.
Red with white face, calved August 13, 1861.

Bred by and the property of the Hon. and Rev. H. Noel Hill, Cronkhill, Salop; got by Byron the Third, dam (Geranium) by Goldsmith (1258), g.d. (Fairy) by Lot (364).

1864, May 13, r.—w.f. H	Silver	Albert 2380	Hon. and Rev. H. N. Hill.

GERTRUDE.
Red with white face, calved June 6, 1859.

Bred by and the property of the Duke of Bedford, Woburn Park, Beds.; got by General (1251), dam (Snowdrop) by The Prince (1092), g.d. (Fidget) by Bonaparte (1152), g.g.d. (Fidget) by Napoleon (1334).

1862, Aug. 7, r.—w.f. B	Steer	Carbonel 1525	Duke of Bedford.
1863, Aug. 29, r.—w.f. B		Victory 2298	do.
1864, Oct. 29, r.—w.f. B		Victory 2296	do.

GIANTESS.
Red with white face, calved July 18, 1860.

Bred by and the property of Mr. P. Turner, The Leen, Pembridge; got by Bertram (1513), dam (Stella) by Andrew the Second (619), g.d. (Sprite) by Sir Walter (352), g.g.d. (Sylph) by Viscount (816), g.g.g.d. (Sylph) by Old Court the Second (1341).

1863, July 16, r.—w.f. B	Steer	Bolingbroke 1883	Mr. Turner.
1864, Nov. 7, r.—w.f. H	Vesta	Grove 2nd 2556	do.

GIANTESS.

Red with white face, of 1857, vol. v., p. 195.

Bred by and the property of Mr. Richard Shirley, Baucott, Church Stretton; got by Marlow (2104), dam (Tasty) by Knockerell, (1630), g.d. (Natty) by Dolluggan (759), g.g.d. — by The Count (2263).

PRODUCE IN	NAME.	BY WHAT BULL.	BY WHOM BRED.
1862, Mar. 23, r.—w.f. B	Steer	Pilot 1036	Mr. Shirley.
1863, April 20, r.—w.f. B	Gigantic 2544	do.	do.

The Steer of 1861 was a winner of the First Prize in his class and Silver Medal at the Birmingham Fat Show; and Second Prize at the meeting of the Smithfield Club, 1863.

GIANTESS THE FOURTH.

Red with white face, calved September 10, 1858.

Bred by Mr. Stedman, Bedstone Hall, Salop, the property of Mr. John Baldwin, Luddington, Stratford-on-Avon; got by Grateful (1260), dam (Giantess the Second) by Bedstone (2411), g.d. (Giantess) by Conrad (1183), g.g.d. (Violet) by Dinedor (395), g.g.g.d. — by Trojan (542).

1862, Dec. 7, r.—w.f. H	Giantess 5th	Kinlet 1293	Mr. Baldwin.
1863, Dec. 30, r.—w.f. H	Giantess 6th	Battersea 1865	do.

GILIA.

Red with white face, calved July 14, 1861.

Bred by and the property of Mr. Naylor, Leighton Hall, Welshpool; got by Royal (2195), dam (Gipsy Lass) by The Knight (185), g.d. (Gipsy) by Big Ben (248), g.g.d. (Grizzle) by Prince (251), g.g.g.d. (Prettymaid) by Tobias (487), g.g.g.g.d. (Cherry) by Red Robin (263).

1864, July 27, r.—w.f. B		Gladstone 2547	Mr. Naylor.

GIPSY.

Red with white face, calved April 3, 1859.

Bred by the late Lord Berwick, Cronkhill, Salop, the property of the Hon. and Rev. Noel Hill, Berrington, Salop; got by Attingham (911), dam (Empress) by Venison (1441), g.d. (Empress) by Dinedor (395).

PRODUCE IN	NAME.	BY WHAT BULL.	BY WHOM BRED.
1862, Nov. 22, r.—w.f. H	Gold	Conqueror 1929	Hon. and Rev. H. Noel Hill.
1863, Oct. 2, r.—w.f. B	Gander	Van Tromp 2291	do.
1864, Sep. 2, r.—w.f. B	Silver King 2756	Albert 2380	do.

GLASBURY.

Red with white face.

Bred by Mr. Edward Williams, Llowes Court, Hay, the property of Mr. E. Drinkwater, Treribble, Ross; got by Quicksilver (353), dam — by Prince (333).

1859, Aug. 8, r.—w.f. B	Treribble 2834	Chieftain 930	Mr. Williams.
1860, Dec. 1, r.—w.f. H	Gipsy	Grove 3rd	Mr. Drinkwater.
1861, Sep. 8, r.—w.f. H	Plum	Portrait 2685	do.
1862, Dec. 29, r.—w.f. B	Steer	do.	do.

GRACE.

Red with white face, calved in the year, 1858.

Bred by Mr. Stedman, Bedstone Hall, Salop, the property of Mr. R. S. Fetherstonhaugh, Rockview, Killucam, Ireland; got by Grateful (1260), dam — by Venison (1441), g.d. — by Dinedor (395).

1862,	r.—w.f. H	Grace 2nd	Kinlet 1293	Mr. Stedman.
1863,	r.—w.f. H	Grace 3rd	do.	do.

GRACE.

Red with white face, calved in the month of September, 1856.

Bred by Mr. J. Hollings, sen., Hillend, the property of Mr. J. A. Hollings, jun., Hillend, Hereford; got by Garrick (1248), dam (Daisy) by Byron (280), g.d. (Cherry the Fourth) by

COWS.

Reveller (2722), g.g.d. (Cherry the Third) by Cornet (1933), g g.g.d. (Cherry the Second) by Young Waterloo (2341), g.g.g.g.d. (Cherry the First) by Foxley, bred by the late Sir Robert Price.

PRODUCE IN	NAME.	BY WHAT BULL.	BY WHOM BRED.
1859, Aug., r.—w.f. H		Noke 1338	Mr. Hollings.
1860, Oct., r.—w.f. H	Grace 2nd	Woodman 1460	do.
1861, Sep., 28, r.—w.f. B	Steer	St. Clement 2201	do.
1862, Oct. 16, r.—w.f. H	Clement's Grace	do.	do.
1863, Nov. 4, r.—w.f. B	Steer	do.	do.
1864, Dec. 1, r.—w.f. B	do.	do.	do.

GRACE THE SECOND.

Red with white face, calved in the month of October, 1860.

Bred by Mr. J. Hollings, sen., Hillend, the property of Mr. J. A. Hollings, Hillend, Hereford; got by Woodman (1460), dam (Grace) by Garrick (1248), g.d. (Daisy) by Byron (380), g.g.d. (Cherry the Fourth) by Reveller (2722), g.g.g.d. (Cherry the Third) by Cornet (1933), g g.g.g.d. (Cherry the Second) by Young Waterloo (2341), — (Cherry the First) by Foxley, bred by the late Sir Robert Price.

1863, May 13, r.—w.f. H	Governess	St. Clement 2201	Mr. Hollings.
1864, Nov. 9, r.—w.f. H	Governess 2nd	do.	do.

GRACE THE THIRD.

Red with white face, calved October 7, 1861.

Bred by Mr. J. Hollings, sen., Hillend, the property of Mr. J. A. Hollings, Hillend, Hereford; got by St. Clement (2201), dam (Hopeful the Second) by Voltigeur (1445), g.d. (Hopeful) by Hope (411), g.g.d. (Silver) by Cornet (1933), g.g.g.d. (Silver) by Young Waterloo (2341).

1864, Oct. 17, r.—w.f. H	Baroness	Chieftain 2nd 1917	Mr. Hollings.

GRACE THE FOURTH.

Red with white face, calved October, 29, 1861.

Bred by Mr. J. Hollings, sen., Hillend, the property of Mr. J. A. Hollings, jun., Hillend, Hereford; got by St. Clement (2201) dam (Hopeful the Third) by Noke (1338), g.d. (Hopeful the Second) by Voltigeur (1445), g.g.d. (Hopeful the First) by Hope (411), g.g.g.d. (Silver) by Cornet (1933), g.g.g.g d. (Silver) by Young Waterloo (2341).

PRODUCE IN	NAME.	BY WHAT BULL	BY WHOM BRED.
1864, Oct. 27, r.—w.f. H	Marchioness	Chieftain 2nd 1917	Mr. Hollings.

GRACE DARLING.

Red with white face, calved in the year 1862.

Bred by and the property of Mr. R. G. Price, M.P., Norton Manor, Presteign; got by Stanage (1742).

1864, Oct. 5, r.—w.f. B		Lord of the Manor 2622	Mr. Price.

GRACEFUL.

Red with white face, calved in the month of September, 1860.

Bred by Mr. J. Hollings, sen., Hillend, the property of Mr. J. A. Hollings, Hillend, Hereford; got by Woodman, (1460), dam (Daisy) by Byron (380), g.d. (Cherry the Fourth) by Reveller (2722), g.g.d. (Cherry the Third) by Cornet (1933), g.g.g.d. (Cherry the Second) by Young Waterloo (2341), g.g.g.g.d. (Cherry the First) by Foxley, bred by the late Sir Robert Price.

1863, Oct. 11, r.—w.f. B	Steer	St. Clement 2201	Mr. Hollings.
1864, Dec. 3, r.—w.f. H	Chieftain's Daisy	Chieftain 2nd 1917	do.

cows.

GRACEFUL.
Red with white face, calved in the year 1858.

Bred by Mr. Stedman, Bedstone Hall, Salop, the property of Mr. W. Powell, Eglwysnunydd, Taibach, Glamorgan; got by Grateful (1260), dam — by Venison (1441), g.d. by Counsellor (422), g g.d. — by Dinedor (395).

PRODUCE IN		NAME.	BY WHAT BULL.	BY WHOM BRED.
1862,	r.—w.f. B	Steer	Kinlet 1293	Mr. Stedman.
1863, Sep.,	r.—w.f. H	Tulip	General 1251	Mr. Powell.

GRACEFUL.
Red with white face, calved November 12, 1860.

Bred by the late Lord Berwick, Cronkhill, Salop, the property of Mr. F. W. Stone, Moreton Lodge, Guelph, Canada West; got by Severn (1382), dam (Lady) by Albert Edward (859), g d. (Zephyr) by Walford (871), g.g d. (Friday the Second) by Wonder (420), g.g.g.d. (Friday by Commerce) (354), g.g.g.g.d. (Prettymaid) by The Sheriff (356), — (Turberville) by Sovereign (404).

1864, April 6,	r.—w.f. H	Graceful 2nd	Patriot 2150	Mr. Stone.

Graceful was a winner of the first prize in her class at the Toronto, Kingston, and Hamilton Meetings of the Canadian Agricultural Society.

GRACEFUL THE SECOND.
Red with white face, calved in the year 1856.

Bred by and the property of Mr. W. H. Oatley, Wroxeter, Salop; got by Walford (871), dam (Graceful) by Charity (375), g.d. (Giantess) by Sovereign (404).

1859,	r.—w.f. H		Julius Cæsar 2054	Mr. Oatley.
1860, April 29,	r.—w.f. B	Claudius Cæsar 1922	Augustus Cæsar 1854	do. do.
1862,	r.—w.f. H	Gertrude	Franky 1243	do.
1863,	r.—w.f. H	Gazelle	Nero 2650	do.
1864,	r.—w.f. B	Steer	do.	do.

GRAND DUCHESS.

Red with white face, of 1855, vol. v., p. 197.

Bred by and the property of Mr. John Monkhouse, The Stow, Hereford; got by Madoc (899), dam (Duchess) by Young Hope (343), g.d. (Duchess) by Chance (348), g.g.d. — bred by the late Mr. D. Williams, Newton, Brecon.

PRODUCE IN	NAME.	BY WHAT BULL.	BY WHOM BRED.
1862, Aug. 12, r.—w.f. B	Steer	Chieftain 930	Mr. Monkhouse.
1863, Aug. 9, r.—m.f. B	do.	do.	do.
1864, Aug. 5, r.—w.f. B	Grandee 2554	do.	do.

Grandee was a winner of the First Prize in his class at the Plymouth Meeting of the Royal Agricultural Society of England.

GREY COUNTESS.

Red with white face, calved in the month of December, 1858.

Bred by Mr. J. Hollings, sen., Hillend, the property of Mr. J. A. Hollings, Hillend, Hereford; got by Woodman (1460), dam (Countess) by Albert Edward (754), g.d. (Countess) by Monarch (504).

1861, Oct. 5, r.—w.f. B	Steer	St. Clement 2201	Mr. Hollings.
1862, Nov. 17, r.—w.f. H	Clement's Countess	do.	do.

GREY OAK-APPLE.

Light Grey, of 1851, vol. v., p. 198.

Bred by the late Lord Berwick, Cronkhill, Salop, the property of Mr. A. R. Boughton Knight, Downton Castle, Ludlow; got by Tom Thumb (243), dam (Oak Apple) by Commerce (354), g.d. (Strawberry) bred by the late Mr. E. Jeffries, by a son of Guinea.

1859, Oct. 21, g.—w.f. H	Apple Blossom	Attingham 911	Lord Berwick.
1860, Aug. 6, r.—w.f. H	Red Oak Apple	do.	do.
1862, June 26, g.—w.f. H	Defiance	Lord Grey 2085	Mr. Knight.
1864, Sep. 9, g.—w.f. H	Grisel	Garibaldi 2003	do.

GRISETTE.
Light Grey, of 1855, vol. v., p. 198.

Bred by the late Lord Berwick, Cronkhill, Salop, the property of Mr. A. R. Boughton Knight, Downton Castle, Ludlow; got by Attingham (911), dam (Dorcas the Third) by Tom Thumb (243), g.d. (Dorcas the Second) by Wonder (420), g.g.d. (Dorcas) by Ashley Moor white bull (870), g.g.g.d. (Old Damsel) by Coleman's bull (1547).

PRODUCE IN	NAME.	BY WHAT BULL.	BY WHOM BRED.
1859 Aug. 26, g.—w.f. H	Gazelle	Sir David 349	Lord Berwick.
1863, June 19, white B	Robin Grey	Lord Grey 2085	Mr. Knight.
1864, Aug. 20, r.—w.f. H	Cherry	Garibaldi 2003	do.

Gazelle, sold to Mr. Ashwood, Langdon Hall, Salop.

GRIZZLE.
Red with white face, calved in the year 1859.

Bred by and the property of Mr. G. T. Forester, High Ercall, Wellington, Salop; got by Discord (1217), dam (Grimy) by Governor (464), g.d. (The Twin) by Hope (439), g.g.d. (Miss Hewer) by Hope (411), g.g.g.d. — bred by Mr. John Hewer.

1862, Aug. 3, r.—w.f. H	Yellow Girl	Severn 1382	Mr. Forester.
1863, July 20, r.—w.f. H	Smudge	do.	do.

GWENNY.
Red with white face, of 1851, vol. iii., p. 170.

Bred by the late Mr. James Rea, Monaughty, the property of Mr. W. Stallard, Brockhampton, Ross; got by Regent (891), dam (Venus the Fifth) by Albert (330), g.d. (Winifred) by Monaughty (220), g.g.d. (Venus the Fourth) by Duke (304), g.g.g.d. (Venus the Third) by Regulator (360), g.g.g.g.d. (Venus the Second) by Noble (238), — (Venus) by Crabstock (303).

1862, Mar., r.—w.f. B	Dead	Sir Oliver 2nd 1733	Mr. Holloway.
1863, June 4, r.—w.f. H	Gertrude	Sir Oliver 3rd 2773	do.
1864, Aug. 24, r.—w.f. H	Minna	do.	do.

Minna, sold to Mr. Stallard, Brockampton.

COWS.

GWENNY THE SECOND.

Red with white face, of 1854, vol. iii., p. 170.

Bred by Mr. James Rea, Monaughty, Knighton, the property of Mr. W. Stallard, Brockhampton, Ross; got by Chieftain (930), dam (Gwenny) by Regent (891), g.d. (Venus the Fifth) by Albert (330), g.g.d. (Winifred) by Monaughty (220), g.g.g.d. (Venus the Fourth) by Duke (304), g.g.g.g.d. (Venus the Third) by Regulator (360), — (Venus the Second) by Noble (238), — — (Venus) by Crabstock (303).

PRODUCE IN	NAME.	BY WHAT BULL.	BY WHOM BRED.
1858, Nov. 25, r.—w.f. H	Gwenny 3rd	Sir Benjamin 1387	Mr. T. Rea.
1859, Sep. 27, r.—w.f. B	Sir Robert 2227	do.	do.
1860, Aug. 10, r.—w.f. H	Gwenny 4th	do.	do.
1861, Sep. 5, r.—w.f. H	Gwenny 5th	Young Hampton 2327	Mr. Stallard.
1862, Sep. 9, r.—w.f. B	Chieftain 3d 2457	Chieftain 2nd 1917	do.
1864, Mar. 5, r.—w.f. B	Subaltern 2794	do.	do.

Chieftain the Third was winner of the Second Prize in his class at the Plymouth Meeting of the Royal Agricultural Society of England.
Gwenny the Fourth, sold to Mr. Stallard, Brockhampton, Ross.

GWENNY THE FOURTH.

Red with white face, calved August 10, 1860.

Bred by Mr. Thomas Rea, Westonbury, Leominster, the property of Mr. W. Stallard, Brockhampton, Ross; got by Sir Benjamin (1387), dam (Gwenny the Second) by Chieftain (930), g.d. (Gwenny) by Regent (891), g.g.d. (Venus the Fifth) by Albert (330), g.g.g.d. (Winifred) by Monaughty (220), g.g.g.g.d. (Venus the Fourth) by Duke (304), — (Venus the Third) by Regulator (360), — (Venus the Second) by Noble (238), — (Venus) by Crabstock (303).

1863, May 7, r.—w.f. H	Bridget	Shamrock 2nd 2210	Mr. Stallard.
1864, Mar. 17, r.—w.f. B	San Sebastian 2741	Chieftain 2nd 1917	do.

HEBE.
Red with white face, calved February 6, 1861.

Bred by Mr. J. Rea, Monaughty, Knighton, the property of Mr. James P. Apperley, Fownhope, Hereford; got by Zealous (2345), dam (Ruth) by Vanguard (1109), g.d. (Fairmaid the Fourth) by Chieftain (930), g.g.d. (Fairmaid) by Cholstrey (217), g.g.g.d. (Fairmaid) by Gallant (239).

PRODUCE IN	NAME.	BY WHAT BULL.	BY WHOM BRED.
1864, Sept. 4, r.—w.f. B		Capt. Perry 2444	Mr. Apperley.

HEBE.
Red with white face, calved September 7, 1861.

Bred by and the property of Mr. P. Turner, The Leen, Leominster; got by Bertram (1513), dam (Spot) by Silurian (1064), g.d. (Gazelle) by Andrew the Second (619), g.g.d. (Vesta) by Sir Walter (352), g.g.g.d. (Myrtle) by Commerce (354), g.g.g.g.d. (Sylph) by Old Court the Second (1341).

1864, Aug. 19, r.—w.f. B		Bolingbroke 1883	Mr. Turner.

HEIRESS.
Red with white face, calved August 10, 1860.

Bred by Lord Berwick, Cronkhill, Salop, the property of Mr. Naylor, Leighton Hall, Montgomeryshire; got by Severn (1382), dam (Young Vic) by Wonder (420), g.d. (Victoria) by Hope (439), g.g.d. (Countess) by Young Chance (449).

1863, July 24, r.—w.f. H	Heiress 2nd	Blondin 1880	Mr. Naylor.
1864, Aug. 26, r.—w.f. B	Steer	Gladstone 2547	do.

HELICAN.
Red with white face, of 1855, vol. iv., p. 133.

Bred by Lord Berwick, Cronkhill, Salop, the property of Mr. R. S. Fetherstonhaugh, Rockview, Killucan, Ireland; got by Albert

Edward (859), dam (Yellow Byron) by Walford (871), g.d. (Miss Byron) by Baron (418), g.g.d. (Little Beauty) bred by Mr. John Hewer.

PRODUCE IN		NAME.	BY WHAT BULL.	BY WHOM BRED.
1861,	B	Dead	Will-o'-the-Wisp 1454	Lord Berwick.
1862,	g.—w.f. B	Handsome	Cropper 1559	Mr. Fetherstonhaugh.
1863,	r.—w.f. B	Steer	Silver Stream 2214	do.
1864,	r.—w.f. B	do.	Leominster 2071	do.

HELIOTROPE.
Red with white face, of 1855, vol. v., p. 202.

Bred by Lord Berwick, Cronkhill, Salop, the property of Mr. R. S. Fetherstonhaugh, Rockview, Killucan, Ireland; got by Attingham (911), dam (Grey Dove) by Wonder (420), g.d. (Dove) by Ashley Moor (791), g.g.d. (Pigeon) by Ashley Moor White Bull (870), g.g.g.d. (Damsel) by Cholstrey (868), &c.

1862, Nov. 29, r.—w.f. B	Holly 2572	Leominster 2071	Mr. Fetherstonhaugh.
1863, r.—w.f. H	Highdrangea	Silver Stream 2214	do.

HER GRACE.
Red with white face, calved July 7, 1858.

Bred by and the property of Mr. John Monkhouse, the Stow, Hereford; got by Formidable (1240), dam (Grand Duchess) by Madoc (899), g.d. (Duchess) by Young Hope (343), g.g.d. (Duchess) by Chance (348).

1862, Sep. 27, r.—w.f. H	Graceful	Chieftain 930	Mr. Monkhouse.
1863, Aug. 21, r.—w.f. H	Grace Darling	do.	do.

HERMIONE.
Red with white face, calved July 30, 1861.

Bred by and the property of Mr. Philip Turner, The Leen, Pembridge; got by Bertram (1513), dam (Venus) by Prince

Charlie (1357), g.d. (Comely) by Confidence (367), g.g.d. (Gaudy) by Defiance (1209), g.g.g.d. (Beauty) by Old Court (306),

PRODUCE IN	NAME.	BY WHAT BULL.	BY WHOM BRED.
1864, Aug. 8, r.—w.f. H	Rosamond	Demetrius 2494	Mr. Turner.

HONEY.
Red with white face, calved August 10, 1860.

Bred by and the property of Mr. T. Olver, Penhallow, Grampound, Cornwall; got by Earl Derby (1979), dam (Honeysuckle) by Duke of Cornwall (1569), g.d. (Honeysuckle) by Woodbine (1120).

PRODUCE IN	NAME.	BY WHAT BULL.	BY WHOM BRED.
1862, Nov. 12, r.—w.f. B	Hercules 2564	Penhallow 2154	Mr. Olver.

HOPBINE.
Red with white face, calved October 17, 1861.

Bred by and the property of Mr. Henry Haywood, Blakemere, Hereford; got by Cholstrey (1918), dam (Hewer the Second) by Blakemere (1151), g.d. (Hewer) by Governor (464), g.g.d. — bred by Mr. J. Hewer, Vern House, Marden, Hereford.

PRODUCE IN	NAME.	BY WHAT BULL.	BY WHOM BRED.
1864, July 12, r.—w.f. H	Hop Duty	Cromwell 1947	Mr. Haywood.

HOPE THE SECOND.
Red with white face, calved in the month of September, 1860.

Bred by and the property of Mr. T. Roberts, Ivington Bury, Leominster; got by Master Butterfly (1313), dam (Hope) by King James (978), g.d. (Rose of the Valley) by Coningsby (718), g.g.d. — by Goldfinder (383).

PRODUCE IN	NAME.	BY WHAT BULL.	BY WHOM BRED.
1864, Mar. 10, r.—w.f. B	Hopeful 2574	Sir Thomas 2228	Mr. Roberts.

COWS.

HOPEFUL THE SECOND.

Red with white face, of 1853, vol. v., p. 205.

Bred by Mr. J. Hollings, sen., Hillend, the property of Mr. J. A. Hollings, jun., Hillend, Hereford; got by Voltigeur (1445), dam (Hopeful) by Hope (411), g.d. (Silver) by Cornet (1933), g.g.d. (Silver) by Young Waterloo (2341).

PRODUCE IN	NAME.	BY WHAT BULL.	BY WHOM BRED.
1862, Nov. 25, r.—w.f. H	Maid of Weston	St. Clement 2201	Mr. Hollings.
1863, Nov. 30, r.—w.f. B	Dead	do.	do.
1864, Dec. 26, r.—w.f. H	Maid of Weston 2nd	do.	do.

HOPEFUL THE THIRD.

Red with white face, of 1855, vol. v., p. 205.

Bred by Mr. J. Hollings, sen., Hillend, the property of Mr. J. A. Hollings, Hillend, Hereford; got by Noke (1338), dam (Hopeful the Second) by Voltigeur (1445), g.d. (Hopeful the First) by Hope (411), g.g.d. (Silver) by Cornet (1933), g.g.g.d. (Silver) by Young Waterloo (2341).

1863, May 27, r.—w.f. H	Clement's Hope	St. Clement 2201	Mr. Hollings.

HOTTENTOT.

Red with white face, of 1852, vol. v., p. 206.

Bred by and the property of Mr. G. T. Forester, High Ercall, Wellington, Salop; got by Wonder (420), dam (Venus) by Hope (439), g.d. (Miss Fitzfavourite) by Young Fitzfavourite g.g.d. — by Fitzfavourite (441).

1862, Aug. 31, r.—w.f. H	Humpy	Severn 1382	Mr. Forester.
1863, Dec. 9, r.—w.f. B	Steer	do.	do.

COWS.

JANE.
Red with white face, calved December 12, 1859.

Bred by and the property of Mr. J. R. Paramore, Dinedor Court, Hereford; got by The General (2817), dam (Jenny) by Upstart (795), g.d. — bred by Dr. Lamb, Henwood, Dilwyn.

PRODUCE IN	NAME.	BY WHAT BULL.	BY WHOM BRED.
1862, May 4, r.—w.f. H		The General 2817	Mr. Paramore.
1863, June 12, r.—w.f. B	Steer	do.	do.
1864, May 28, r.—w.f. B	do.	Shamrock 2nd 2210	do.

JENNY.
Red with white face, calved June 9, 1853.

Bred by and the property of Mr. J. R. Paramore, Dinedor Court, Hereford; got by Upstart (795), dam (Jenny) bred by Dr. Lamb, Henwood, Dilwyn.

1856, r.—w.f. B	Steer	Hotspur 972	Mr. Paramore.
1857, r.—w.f. B	do.	do.	do.
1858, r.—w.f. B	do.	do.	do.
1859, Dec. 12, r.—w.f. H	Jane	The General 2817	do.
1860, r.—w.f. B	Steer	do.	do.
1861, r.—w.f. B	do.	do.	do.
1862, r.—w.f. B	do.	do.	do.
1863, April 7, r.—w.f. B	do.	do.	do.
1864, Mar. 11, r.—w.f. H	Lovely	The Jew 2266	do.

JENNY.
Red with white face, calved September 5, 1858.

Bred by the late Lord Berwick, Cronkhill, Salop, the property of Mr. J. J. Stone, Scyborwen, Llantrissent, Monmouth; got by Attingham (911), dam (Barbara), by Albert Edward (859), g.d. (Big Damsel) by The Count (351), g g.d. (Prettymaid) by Cholstrey (868), g.g.g.d. (Old Damsel) by Coleman's bull (1547), g.g.g.g.d. (Old Daisy) by Chancellor (156), — (Cherry the Second) by Thickset (1769).

1861, Sep. 23, r.—w.f. B	Lord Berwick 2082	Canning 1522	Mr. Stone.
1863, July 27, r.—w.f. B	Wentwood 2872	Prince of Wales 2172	do.
1864, Oct. 2, r.—w.f. B	Scyborwen 2743	Mountain Chief 2645	do.

COWS.

JESSAMINE.

Red with white face, of 1855, vols. iv. and v., pp. 134, 209.

Bred by the late Lord Berwick, Cronkhill, Salop, the property of Mr. J. O. G. Pollock, Mountainstown, Navan, Ireland; got by Attingham (911), dam (Becky) by Young Byron (832), g.d. (Rebecca) by Governor (464), g.g.d. (Old Prettymaid) by Young Sovereign (1472), g.g.g.d. — by (Whitenob) (345), g.g.g.g.d. — by Young Wellington (505).

PRODUCE IN		NAME.	BY WHAT BULL.	BY WHOM BRED.
1862, Sept.,	r.—w.f. B	Jolly Boy 2580	Sir Robert 2227	Mr. Pollock.
1863, Aug.,	r.—w.f. H	Jessy	Master Willie 2637	do.

JESSIE.

Red with white face, of 1855, vols. iii. and v., p.p. 174, 209.

Bred by Mr. Johnstone, Broncroft Castle, the property of Mr. Hinckesman, The Poles, Ludlow; got by Byron (559), dam — bred by the late Mr. Tully.

1862, April 21, r.—w.f. H	Jessica	The Friar 1085	Mr. Hinckesman.
1863, Sept. 9, r.—w.f. H	Jessamine	Forge	do.
1864, Oct. 21, r.—w.f. B	Oakley 2657	Salopian 2739	do.

JEWEL (A TWIN).

Red with white face, calved July 14, 1858.

Bred by and the property of Mr. Philip Turner, The Leen, Pembridge; got by Felix (953), dam (Brilliant) by Andrew the Second (619), g.d. (Gem) by Sir Walter (352), g.g.d. (Jewel) by Commerce (354), g.g.g.d. (Mischief) by a bull bred by Mr. Bowen, Monkland.

1861, May 9, r.—w.f. H		Logic 2079	Mr. Turner.
1862, May 10, r.—w.f. H	Sapphire	Bolingbroke 1883	do.

Jewel was one of a pair winners of the Second Prize in their class at the Worcester Meeting of the Royal Agricultural Society of England.

COWS.

JEWESS.
Red with white face, calved in the year 1860.

Bred by Capt. Power, Hill Court, Ross, the property of Mr. William Jones, Hill of Eaton, Ross; got by The Jew (2266), dam — by Uncle Tom (1108), g.d. — bred by the late Mr. Phillipps, Bryngwyn.

PRODUCE IN	NAME.	BY WHAT BULL.	BY WHOM BRED.
1863, Dec. 1, r.—w.f. B	Steer	Esperus	Mr. Jones.
1864, Nov. 8, r.—w.f. B		Chieftain 3rd 2457	do.

JEWESS.
Red with white face, calved in the month of July, 1860.

Bred by the late Lord Berwick, Cronkhill, Salop, the property of the Hon. and Rev. Noel Hill, Berrington, Salop; got by Attingham (911), dam (Rebecca) by Governor (464).

1863, July 6, r.—w.f. B	Shylock 2754	Van Tromp 2291	Hon. and Rev. H. Noel Hill.
1864, June 24, r.—w.f. H	Rebecca	Conqueror 1929	do.

JUNIPER.
Red with white face, of 1855, vols. iv. and v., p.p. 135, 210.

Bred by the late Lord Berwick, Cronkhill, Salop, the property of Mr. J. O. G. Pollock, Mountainstown, Navan, Ireland; got by Walford (871), dam (Rebecca) by Governor (464), g.d. (Old Prettymaid) by Young Sovereign (1472), g.g.d. — by Whitenob (345), g.g.g.d. — by Young Wellington (505).

1862, Sep., r.—w.f. H	Jonquil	Sir Robert 2227	Mr. Pollock.
1863, July, r.—w.f. B	Gin 2545	Master Willie 2637	do.
1864, July 25, r.—w.f. H	Juicy	Reindeer 2717	do.

JUNO.
Red with white face, calved in the month of September, 1860.

Bred by Mr. Stedman, Bedstone Hall, Salop, the property of Mr. John Barton, Coln, Fairford; got by Young Perfection, dam (Venus the Second) by Bedstone (2411), g.d. (Venus) by

Venison (1441), g.g.d (Venus) by Cotmore the Second (1191), g.g.g.d. — by Dinedor (395), g.g g.g.d — by Trojan (542).

PRODUCE IN	NAME.	BY WHAT BULL.	BY WHOM BRED.
1863, Jan. 3, r.—w.f. H	Jenny Lind	Kinlet 1293	Mr. Barton.
1864, Mar. 10, r.—w.f. B	Steer	Young Cardinal Wiseman 2882	do.

JUNO.

Red with white face, calved February 24, 1860.

Bred by H.R.H. the Prince Consort, Windsor Castle, the property of Major-General the Hon. A. N. Hood, Cumberland Lodge, Windsor; got by Brecon (918), dam (Fanciful) by Phantom (1035), g.d. (Fancy the First).

1862, Sep., 2, r.—w.f. H	Crown Princess	Ajax 1843	Hon. A. N. Hood.
1863, July 10, r.—w.f. H	Lady Mary	do.	do.

KATE THE SECOND.

Red with white face, of 1855, vol. v., p. 211.

Bred by the late Mr. T. Rea, Westonbury, Leominster, the property of Mr. J. H. Whitehouse, Ipsley Court, Redditch; got by Sir Benjamin (1387), dam (Kate) by Chieftain (930), g.d. (Venus the Fifth) by Albert (330), g.g.d. (Winifred) by Monaughty (220), g.g.g.d. (Venus the Fourth) by Duke (304), g.g.g.g.d. (Venus the Third) by Regulator (360). — (Venus the Second) by Noble (238), — (Venus) by Crabstock (303).

1861, Aug. 9, r.—w.f. B	Steer	Lord Nelson 2088	Mr. T. Rea.
1863, Sep. 19, r.—w.f. B	Earl Berkeley	Artful 1852	do.
1864, Nov. 1, r.—w.f. { B / H }	Twins	Zeno 1825	Mr. Whitehouse.

Kate the Second was a winner of the Second Prize in her class at the Worcester and Newcastle Meetings of the Royal Agricultural Society of England; and one of seven winners of First Prizes at the meetings of the Leominster and Herefordshire Agricultural Societies, 1863.

cows.

KATE THE THIRD.

Red with white face, calved September 20, 1861.

Bred by and the property of Mr. J. Wigmore, Bickerton Court, Dymock; got by Surprise (2250), dam (Kate) by Chieftain (930), g.d. (Venus), by Albert (330), g.g.d. (Winifred) by Monaughty (220), g.g.g.d. (Venus the Fourth) by Duke (304), g.g.g.g.d. (Venus the Third) by Regulator (360).

PRODUCE IN	NAME.	BY WHAT BULL.	BY WHOM BRED.
1863, Dec. 27, r.—w.f. B	Steer	Speculator 2240	Mr. Wigmore.
1864, Dec. 21, r.—w.f. H	Kate 4th	do.	do.

LABURNUM.

Red with white face, calved in the month of January, 1858.

Bred by Mr. H. E. Powell, Great Brampton, Madley, the property of Mr. J. H. Whitehouse, Ipsley Court, Redditch; got by Mussulman (1333), dam (Lady) by Young Sovereign (379), g.d. (Miss Cotmore the Second) by Lottery the Second (408), g.g d. (Miss Cotmore) by Cotmore (376), g.g.g.d. — by Conqueror (412).

1864, Jan., r.—w.f. B	Golden-chain 2552	Vincent 2858	Mr. Powell.

LADY.

Red with white face, calved October 20, 1860.

Bred by and the property of Mr. T. Olver, Penhallow, Grampound, Cornwall; got by Conservative (1931), dam (Strawberry) by Attingham (911), g.d, (Young Oak-apple) by Tom Thumb (243), g.g.d. (Oak-apple) by Commerce (354), g.g.g.d. (Strawberry) bred by Mr. Jefferies, The Grove.

1863, June 22, r.—w.f. H	Lucy	Sir Hugh 2223	Mr. T. Olver
1864, Sept. 19, r.—w.f. B	Leotard 2595	Zippor 2354	do.

COWS.

LADY.

Red with white face, calved January 6, 1861.

Bred by and the property of Mr. Thomas Thomas, St. Hilary, Cowbridge, Glamorganshire; got by Goldfinder the Second (959), dam (Comely) by Young Royal (1469).

PRODUCE IN	NAME.	BY WHAT BULL.	BY WHOM BRED.
1863, Nov. 15, r.—w.f. H	Lucy	Goldfinder 2nd 959	Mr. Thomas.
1864, June 28, r.—w.f. H		do.	do.

Lady was one of a pair, winners of the First Prize in their class at the meeting of the Carmarthenshire Agricultural Society, 1861; she was also a winner of a First Prize at Cowbridge, 1862, 1863, and 1864; First at Newport, 1862, 1863, and 1864; First at Carmarthen, 1864; and Third at the Bristol meeting of the Bath and West of England Society.

LADY.

Red with white face, of 1857, vol. v., p. 213.

Bred by and the property of Mr. C. H. Hinckesman, The Poles, Ludlow; got by Old Jim (1026), dam (Duchess) by Nelson (1021), g.d. — bred by Mr. Jellicoe, by Lundyfoot (88), g.g.d. bred by Mr. B. Tomkins, g.g.g.d. bred by Mr. Turner, of Noke.

1862, Mar. 23, r.—w.f. H	Gazelle	The Friar 1035	Mr. Hinkesman.
1863, Mar. 24, r.—w.f. H	Alexandra	The Abbot 2258	do.
1864, July 18, r.—w.f. B	Steer	Salopian 2739	do.

LADY.

Red with white face, calved December 14, 1859.

Bred by and the property of Mr J. M. Read, Elkstone, Cheltenham; got by Sebastopol (1381), dam (Toilette) by the Duke (493), g.d. (Miss Talbot) by Prince (251), g.g.d. (Lady Ingestre) bred by Lord Radnor.

1863, June 3, r.—w.f. B	Steer	Caliban 1163	Mr. Read.
1864, April 12, r.—w.f. H	Lady 2nd	do.	do.

cows.

LADY.
Red with white face, of 1854, vol. iv., p. 138.

Bred by and the property of the Rev. Archer Clive, Whitfield. Hereford; got by Trader (1101) dam (Lively) by Andrew the Second (619).

PRODUCE IN	NAME.	BY WHAT BULL.	BY WHOM BRED.
1862, Sep. 16, r.—w.f. H	Luna	Ballarat 1858	Rev. A. Clive.
1863, Oct. 2, r.—w.f. B	Steer	Bertram 1513	do.

LADY.
Red with white face, calved in the year 1848.

Bred by the late Mr. J. Powell, Great Brampton, Hereford; got by Young Sovereign (379), dam (Miss Cotmore the Second) by Lottery the Second (408) g.d. (Miss Cotmore) by Cotmore (376), g.g.d. — by Conqueror (412).

1858, Jan., r.—w.f. H	Laburnum	Mussulman 1333	Mr. Powell.
1859, Sep., r.—w.f. B	Steer	do.	do.
1859, Dec., r.—w.f. H	Leonora	Dutiful 1978	do.
1861, Oct. 25, r.—w.f. H	Lola	Mentor 2112	do.
1862, Oct. 6, r.—w.f. H	Lurline	Doctor 1964	do.

Laburnam, sold to Mr. J. H. Whitehouse, Ipsley Court, Redditch.

LADY.
Red with white face, calved in the year 1861.

Bred by Mr. John Rogers, Letchmoor, Presteign; got by Young David (2884), dam — by Young Royal (1470), g.d. — by Gaylad the Second (1589), g.g.d. — by a bull bred by the late Mr. Powis, Heartsease, g.g.g.d — by Portrait (372).

1864, June, r.—w.f. B		Matchless 2110	Mr. Rogers.

LADY ANN.
Red with white face, of 1857, vol. v., p. 213.

Bred by and the property of Mr. T. Roberts, Ivington Bury,

Leominster; got by Arthur Napoleon (910), dam (Lady Jane) by Cholstrey (217).

PRODUCE IN	NAME.	BY WHAT BULL.	BY WHOM BRED.
1862, June 30, r.—w.f. B	Steer	Master Butterfly 1313	Mr. T. Roberts.
1863, July 10, r.—w.f. B	Oliver Cromwell 2658	Sir Oliver 2nd 1733	do.
1864, July 11, r.—w.f. B	Garrick 3rd 2535	Garrick 2nd 2533	do.

LADY ASHFORD.
Red with white face, calved December 26, 1860.

Bred by Mr. Wm. Tudge, Adforton, Leintwardine, the property of Mr. John Baldwin, Luddington, Stratford-on-Avon; got by Carbonel (1525), dam (Lady) by Orleton (901), g.d. (Stout) by Nelson (1021), g.g.d. (Prettymaid) by Turpin (300).

1863, Sep. 2, r.—w.f. H	Lady Adforton	Pilot 2156	Mr. Tudge.
1864, July 27, r.—w.f. H	Lucy Ashford	do.	do.

Lady Ashford was a winner of the First Prize in her class at the Worcester Meeting of the Royal Agricultural Society of England, and at the Tredegar Meeting, Newport, 1863; also, a First at the Bristol Meeting of the Bath and West of England Society, First at Smithfield, and Second at Birmingham, 1864.

LADY ASHTON.
Red with white face.

Bred by Mr. Manwaring, Berrington, Leominster, the property of Mr. Thomas Roberts, Ivington Bury, Leominster; got by Ashton (1500), g.d. — by The Knight (185).

1863, Oct. 29, r.—w.f. B	Berrington 2414	Sir Thomas 2228	Mr. Roberts.
1864, Nov. 4, r.—w.f. B	Lord Rodney 2623	do.	do.

LADY BERWICK.
Red with white face, calved in the month of March, 1860.

Bred by and the property of Mr. William Lort, King's Norton, Worcestershire; got by Gambler (1247), dam (Woodpigeon) by The Count (351), g.d. (Pigeon the Second) by Young Trueboy (1475), g.g.d. (Pigeon) by Ashley Moor White Bull (870), g.g.g.d. (Damsel) by Cholstrey (868).

1863, Sep. 7, r.—w.f. H	Finella	Cæsar 1894	Mr. Lort.

COWS.

LADYBIRD.

Red with white face, calved December 20, 1860.

Bred by and the property of Mr. P. Turner, The Leen, Pembridge; got by Bertram (1513), dam (Countess) by Silurian (1064), g.d. (Princess) by Andrew the Second (619), g.g.d. (Brenda) by Viscount (816), g g.g.d. (Rarity) by Cupid (1950).

PRODUCE IN	NAME.	BY WHAT BULL.	BY WHOM BRED.
1863, July 17, r.—w.f. B	Ambassador 2388	Bolingbroke 1883	Mr. Turner.
1864, July 4, r.—w.f. H	Dowager	do.	do.

LADY BYRON.

Red with white face, calved in the year 1861.

Bred by and the property of Mr. T. Morris, Therrow, Llyswen, Hay; got by Druid (1220), dam (Miss Byron the Second) by Prior (1359), g.d. (Miss Byron) by Young Byron (832), g.g.d. (Miss Hope) by Young Hope (343), g.g.g.d. (Beauty the Second) by Counsellor (422).

1863, Sep., r.—w.f. H	Ada	Don Salisbury 1969	Mr. Morris.
1864, Aug., r.—w.f. H	Medora	Prince Imperial 2171	do.

LADY GREY.

Grey, calved June 25, 1854.

Bred by and the property of Mr. James Bourn, Mawley Town Farm, Cleobury Mortimer; got by Regent (1705), dam (Daisy) by Bowlegs (1517), g.d. (Damsel).

1859, Dec. 5, r.—w.f. H		Wigmore 1800	Mr. Bourn.
1860, Oct. 2, r.—w.f. {B/B} Twins		Cardinal 1526	do.
1861, Aug. 27 {r.—w.f. B / grey H} Twins		do.	do.
1862, Sep. 22, r.—w.f. H	Linnet	do.	do.
1863, Sep. 3, r.—w.f. H	Lady Jane	do.	do.

Lady Jane, sold to Mr. Vincent Gosford, Tanylan, Holywell.

COWS.

LADY HASTINGS.
Red with white face, of 1858, vol. v., p. 214.

Bred by and the property of Mr. T. Roberts, Ivington Bury, Leominster; got by Master Butterfly (1313), dam (Prima Donna) by King James (978), g.d. (Long Horns) by Andrew the Second (619), g.g.d. (Pigeon) by Prince by Dayhouse (299.)

PRODUCE IN	NAME.	BY WHAT BULL.	BY WHOM BRED.
1862, July 13, r.—w.f. H	MissHastings2nd	Sir Thomas 2228	Mr. Roberts.
1863, Sep. 15, r.—w.f. B	Lord Hastings 2616	do.	do.
1864, Aug., r.—w.f. H		do.	do.

LADY HEREFORD.
Red with white face.

Bred by Capt. Power, Hill Court, Ross, the property of Mr. William Jones, Hill of Eaton, Ross; got by Uncle Tom (1108), dam — by Vanguard (1109), g.d. — bred by the late Viscount Hereford.

PRODUCE IN	NAME.	BY WHAT BULL.	BY WHOM BRED.
1862, r.—w.f. H		Geologist 2012	Mr. Wm. Jones.
1863, June 2, r.—w.f. H		do.	do.
1864, May 3, r.—w.f. B	Steer	Esperus	do.

LADY HONEYBOURNE.
Red with white face, calved in the month of May, 1859.

Bred by and the property of Mr. J. Prosser, Honeybourne Grounds Worcester; got by Medallist (1009).

PRODUCE IN	NAME.	BY WHAT BULL.	BY WHOM BRED.
1863, Jan., r.—w.f. H	Lady of the Grounds	The Jew 2266	Mr. J. Prosser.
1864, Mar. 10, r.—w.f. H	Weston Lass	do.	do.

LADY JANE.
Red with white face, calved in the year 1860.

Bred by Mr. John Rogers, Boultibrook, Presteign, the property of the Rev. A. Clive, Whitfield; got by Mameluke (1307), dam

COWS.

— by Sir David the Second (1065), g.d. — by Albert (330), g.g.d. — by Prince (251), g.g.g.d. — Young Old Court (1341).

PRODUCE IN		NAME.	BY WHAT BULL.	BY WHOM BRED.
1862, July,	r.—w.f. H	Janet	Ballarat 1858	Rev. A. Clive.
1863, July 20,	r.—w.f. B	Steer	Bertram 1513	do.
1864, Oct. 14,	r.—w.f. H	Linnet	Plato 2160	do.

LADY LUCY.

Red with white face, of 1858, vol. v., p. 214.

Bred by and the property of Mr. T. Roberts, Ivington Bury, Leominster; got by Arthur Napoleon (910), dam (Lady Jane) by Cholstrey (217).

PRODUCE IN		NAME.	BY WHAT BULL.	BY WHOM BRED.
1862, Aug.,	r.—w.f. B	Steer	Master Butterfly 1313	Mr. Roberts.
1863, Sep. 9,	r.—w.f. B	Lord Lucy 2619	Sir Thomas 2228	do.
1864, Sep. 3,	r.—w.f. H	Lady Lucy 2nd	Sir George 2763	do.

LADY NOBLE.

Red with white face, of 1851, vol. v., p. 215.

Bred by Mr. T. Morris, Therrow, Llyswen, Hay, the property of Mr. E. Wright, Halston Hall, Oswestry; got by Young Byron (832), dam (Miss Noble) by Noble (543), g.d. (Favourite) by a son of Sovereign (404), g.g.d. (Damsel) by Young Wellington (505).

PRODUCE IN		NAME.	BY WHAT BULL.	BY WHOM BRED.
1863, Aug. 11,	r.—w.f. H	Nobless	Magnet 2nd 989	Mr. Wright.
1864, Oct. 25,	r.—w.f. B	Baron 2401	Hero 2039	do.

LADY THE SECOND.

Red with white face, of 1856, vol. v., p. 215.

Bred by and the property of Mr. T. Edwards, Wintercott, Leominster; got by Croft (937) dam (Lady) by Paddock (773), g.d. (Lovely) by Coningsby the Second (1552).

PRODUCE IN		NAME.	BY WHAT BULL.	BY WHOM BRED.
1862, Oct. 7,	r.—w.f. H	Ladylift	Sir Newton 1731	Mr. Edwards.
1863, Aug. 14,	r.—w.f. B	Adforton 3rd 2372	Adforton 1839	do.
1864, Aug. 16,	r.—w.f. B	Young Grove 2888	do.	do.

cows.

LADY THE THIRD.
Red with white face, calved June 12, 1860.

Bred by and the property of Mr. Naylor, Leighton Hall, Montgomeryshire; got by Tom of Lincoln (1099), dam (Lady the Second) by Ashford (1499), g.d. (Lady) by Attraction (892), g.g.g.d. (Lady Beddoes).

PRODUCE IN	NAME.	BY WHAT BULL.	BY WHOM BRED.
1863, July 23, r.—w.f. H	Lady 4th	Salisbury 2204	Mr. Naylor.

LADY WISEMAN.
Red with white face, calved July 30, 1859.

Bred by and the property of Mr. Edward Bowen, Corfton, Ludlow; got by Cardinal Wiseman (1168), dam — by Governor (464), g.d. — by Mercury (361), g.g.d. — by Corfton (1188), g.g.g.d. — bred by the late Lord Rodney.

1862, July 24, r.—w.f. B	Don Juan 2502	Cardinal Wiseman 1168	Mr. Bowen.
1863, Aug. r.—w.f. B	Steer	do.	do.
1864, Aug. 10, r —w.f. B	Macaroni 2627	Oxenbold 2145	do.

LARK.
Red with white face, of 1860, vol. v., p. 239.

Bred by and the property of Mr. E. Tanner, Hopton Castle Clun, Salop; got by Buckton (1891), dam (Moorhen) by Northampton (600), g.d. (Lottery.)

1863, Oct. 5, r.—w.f. B	The Digger 2813	The Doctor 1083	Mr. Tanner.
1864, Nov. 14, r.-w.f. { B / H }	Twins	do.	do.

LARKSPUR.
Red with white face, calved April 2, 1861.

Bred by and the property of Mr. C. H. Hinckesman, the Poles, Ludlow; got by The Friar (1085), dam (Lady) by Old Jim

(1026) g.d. (Duchess) by Nelson (1021), g.g.d. — bred by Mr. W. Jellicoe, by Lundyfoot (88), g.g.g.d. — bred by Mr. B. Tomkins, g.g.g.g.d. — bred by Mr. Turner, Noke.

PRODUCE IN	NAME.	BY WHAT BULL.	BY WHOM BRED.
1864, Sep. 10, r.—w.f. B	Bromfield 2432	Berwick 1874	Mr. Hinckesman.

LARKSPUR.
Red with white face, calved March 7, 1860.

Bred by and the property of Mr. John Monkhouse, The Stow, Hereford; got by Chieftain (930)), dam (Dahlia) by Madoc (899), g.d. (Dahlia) by Guy Fawkes (581), g.g.d. (Daffodil) by Charity (375), g.g.g.d. (Tulip) by Sir Andrew (183).

PRODUCE IN	NAME.	BY WHAT BULL.	BY WHOM BRED.
1863, June 24, r.—w.f. H	Skylark	Sir Richard 1734	Mr. Monkhouse.

LARKSPUR.
Red with mottled face, calved September 2, 1860.

Bred by and the property of Capt. Peploe, Garnstone, Weobley; got by The Twin (1420), dam (Larkspur) by Musician (725), g.d. (Larkspur) by Victory (2297), g.g.d. (Larkspur) by Semplon (58). g.g.g.d. (Twin the Second) by George (2013), g.g.g.g.d. (Twin) by Phœnix (55).

PRODUCE IN	NAME.	BY WHAT BULL.	BY WHOM BRED.
1864, Jan. 23, r.—m.f. H	Daisy	Leo 2070	Capt. Peploe.

·LAUNDRESS.
Red with white face, of 1858, vol. v., p. 217.

Bred by and the property of Mr. John Partridge, Bishop's Wood, Ross; got by General (1251), dam (Lady the Second) by Brecon (918), g.d. (Ladyday) by Newton (344), g.g.d. (Cowslip) by Young Hope (343),

PRODUCE IN	NAME.	BY WHAT BULL.	BY WHOM BRED.
1862, May 13, r.—w.f. B	Steer	Noble Tom 2135	Mr. Partridge.
1863, April 10, r.—w.f. B	do.	do.	do.
1864, July 5, r.—w.f. H	Misss Garway	Garway 2536	do.

COWS.

LAURA.

Red with white face, calved July 1, 1861.

Bred by and the property of Mr. T. S. Bradstock, Cobrey Park, Ross; got by The Jew (2266), dam (Margaret) by Daniel (1201), g.d. (Stately) by Deluge (1210) g.g.d. (Lofty) by Young Gaylad (1463).

PRODUCE IN	NAME.	BY WHAT BULL.	BY WHOM BRED.
1863, Aug. 13, r.—w.f. B	Lord Lincoln 2618	Young Ramber 2335	Mr. Bradstock.
1864, Aug. 2, r.—w.f. H	Lydia	do.	do.

LAURA.

Red with white face, calved January 2, 1860.

Bred by and the property of Mr. John Jones, Llwyn-y-Gaer, Raglan; got by Chancellor (1172), dam (Lolla) by Patron (803), g.d. (Lily).

1863, May, 1, r.—w.f. B	Steer	Bold David 1881	Mr. Jones.
1864, Jan., 1, r.—w.f.	Dead	do.	do.
1864, Dec. 6, r.—w.f. B	Ragman 2709	Trusty 2846	do.

Laura with her sire and dam were winners of a prize at the meetings of the Abergavenny and Tredegar Agricultural Societies, 1860; she was also a winner at Tredegar, 1861; and as one of eight at Monmouth, 1862.

LAURA.

Red with white face, calved March 1, 1856.

Bred by and the property of Mr. J. R. Paramore, Dinedor Court, Hereford; got by a Bull bred by the late Mr. J. Turner, Noke Court, dam — bred by Dr. Lamb, Henwood, Dilwyn.

1858, May 3, r.—w.f. H	Ada	Hotspur 972	Mr. Paramore.
1859, June 5, r.—w.f. H	Laura 2nd	The General 2817	do.
1860, June 18, r.—w.f. H	Laura 3rd	do.	do.
1861, r.—w.f. B	Dead	do.	do.
1862, r.—w.f. B	Steer	do.	do.
1863, Aug. 24, r.—w.f. H	Laura 4th	The Jew 2266	do.

COWS.

LAURA.

Red with white face, calved December 15, 1858.

Bred by and the property of Mr. James Bourn, Mawley Town Farm, Cleobury Mortimer; got by Wigmore (1800), dam (Lady Grey) by Regent (1705), g d. (Daisy) by Bowlegs (1517). g.g.d. (Damsel).

PRODUCE IN	NAME.	BY WHAT BULL.	BY WHOM BRED.
1861, Aug. 18, r.—w.f. H	Lovely	Cardinal (1526)	Mr. Bourn.
1862, July 27, r.—w.f. H	Luna	do.	do.
1863, July 5, r.—w.f. B	Steer	do.	do.
1864, July 13, r.—w.f. B	Sweetmeat 2798	do.	do.

LAURA THE SECOND.

Red with white face, calved June 5, 1859.

Bred by and the property of Mr. J. R. Paramore, Dinedor Court, Hereford; got by The General (2817), dam (Laura) by a Bull bred by the late Mr. Turner, Noke Court, g. d. — bred by Dr. Lamb, Henwood, Dilwyn.

1861, July 11, r.—w.f. B	Steer	The General 2817	Mr. Paramore.
1863, Aug. 26, r.—w.f. H	Blossom	The Jew 2266	do.

LAURA THE THIRD.

Red with white face, calved June 18, 1860.

Bred by and the property of Mr. J. R. Paramore, Dinedor Court, Hereford; got by The General (2817), dam (Laura) by a Bull bred by the late Mr. Turner, Noke Court, g.d. — bred by Dr. Lamb, Henwood, Dilwyn,

1863, Aug. 24, r.—w.f. H	Damsel	The Jew 2266	Mr. Paramore.
1864, Aug. 2, r.—w.f. H		Portly 2165	do.

COWS.

LEIGHTON PIGEON.
Red with white face, calved November 11, 1856.

Bred by and the property of Mr. P. Ballard, Leighton Court, Bromyard; got by Young Sir David (1137), dam (Thingehill Pigeon) by Reform (508), g.d. (Hampton Pigeon) by Young Sovereign (506), g.g.d. (Sylph) by Chance (355).

PRODUCE IN	NAME.	BY WHAT BULL.	BY WHOM BRED.
1860, Dec. 5, r.—w.f. B	Leighton 2069	Tell-tale 1757	Mr. Ballard.
1862, Sep. 14, r.—w.f. B	Banjo 2394	Tambarine 2254	do.

LILIUM.
Red with white face, calved July 7, 1860.

Bred by and the property of Mr. Evan Davies, Patton, Wenlock, Salop; got by Franky (1243), dam (Lilac) by Tyrant (1784).

1863, July 26, r.—w.f. B	Dead	Lincoln 2076	Mr. Evan Davies.
1864, June 10, r.—w.f. H	Lufra	do.	do.

LILY.
Red with white face, calved in the year 1858.

Bred by Mr. H. Gibbons, Hampton Bishop, Hereford, the property of Mr. J. R. Paramore, Dinedor Court; got by Medallist (1009), dam (Lily) by Young Gaylad (1463), g d. (Lily) by Zephyr (1826), g.g.d. (Silk) by Dewshall (358).

1861, Oct. 30, r.—w.f. B	Steer	Shamrock 2nd 2210	Mr. Gibbons.
1862, Sep., r.—w.f. H		do.	do.
1863, Aug. 10, r.—w.f B	Steer	do.	Mr. Paramore.
1864, Sep., 22, r.—w.f. B		Grateful 1596	do.

LILY.
Red with white face, calved in the year, 1858.

Bred by and the property of Mr. John Rogers, Letchmoor, Presteign; got by Trusty (2847), dam — by Gaylad the Second (1589).

1861, r.—w.f. B	Steer	Young David 2884	Mr. Rogers.
1862, r.—w.f. H	Lily 2nd	Matchless 2110	do.

LILY.

Red with white face, calved August 7, 1858.

Bred by and the properey of Mr. T. Olver, Penhallow, Grampound, Cornwall; got by the Duke of Cornwall (1569), dam (Lily) by Rory o'More (1711).

PRODUCE IN	NAME.	BY WHAT BULL.	BY WHOM BRED.
1862, Nov. 1, r.—w.f. B	Steer	Volunteer 2300	Mr. Olver.
1863, Nov. 19, r.—w.f. B		Sir Hugh 2223	do.

LINNET.

Red with white face, calved in the month of October, 1859.

Bred by the Rev. A. Clive, Whitfield, Hereford, the property of Mr. J. R. Paramore, Dinedor Court, Hereford; got by Alma (1144), dam (Lark) by Royal (331).

1863, June 12, r.—w.f. B	Sovereign 2786	Bertram 1513	Mr. Paramore.
1864, May 1, r.—w.f. H	Queen of May	The Priest 1416	do.

LIONESS.

Red with white face, calved March 16, 1861.

Bred by and the property of Mr. W. D. Turner, Lynch Court, Leominster; got by Logic (2079), dam (Little Lady) by The Doctor (1083), g.d. (Rebecca) by Burton (1159), g.g.d. (Vesta) by Andrew the Second (619), g.g.g.d. — bred by the late Mr. Turner, Aymestry.

1863, Dec. 5, r.—w.f. B		Logic 2079	Mr. Turner.
1864, Nov. 15, r.—w.f. B		Stockwell 2793	do.

LIONESS.

Red with white face, of 1857, vol. v., p. 220.

Bred by Lord Bateman, Shobdon Court, Leominster, the property of Mr. Wright, Halston Hall, Oswestry; got by Carlisle (923), dam (Lofty) by Albert Edward (859).

1862, July 7, r.—w.f. B	Lion 2600	Magnet 2nd 989	Mr. Wright.
1863, Nov. 24, r.—w.f. B	Lion 2nd 2601	do.	do.
1864, Oct. 5, r.—w.f. H	Lofty	Prince 2691	do.

LITTLE POLLY.
Red with white face, calved in the year 1861.

Bred by Mr. Sheriff, Coxall, Ludlow, the property of Mr. R. H. Capper, The Northgate, Ross; got by Sir Colin (2216), dam (Miss Noble) by Brilliant (1518),

PRODUCE IN	NAME.	BY WHAT BULL.	BY WHOM BRED.
1864, July 30, r.—w.f. H	Polly	Garibaldi 2008	Mr. Capper.

LIVELY.
Red with white face, calved October 22, 1859.

Bred by and the property of Mr. T. Edwards, Wintercott, Leominster; got by Leominster (1634), dam (Lady the Second) by Croft (937), g.d. (Lady) by Paddock (773), g.g.d. (Lovely) by Coningsby the Second (1552).

1862, July 22, r.—w.f. H	Young Lively	Ben 1870	Mr. Edwards.
1863, July 18, r.—w.f. B	Comet 2469	Sir William 2233	do.
1864, July 18, r.—w.f. H		Adforton 1839	do.

LIZZIE.
Red with white face, calved in the month of April, 1860.

Bred by and the property of Mr. Rees Keene, Pencraig, Caerleon, Monmouth; got by General Wyndham (1590), dam (Bonny) by Prince Albert (2168), g.d. (Bonny) by Young David (2325), g.g.d. (Beauty) by Foxhall (2520).

1863, March, r.—w.f. B	Steer	Dolward 1966	Mr. Keene.
1864, March, r.—w.f. H	Bella	Cholstrey 2nd 1919	do.

LOFTY.
Red with white face, of 1855, vol. iv., p. 144.

Bred by Lord Berwick, Cronkhill, Salop, the property of Mr. W. Newbery, Fernhill, Kenilworth; got by Attingham (911), dam (Grey Oak-apple) by Tom Thumb (243), g.d. (Oak-apple) by

Commerce (354), g.g.d. (Strawberry) bred by the late Mr. Jeffries by a son of Guinea.

PRODUCE IN	NAME.	BY WHAT BULL.	BY WHOM BRED.
1862, Dec. 20, r.—w.f. H	Clara	Comus 2477	Mr. Newbery.
1864, Jan. 14, r.—w.f. B	Albert Victor 2381	do.	do.

LOFTY.

Red with white face, calved November 30, 1855.

Bred by and the property of Mr. H. Gibbons, Hampton Bishop, Hereford; got by The Admiral (1078), dam (Lofty) by Young Gaylad (1463).

1858, Oct. 13, r.—w.f. H		Defence 2nd 1208	Mr. Gibbons.
1859, Sep. 23, r.—w.f. B	Steer	Woodman 2nd 1459	do.
1860, Oct. 15, r.—w.f. B	do.	do.	do.
1861, Oct. 24, r.—w.f. H	Jessie	Shamrock 2nd 2210	do.
1862, Sep. 1, r.—w.f. B	dead	do.	do.
1863, Aug. 28, r.—w.f. B		do.	do.
1864, Sep. 12, r.—w.f. B	Steer	do.	do.

LOFTY THE SECOND.

Red with white face, calved November 3, 1857.

Bred by the late Mr. James Rea, Monaughty, Knighton, the property of Mr. P. J. Kearney, Miltown House, Clonmellon, Ireland; got by Treasurer (1105), dam (Lofty) by Regent (891), g.d. (Lofty) by Barrister (658), g.g.d. (Lofty) by Mask (307), g.g.g.d. — by Duke (304).

1864, Jan. 10, r.—w.f. B	Longbow 2610	Sir Benjamin 1387	Mr. Kearney.

LONG THE SECOND.

Red with white face, calved in the month of July, 1861.

Bred by and the property of Mr. J. Jones, Llwyn-y-gaer, Raglan; got by Chancellor (1172), dam (Lucy Long) by

COWS.

Patron the Third (2151), g.d. (Long) by a son of Garway (2009), g.g.d. — by Chance's son (1911).

PRODUCE IN	NAME.	BY WHAT BULL.	BY WHOM BRED.
1864, Sep. r.—w.f. B	Reliance 2713	Red Man 2715	Mr. Jones.

Long the Second, with sire and dam were winners of the First Prize in their class at the meetings of the Abergavenny and Monmouthshire Agricultural Societies, and Second at Tredegar, 1861; she was also one of a pair winners of the First Prize at Abergavenny, and Second at Monmouth, 1862; and First at Monmouth, 1863.

LONG HORNS.

Red with white face, calved in the year 1849.

Bred by and the property of Mr. W. H. Oatley, Wroxeter, Salop; got by Surprise (779), dam (Princess Royal) by Emigrant (1980), g.d. (Sovereign Cow) by Sovereign (404).

1852, Oct. r.—w.f. B	Steer		Mr. Oatley.
1853, Oct. r.—w.f. B		Walford, 871	do.
1854, r.—w.f. B	Steer	Julius Cæsar 2054	do.
1855, r.—w.f. B	do.	do.	do.
1856, r.—w.f. B	do.	do.	do.
1857, Dec. r.—w.f. H	Agrippina	do.	do.
1859, Jan. r.—w.f. H	Agnes	do.	do.
1859, Dec. r.—w.f. H	Augusta	Carausius 2446	do.
1860, Nov. r.—w.f. B	Steer	do.	do.
1861, Oct. r.—w.f. H			do.
1862, July 24, r.—w.f. H	Angelica	Franky 1243	do.
1863, July 3, r.—w.f. B	Steer	Claudius Cæsar 1922	do.
1864, Aug. 5, r.—w.f. H	Agnes	Nero 2650	do.

LOO.

Red with white face, calved in the year 1861.

Bred by and the property of Mr. W. Lort, The Cotteridge, King's Norton; got by Gambler (1247), dam (Silk) by Clyro, g.d. (Duchess) by Quicksilver (858), g.g.d. (Fairmaid) bred by Mr. J. Hewer, Vern House, Marden.

1864, July, r.—w.f. H	Nola	Nabob 2648	Mr. Lort

LOVELY.
Red with white face, calved September 2, 1862.

Bred by and the property of Mr. T. Olver, Penhallow, Grampound, Cornwall; got by Earl Derby (1979), dam (Lady) by The Earl (1761), g.d. — bred by Mr. P. Turner, The Leen, Leominster.

PRODUCE IN	NAME.	BY WHAT BULL.	BY WHOM BRED.
1864, Sep. 12, r.—w.f. H	Laura	Zippor 2354	Mr. T. Olver.

LOVELY.
Red with white face, calved in the month of September, 1860.

Bred by Mr. John Hollings, Hillend, Hereford, the property of Mr. J. A. Hollings, Hillend, Hereford; got by Woodman (1460), dam (Gentle) by Noke (1338), g.d. (Daisy) by Byron (380), g.g.d. (Cherry the Fourth) by Reveller (2722), g.g.g.d. (Cherry the Third) by Cornet (1933), g.g.g.g.d. (Cherry the Second) by Young Waterloo (2341), — (Cherry the First) by Foxley.

1863, Oct. 22, r.—w.f. H	Lovely	St. Clement 2201	Mr. Hollings
1864, Dec. 8, r.—w.f. B	Steer	Chieftain 2nd 1917	do.

LOVELY.
Red with mottled face, of 1851, vols. iii. and iv., pp. 186, 146.

Bred by Mr. E. Price, Court House, Pembridge, the property of Mr. W. R. Grose, Penpont, Wadebridge, Cornwall; got by Pembridge (721), dam (Luck's All) by Prince Dangerous (362), g.d. (Mottle) by The Sheriff (356), g.g.d. (Lady) by Forester (398), g.g.g.d. (Pink) by Crabstock (303).

1860, Oct. 4, r.—m.f. B	Steer	Volunteer 2300	Mr. Grose
1861, Sep. 25, r.—w.f. H	Barbara	do.	do.
1862, Aug. 21, r.—m.f. B	Dick the Dustman 2495	Conservative 1931	do.

COWS.

LOVELY.

Red with white face, calved August 22, 1861.

Bred by and the property of Mr. H. Gibbons, Hampton Bishop, Hereford; got by Shamrock the Second (2210), dam (Lovely) by Defence the Second (1208), g.d. (Spot) by The Admiral (1078), g.g.d. (Old Spot) by Young Gaylad (1463).

PRODUCE IN	NAME.	BY WHAT BULL.	BY WHOM BRED.
1864, June 22, r.—w.f. B		Trumpeter 2282	Mr. Gibbons

LOVELY.

Red with white face, of 1857, vol. v., p. 225.

Bred by and the property of Mr. B. Hawkins, Orleton, Ludlow; got by Jerry (976), dam (Dainty) by Emperor (221), g.d. (Dainty) by Quality (703).

1862, July 22, r.—w.f. B	Steer	The Grove 1764	Mr. Hawkins.
1863, May 22, r.—w.f. H	Lily	do	do.
1864, May 17, r.—w.t. H	Lovely 2nd	Noble Boy 1337	do.

LOVELY.

Red with white face, calved November 9, 1860.

Bred by the late Lord Berwick, Cronkhill, Salop; the property of the Hon. and Rev. Noel Hill, Berrington, Salop; got by Severn (1382), dam (Bedstone Prettymaid) by Conrad (1183), g.d. (Prettymaid) by Dinedor (395), g.g.d. — bred by Mr. Stedman, Bedstone Hall, Salop.

1863, Sep. 5, r.—w.f. B	Steer	Van Tromp 2291	Hon. and Rev. Noel Hill.
1864, Sep. 11, r.—w.f. B		Conqueror 1929	do.

LOVELY THE SECOND.

Red with white face, calved August 29, 1859.

Bred by and the property of Mr. James Taylor, Stretford Court,

Leominster; got by St. Oswall (1378), dam (Lovely) by Young Emperor (1812).

PRODUCE IN	NAME.	BY WHAT BULL.	BY WHOM BRED.
1863, Jan. 25, r.—w.f. B	Steer	Croft 937	Mr. Taylor.
1864, May 12, r.—w.f. H	Hebe	Rambler 1046	do.

Lovely the Second was one of seven winners of the First Prize in their class at the meetings of the Herefordshire and the Leominster Agricultural Societies, 1864.

LOVELY THE THIRD.

Red with white face, calved September 9, 1860.

Bred by and the property of Mr. James Taylor, Stretford Court, Leominster; got by Croft (937), dam (Lovely) by Young Emperor (1812).

1863, Oct. 14, r.—w.f. H	Lovely 4th	Trustful 2845	Mr. Taylor.
1864, Oct. 12, r.—w.f. B	Steer	do.	do.

Lovely the Third was one of four winners of the First Prize in their class at the meeting of the Ludlow Agricultural Society, and one of six winners of a First Prize at Hereford, 1862; she was also one of four at Leominster, and one of seven at Hereford, winners of First Prizes, 1864.

LUCY.

Red with white face, calved in the year 1859.

Bred by Mr. J. B. Green, Marlow, Salop, the property of Capt. Crawshay, Danypark, Crickhowell; got by Vanguard (1109), dam — by Beefy Ben (1869), g.d. — by Cholstrey (217), g.g.d. — by Zest of Oxford (2352), g.g.g.d. — by Discount (339).

1862,	r.—w.f. B	Steer	Zealot 2344	Mr. Green.
1863,	r.—w.f. H	Lucena	do.	do.
1864,	r.—w.f. B		Zealous 2349	Capt. Crawshay

LUCY.

Red with white face, of 1855, vols. iv. and v., pp. 148, 227.

Bred by the late Mr. James Rea, Monaughty, Knighton, the property of Mr. E. Lewis, Breinton, Hereford; got by Chieftain

COWS.

(930), dam (Fairmaid) by Regent (891)) g.d. (Fairmaid) by Barrister (658), g.g.d. (Fairmaid) by Old Court (306).

PRODUCE IN		NAME.	BY WHAT BULL.	BY WHOM BRED.
1862, Sep. 20,	r.—w.f. H	Lucy 5th	Lord Nelson 2088	Mr. Thos. Rea.
1864, Mar. 16,	r.—w.f. B	Triumph 2836	Sir Benjamin 1387	do.

LUCY.
Red with white face, calved July 8, 1860.

Bred by and the property of Mr. Naylor, Leighton Hall, Montgomeryshire; got by Admiral (1481), dam (Purity) by Silvester (797), g.d. (Wren) by Big Ben (248)) g.g.d (Tidy), bred by Mr. Rea, Monaughty, g.g.g.d. — by Old Court (306).

1863, June 18th, r.—w.f. H	Lucella	Salisbury 2204	Mr. Naylor.

LUCY LONG.
Red with white face, calved July 10, 1857.

Bred by and the property of Mr. T. Morris, Therrow, Llyswen, Hay; got by Telegraph (1404), dam (Lucy) by Newton (344), g.d. (Lucy) by Young Hope (343), g.g.d. (Lady), by Prince Llewellyn (713), g.g.g.d. (Lily) by Charity the Second (516).

1860,	r.—w.f. B	Steer	Druid 1220	Mr. Morris
1861,	r.—w.f. B	do.	do.	do.
1862,	r.—w.f. B	do.	do.	do.
1863,	r.—w.f. B	do.	Prince Imperial 2171	do.
1864,	r.—w.f. H	Lovely	do.	do.

LUCY THE SECOND.
Red with white face, calved June 14, 1860.

Bred by and the property of Mr. H. R. Evans, jun., Swanstone Court, Leominster; got by Sir Franklin (1068), dam (Lucy) by Rambler (1046), g.d. (Young Lovely) by Emperor (373), g.g.d. (Lovely) by Young Trueboy (1475), g.g.g.d. (Lovely) by Ashley Moor White Bull (870), g.g.g.g d (Old Damsel) by Coleman's Bull (1547), — (Old Daisy) by Chancellor (156).

1863, April 27, r.—w.f. H	Hope	Chatham 1914	Mr. Evans.

Hope was one of four winners of the First Prize in their class at the meeting of the Leominster Agricultural Society, 1863.

LUCY THE SECOND.
Red with white face, calved February 5, 1860.

Bred by the late Mr. Thomas Rea, Westonbury, Leominster, the property of Lord Wenlock, Salop; got by Sir Benjamin (1387), dam (Lucy) by Chieftain (930), g.d. (Fairmaid) by Regent (891), g.g.d. (Fairmaid) by Barrister (658), g.g.g.d. (Fairmaid) by Old Court (306).

PRODUCE IN	NAME.	BY WHAT BULL.	BY WHOM BRED.
1862, Feb. 16, r.—w.f. H		Sir Richard 1734	Mr. Rea.
1862, Dec. 29, r.—w.f. H	Lucy 6th	Whitfield	do.

Lucy the Sixth, sold to Mr. Thomas.

LUCY THE THIRD.
Red with white face, of 1860, vol. v., p. 227.

Bred by the late Mr. Thomas Rea, Westonbury, Leominster, the property of Mr. John Burlton, Luntley Court, Leominster; got by Sambo (1720), dam (Lucy) by Chieftain (930), g.d. (Fairmaid) by Regent (891), g.g.d. (Fairmaid) by Barrister (658), g.g.g.d. (Fairmaid) by Old Court (306).

1864, July 24, r.—w.f. H	Lola	Sir Richard 1734	Mr. Rea.

LUNA.
Red with white face, calved in the month of February, 1857.

Bred by and the property of Mr. G. T. Forester, High Ercall, Wellington, Salop; got by Darling (1202), dam (Hottentot) by Wonder (420), g.d. (Venus) by Hope (439), g.g.d. (Miss Fitzfavourite) by Young Fitzfavourite, g.g.g.d. — bred by the late Mr. Jeffries, Grove, by Fitzfavourite (441).

1859, Nov. r.—w.f. H	Diana	Discord 1217	Mr. Forester.
1860, Sep. r.—w.f. B	Steer	do.	do.
1861, July 2, r.—w.f. H	Darky	do.	do.
1862, Aug. 4, r.—w.f. H	Selene	Severn 1382	do.
1863, July 10, r.—w.f. H	Cynthia	do.	do.
1864, July 1, r.—w.f. B	Steer	do.	do.

COWS.

LUNA.
Red with white face, calved January 28, 1861.

Bred by the Rev. Archer Clive, Whitfield, Hereford, the property of Mr. William Stallard, Brockhampton, Ross; got by Baron of Noke (1862), dam (Lofty) by Sir David (349), g.d. (Lofty) by Andrew the Second (619), g.d. — by Monarch (504).

PRODUCE IN	NAME.	BY WHAT BULL.	BY WHOM BRED.
1863, Aug. 6, r.—w.f. B	Sol 2784	Chieftain 2nd 1917	Mr. Stallard.
1864, Aug. 16, r.—w.f. B	Steer	do.	do.

LUPIN.
Red with white face, calved December 25, 1858.

Bred by and the property of Mr. W. Lane, Compton Casey; got by Tyro (1786), dam (Broad) by Planet (1690).

1862, Jan. 18, r.—w.f. H	Larkspur	Hardy 2027	Mr. Lane.

LYDIA.
Red with white face, of 1857, vol. v., p. 229.

Bred by Mr. Rea, Monaughty, Knighton, the property of Mr. J. P. Apperley, Fownhope, Hereford; got by Rufus (1058), dam (Gwenny) by Regent (891), g.d. (Venus the Fifth) by Albert (330), g.g d. (Winifred) by Monaughty (220), g.g.g.d. (Venus the Fourth) by Duke (304), &c.

1863, Oct. 22, r.—w.f. B	Steer	Sir Benjamin 1387	Mr. Apperley.
1864, Dec. 10, r.—w.f. B		Captain Perry 2444	do.

MABEL.
Red with white face, calved February 13, 1862.

Bred by and the property of Mr. R. Davey, M.P., Polsue House, Grampound; got by (Penhallow) (2154), dam (Ringdove) by Great Eastern (1598), g.d. (Young Pigeon) by a son of Confidence (367).

PRODUCE IN	NAME.	BY WHAT BULL.	BY WHOM BRED.
1864, Feb. 23, r.—w.f. B		Zippor 2354	Mr. Davey.

MAID.
Red with white face, calved July 24, 1861.

Bred by and the property of Mr. Naylor, Leighton Hall, Montgomeryshire; got by Admiral (1481), dam (Milkmaid) by Tom of Lincoln (1099), g.d. (Miriam) by Silvester (797), g.g.d. (Lovely) by Big Ben (248), g.g.g.d. (Strapper) bred by Mr. Rea, Monaughty, g.g.g.g.d. — by Old Court (306).

1864, Oct. 13, r.—w.f. H	Maiden	Salisbury 2204	Mr. Naylor.

MAIDEN.
Red with white face, of 1854, vol. v., p. 229.

Bred by and the property of Mr. John Monkhouse, The Stow, Hereford; got by Madoc (899), dam (Maiden) by Guy Fawkes (581), g.d. (Damsel) by Sir Andrew (183), g.g.d. (Virgin) by Grove (370), g.g.g d. — by Solon (92).

1862, Sep. 10, r.—w.f. B	Steer	Chieftain 930	Mr. Monkhouse.
1883, Sep. 2, r.—w.f. B	Steer	do.	do.
1864, Aug. 20, r.—w.f. B		do.	do.

MAJESTY.
Red with white face, calved August 8, 1859.

Bred by the late Mr. Sobey, Tencreek, Liskeard, the property of Mr. W. R. Grose, Penpont, Wadebridge, Cornwall; got by Big Ben (1875), dam Miss (Coningsby) by Coningsby (718), g.d.. (Nockerell, late Lilac) by Berrington (435).

1861, Dec. 21, r.—w.f. H	Bessie	Penhallow 2154	Mr. Grose.
1864, April 5, r.—w.f. B	Fanny	Sir Hugh 2223	do.

COWS.

MARCHIONESS.

Red with white face, calved September 19, 1860.

Bred by and the property of Mr. T. Duckham, Baysham Court, Ross; got by Cronkhill (1558), dam (Duchess) by Uncle Tom (1108), g.d. (Countess) by Colossus (591), g.g.d. (Duchess) by Woodman (402), g.g.g.d. (Dainty) by Sheriff (397), g.g.g.g.d. (Duchess) bred by Mr. Corbett, The Sheriff's, Leominster.

PRODUCE IN	NAME.	BY WHAT BULL.	BY WHOM BRED.
1863, June 16, r.—w.f. B	Steer	Castor 1900	Mr. Duckham.
1864, April 25, r.—w.f. H	Lucy	Cato 1902	do.

MARCHIONESS.

Red with white face, calved October, 24, 1860.

Bred by and the property of Mr. P. Turner, The Leen, Pembridge; got by Bertram (1513), dam (Luna) by Sir David (349), g.d. (Stella) by Andrew the Second (619), g.g.d. (Sprite) by Sir Walter (352), g.g.g.d. (Sylph) by Viscount (816).

1863, July 11, r.—w.f. H	Queen of the Vale	Bolingbroke 1883	Mr. Turner.
1864, July 21, r.—w.f. H	Primula	do.	do.

MARGARET.

Red with white face, calved in the month of October, 1857.

Bred by and the property of Mr. T. S. Bradstock, Cobrey Park, Ross; got by Daniel (1201), dam (Stately) by Deluge (1210), g.d. (Lofty) by Young Gaylad (1403).

1860, July 29, r.—w.f. B	Bluster	Uncle Tom 1108	Mr. Bradstock.
1861, July 1, r.—w.f. H	Lana	The Jew 2266	do.
1862, Aug. 3, r.—w.f. H	Maple	Young Rambler 2335	do.
1863, July 20, r.—w.f. B	Ironmaster 2576	do.	do.
1864, July 8, r.—w.f. H	Modesty	do.	do.

COWS.

MARGERY.
Red with white face, calved September 17, 1860.

Bred by and the property of Mr. E. Tanner, Hopton Castle, Clun, Salop; got by a son of Buckton (1891), dam (Bountiful) by Young Walford, g.d. (Beauty) by Northampton (600).

PRODUCE IN	NAME.	BY WHAT BULL	BY WHOM BRED.
1863, Oct. 12, r.—w.f. B	Confederate 2478	The Doctor 1083	Mr. Tanner.
1864, Sep. 15, r.—w.f. B		Young Buckton 2881	do.

MARIA.
Red with white face, calved December 29, 1860.

Bred by and the property of Mr. T. S. Bradstock, Cobrey Park, Ross; got by The Jew, (2266), dam (Fairy) by Daniel (1201), g.d. (Freckle) by Deluge (1210), g.g.d. (Blossom) by Young Royal (1468).

1864, Jan. 9, r.—w.f. B	West of England 2873	Young Rambler 2335	Mr. Bradstock.

MARIANNE.
Red with white face, calved August 3, 1858.

Bred by Mr. John Hewer, Vern House, Hereford, the property of Mr. John Haynes, Llanrothall, Monmouth; got by Meteor (1319), dam (Promise) by Lot (364), g.d. Fanny, by Young Sovereign (506), g.g.d. (Old Fanny) by Fitzfavourite (441).

1862, July, r.—w.f. B	Pool Wharf 2684	Abdel Kader 1837	Mr. Haynes.
1863, June 4, r.—w.f. B	Rector 2714	The Priest 1416	do.
1864, July 1, r.—w.f. B	Steer	Skenchill 2234	do.

MARIGOLD.
Red with white face, calved in the month of February, 1861.

Bred by Mr. H. E. Powell, Great Brampton, Hereford, the property of Mr. J. Prosser, Honeybourne Gounds, Worcester; got by Dutiful (1978), dam (Mulberry) by Mussulman (1333), g.d. (Mayoress) by Prince (524), g.d. (Marchioness) by Young Sovereign (379), g.g.g.d. (Miss Cotmore the Second) by Lottery the Second (408), g.g.g.g.d. (Miss Cotmore) by Cotmore (376), — by Conqueror (412).

PRODUCE IN	NAME.	BY WHAT BULL.	BY WHOM BRED.
1863, Oct. 22, r.—w.f. B	Steer	Mentor 2112	Mr. Powell.
1864, Dec. 24, r.—w.f. H	Mulberry	The Jew 2266	Mr. Prosser.

MARQUISE.
Red with mottled face, calved August 13, 1860.

Bred by and the property of Mr. John Monkhouse, The Stow, Hereford; got by Madoc (899), dam (Marchioness) by Cantab (717), g.d. (Marchioness of Northampton) by Badger (202).

1863, June, 24, r.—w.f. H	Marcia	Chieftain 930	Mr. Monkhouse.
1864, July 25, r.—w.f. H	Maria	do.	do.

MARSTOW PIGEON THE SECOND.
Red with white face, calved June 3, 1858.

Bred by and the property of Mr. John Partridge, Bishop's Wood, Ross; got by Uncle Tom (1108), dam (Miss Thingehill the Fifth), by Young Sir David (1137), g.d. (Miss Thingehill) by Thingehill (546), g.g.d. (Thingehill Pigeon) by Reform (508), g.g.g d. (Hampton Pigeon) by Young Sovereign (506), g.g.g.g.d. (Sylph) by Chance (355).

1861, April 2, r.—w.f. B	Steer	Walford 1792	Mr. Partridge.
1862, May 10, r.—w.f. B	Steer	Noble Tom 2135	do.
1863, Mar. 4, r.—w.f. H	Marstow Pigeon 7th	Geologist 2012	do.

MARSTOW PIGEON THE THIRD.
Red with white face, calved July 4, 1859.

Bred by and the property of Mr. John Partridge, Bishop's

...ood, Ross; got by The Jew (2266), dam (Miss Thing Fifth) by Young Sir David (1137), g.d. (Miss Thingehill) by Thingehill (546), g.g.d. (Thingehill Pigeon) by Reform (508), g.g.g.d. (Hampton Pigeon) by Young Sovereign (506), &c.

PRODUCE IN	NAME.	BY WHAT BULL.	BY WHOM BRED.
1862, June 10, r.—w.f. H		Noble Tom 2135	Mr. Partridge.
1863, June 20, r.—w.f. H	Pheasant	do.	do.
1864, June 29, r.—w.f. B	Steer	Garway 2536	do.

MARSTOW PIGEON THE FOURTH.

Red with white face, calved September 20, 1861.

Bred by and the property of Mr. John Partridge, Bishop's Wood, Ross; got by Noble Tom (2135), dam (Miss Thingehill the Fifth) by Young Sir David (1137), g.d. (Miss Thingehill) by Thingehill (546), g.g.d. (Thingehill Pigeon) by Reform (508), g.g.g.d. (Hampton Pigeon) by Young Sovereign (506).

1864, June 10, r.—w.f. H	Pleasant	Garway 2536	Mr. Partridge.

MARY ANN.

Red with white face of 1858, vol. v., p. 232.

Bred by and the property of Mr. Naylor, Leighton Hall, Montgomeryshire; got by Tom of Lincoln (1099), dam (Mary) by Silvester (797), g.d. (Lady Elinor) by Big Ben (248), g.g.d. (Butterfly) by Prince (251), g.g.g.d. — by a son of Sir Andrew (183).

1862, July 4, r.—w.f. B	Tom King 2829	Volunteer 2299	Mr. Naylor.
1863, June 24, r.—w.f. H	Mary Ann 2nd	Salisbury 2204	do.
1864, July 11, r.—w.f. B	Steer	do.	do.

MARY GREY.

Grey with white face, calved July 14, 1859.

Bred by the late Mr. Sobey, Tencreek, Liskeard, the property of Mr. W. R. Grose, Penpont, Wadebridge, Cornwall; got by Great Eastern (1598), dam (Lady Grey) by a bull bred by Mr. Knight, g.d. (Lady Grey) bred by Mr. Thomas Longmore, Buckton, Salop.

PRODUCE IN	NAME.	BY WHAT BULL.	BY WHOM BRED.
1862, Feb. 22, r.—w.f. H	Polly	Penhallow 2154	Mr. Grose.
1863, Nov. 29, r.—w.f. H	Grossie	Sir Hugh 2223	do.
1864, Sept. 20, r.—w.f. B	Steer	Dick the Dustman 2495	do.

MAUDE.

Red with white face, calved in the year 1858.

Bred by Mr. Turner, Court of Noke, Leominster, the property of Mr. J. P. Apperley, Fownhope, Hereford; got by Sir Roderick (1393), dam (Church House) by Andrew the Second (619), g.d. (Church House) by Cotmore (376).

1860, Nov. 29, r.—w.f. B	Steer	Baron of Noke 1862	Mr. Apperley.
1861, Dec. 5, r.—w.f. H	Miss Churchhouse	Cornet 1934	do.
1862, Oct. 19, r.—w.f. H	Nell	do.	do.
1863, Oct. 6, r.—w.f. H	Lofty	do.	do.
1864, Oct. 26, r.—w.f. B	Steer	Abbot 2367	do.

MAUDE.

Red with white face, calved August 12, 1859.

Bred by and the property of Mr. Naylor, Leighton Hall, Montgomeryshire; got by Tom of Lincoln (1099), dam (Miriam) by Silvester (797), g.d. (Lovely) by Big Ben (248), g.g.d. (Strapper) bred by Mr. Rea, Monaughty, g.g.g.d. — by Old Court (306).

1862, July 26, r.—w.f. H	Maude 2nd	Volunteer 2299	Mr. Naylor.
1863, June 13, r.—w.f. H	Maude 3rd	Blondin 1880	do.

COWS.

MAUDE.

Red with white face, calved September 26, 1859.

Bred by and the property of Mr. W. Tudge, Adforton, Leintwardine, Ludlow; got by The Doctor (1083), dam (Morella) by Young Walford (1820), g.d. (Cherry) by Nelson (1021), g.g.d. (Cherry) by Turpin (300), g.g.g.d. (Cherry) by a Tully bull.

PRODUCE IN	NAME.	BY WHAT BULL.	BY WHOM BRED.
1862, Dec. 1, r.—w.f. B	Orleans 2601	Magnum Bonum 2097	Mr. Tudge.
1863, Oct. 20, r.—w.f. H	Minna	Pilot 2156	do.
1864, Oct. 3, r.—w.f. B	Steer	do.	do.

MAUDE.

Red with white face, calved March 22, 1861.

Bred by H.R.H. the Prince Consort, the property of Major-General the Hon. A. N. Hood, Cumberland Lodge, Windsor; got by Windsor (1456), dam (Superb) by Carlisle (923), g.d. (Stella) by Venison the Second (1442).

1863, Aug. 17, r.—w.f. H	Lady Emily	Ajax 1843	Gen. the Hon. A. N. Hood.

MAUDE.

Red with white face, calved in the year 1857.

Bred by the Rev. Archer Clive, Whitfield, Hereford, the property of Lieut-Col. Feilden, Dulas Court, Hereford; got by Mameluke (1307), dam (Miss Coningsby) by Sir David the Second (1065), g.d. — by Young Royal (1469), g.g.d. — by Coningsby (718), g.g.g.d. — by Portrait (372).

1863, July 20, r.—w.f. B	Steer	Bertram 1513	Lt.-Col. Feilden.
May 25, r.—w.f. H	Young Maude	Gift 1254	do.

cows.

MAYFLY.

Red with white face, calved in the year 1858.

Bred by and the property of Mr. Evan Davies, Patton, Wenlock, Salop; got by Tyrant (1784), dam (Young Mottle) by Patton (1679).

PRODUCE IN	NAME.	BY WHAT BULL.	BY WHOM BRED.
1861, June 29, r.—w.f. H	Mayflower	Cound 1193	Mr. Evan Davies.
1862, Nov. 1, r.—w.f. H	Mayday	Lincoln 2076	do.
1863, Sep. 16, r.—w.f. B	Steer	do.	do.
1864, July 29, r.—w.f. B	Steer	do.	do.

MELODY

Red with white face, of 1855, vol. v., p. 233.

Bred by the late Mr. James Rea, Monaughty, Knighton, the property of Mr. J. Taylor, Stretford Court, Leominster; got by Chieftain (930), dam (Eva), by Regent (891), g.d. (Spot) by Caractacus (659), g.g.d. (Spot) by Hope (439).

1862, July 24, r.—w.f. H	Melody 3rd	Unity 2287	Mr. Taylor.
1863, Sep. 20, r.—w.f. B	Steer	Trustful 2845	do.
1864, Oct. 2. r.—w.f. B	Pompey 2683	do.	do.

Melody was one of six winners of the First Prize in their class at the meeting of the Herefordshire Agricultural Society, 1862; also one of five winners of a First Prize at Leominster, 1863; and an Extra Stock Prize at Leominster and Hereford, 1864.

MERRY LASS.

Red with white face, calved September 1, 1860.

Bred by Mr. T. Rea, Westonbury, Pembridge, the property of Mr. J. Monkhouse, The Stow. Hereford; got by Sir Benjamin (1387), dam (Gaiety) by Regent (891), g.d. (Lively the Second) by Barrister (658), g.g.d. (Pert) by Old Court (306).

1863, May 4, r.—w.f. H	Merry Thought	Chieftain 930	Mr. Monkhouse.
1864, Aug. 7, r.—w.f. H	Minie ha ha	do.	do.

cows.

MILKMAID.

Red with white face, calved in the month of June 1857.

Bred by Captain Power, Hill Court, Ross, the property of Mr. John Wigmore, Bickerton Court, Dymock; got by Uncle Tom (1108).

PRODUCE IN	NAME.	BY WHAT BULL.	BY WHOM BRED.
1864, Feb. r.—w.f. B		Speculator 2240	Mr. Wigmore.

MILLER'S MAID.

Red with white face, calved November 15, 1861.

Bred by and the property of Mr. J. R. Paramore, Dinedor Court, Hereford; got by The General (2817), dam (Greylass) bred by Mr. Hiles, Dinedor.

1864, March 3, r.—w.f. H	Fairmaid	Portly 2165	Mr. Paramore.

MILTON.

Red with white face, of 1850, vol. v., p. 234.

Bred by the late Mr. Longmore, Orleton, Ludlow, the property of Mr. J. Merryman, Heyfields, Baltimore, Maryland, America; got by Wonder (420), dam (Old Milton) by Milton.

1863, July 3, r.—w.f. H	Marion	Curly 801	Mr. Merryman.

MINNA.

Red with white face, calved in the year 1859.

Bred by and the property of the Rev. A. Clive, Whitfield, Hereford; got by Treville (1432), dam (Miss Lawrence) by a son of Governor (464).

1862, July 10, r.—w.f. H	Myrtle	Ballarat 1858	Rev. A. Clive.

COWS.

MINNIE.
Red with white face of 1857, vol. v., p. 234.

Bred by the late Mr. Thomas Rea, Westonbury, Leominster, the property of Mr. J. H. Arkwright, Hampton Court, Leominster; got by Sir Benjamin (1387), dam (Gwenny the Second) by Chieftain (930), g.d. (Gwenny) by Regent (891), g.g.d. (Venus the Fifth) by Albert (330), g.g.g.d. (Winifred) by Monaughty (220), g.g.g.g.d. (Venus the Fourth) by Duke (304), — (Venus the Third) by Regulator (360), — (Venus the Second) by Noble (238), — (Venus) by Crabstock (303).

PRODUCE IN	NAME.	BY WHAT BULL.	BY WHOM BRED.
1862, Oct. 14, r.—w.f. H	Minnie 2nd	Lord Nelson 2033	Mr. Rea.
1864, Mar. 23, r.—w.f. B		Artful 2391	do.

Minnie the Second, sold to Mr. Lumsden, Auchry House, Aberdeenshire.

MIRIAM.
Red with white face, of 1855, vol. iv., p. 152.

Bred by and the property of Mr. Naylor, Leighton Hall, Montgomeryshire; got by Silvester (797), dam (Lovely) by Big Ben (248), g.d. (Strapper), bred by Mr. Rea, Monaughty, g.g.d. — by Old Court (306).

1859, Aug. 12, r.—w.f. H	Maude	Tom of Lincoln 1099	Mr. Naylor.
1860, July 18, r.—w.f. B	Steer	Admiral 1481	do.
1861, July 29, r.—w.f. B	Steer	Tom of Lincoln 1099	do.
1862, July 2, r.—w.f. B	Gladstone 2547	Salisbury 2204	do.
1863, June 18, r.—w.f. B	Steer	do.	do.

MISS BRONCROFT.
Red with white face, calved June 28, 1859.

Bred by and the property of Mr. Richard Shirley, Baucott Munslow, Church Stretton, Salop; got by Pilot (1036), dam (Primrose) by Pipton, g.d. (Old Rosy) by a Tully bull.

1862, March 21, r.—w.f. H	Miss Palmerston	Prime Minister 1696	Mr. Shirley.
1863, April 9, r.—w.f. B	Steer	Zoar 2355	do.
1864, Aug. 23, r.—w.f. H	Dead	do.	do.

COWS.

MISS BURRINGTON.
Red with white face, calved in the month of December, 1860.

Bred by the late Mr. Sherriff, Burrington, Salop, the property of Mr. J. Taylor, Stretford Court; got by Ben Bolt (1871), dam (Countess) by Confidence (367), g.d. — by Emperor (221), g.g.d. — by Byron (440),

PRODUCE IN	NAME.	BY WHAT BULL.	BY WHOM BRED.
1863, Aug. 6. r.—w.f. B	Steer	Trustful 2845	Mr. Taylor.
1864, July 29, r.—w.f. B		do.	do.

The Steer of 1863, was one of four winners of the First Prize in their class at the meeting of the Ludlow Agricultural Society, 1864; and one of a pair winners of a Second Prize at Hereford, the same year.

MISS BURRY.
Red with white face, calved October 25, 1860.

Bred by the late Mr. Josiah Davies, Ivington, Leominster, the property of Mrs. Ann Davies, Ivington, Leominster; got by Master Butterfly (1313), dam (Snowdrop) by Sutton (1752).

1863, Aug. 27, r.—w.f. B	Dead	Conqueror	Mr. Davies.
1864, July 26. r.—w.f. B		do.	do.

MISS BYRON THE FIFTH.
Red with white face, calved in the year 1860.

Bred by and the property of Mr. Thos. Morris, Therrew, Llyswen, Hay; got by Druid (1220), dam (Miss Byron the Second) by Prior (1359), g.d. (Miss Byron) by Young Byron (832), g.g.d. (Miss Hope) by Young Hope (343), g.g.g.d. (Beauty the Second) by Counseller (422).

1863, Aug. r.—w.f. H	Boadicea	Don Salisbury 1969	Mr. Morris.
1864, Aug, 22, r.—w.f. H	Cornelia	Prince Imperial 2171	do

MISS CHANCE.
Red with white face, calved in the year 1858.

Bred by the late Mr. Thos. Longmore, Buckton, Salop, the property of Mr. P. R. Jackson, Blackbrook, Skenfrith, Monmouth;

got by Young Sir David (1818), dam (Cherry) by Young Walford (1820).

PRODUCE IN	NAME.	BY WHAT BULL.	BY WHOM BRED.
1862, May 10, r.—w.f. H	Cactus	Carlisle 923	Mr. Jackson.
1864, Jan. 21, r.—w.f. H	Clarence 2461	do.	do.

MISS CHANCE.
Red with white face, calved November 1, 1861.

Bred by Mr. John Smith, Sevenhampton, Cheltenham, the property of Mr. John Barton, Coln, Fairford; got by St. Michael (1718), dam (Chance) by Vanquish (1439), g.d. (Matchless) by Phantom (1035), g.g.d. (Old Stately), bred by Lord Hereford.

1864, July r.—w.f. H		Young Cardinal Wiseman 2882	Mr. Barton.

MISS CHURCHHOUSE.
Red with white face, calved December 5, 1861.

Bred by and the property of Mr. J. P. Apperley, Fownhope, Hereford; got by Cornet (1934), dam (Maud) by Sir Roderick (1393), g.d. (Churchhouse) by Andrew the Second (619), g.g.d. (Churchhouse) by Cotmore (376).

1864, May 21, r.—w.f. H	Rose	Capt. Perry 2444	Mr. Apperley.

MISS CONINGSBY.
Red with white face, calved in the year 1859.

Bred by Mr. Sobey, Tencreek, Liskeard, the property of Mr. W. R. Grose, Penpont, Wadebridge, Cornwall; got by Big Ben (1875), dam — by Coningsby (718), g.d. — bred by Mr. Goode, Felton Court, Bromyard.

1862, Jan. 4, r.—w.f. H	Miss Hopeful	Penhallow	Mr. Grose.

MISS COPPICE THE SECOND.
Red with white face, calved in the month of April, 1859.

Bred by and the property of Mr. John Partridge, Bishop's

Wood, Ross; got by Uncle Tom (1108), dam (Young Miss Chance the Sixth) by Young Sir David (1137), g.d. (Miss Chance) by Confidence (367), g.g.d. (Miss Chance) by Chance (355), g.g.g.d. (Miss Chance) bred by the late Mr. T. Jeffries, the Grove.

PRODUCE IN	NAME.	BY WHAT BULL.	BY WHOM BRED.
1862, April 10, r.—w.f. H	Dead	Noble Tom 2135	Mr. Partridge.
1863, May 4, r.—w.f. H	Miss Coppice 3rd	Geologist 2012	do.

MISS COTMORE.

Red with white face, calved September 14, 1861.

Bred by and the property Mr. C. H. Hinckesman, The Poles, Ludlow; got by Sir Oliver the Second (1733), dam (Churchhouse) by Andrew the Second (619), g.d. (Churchhouse) by Cotmore (376),

1864, Aug. 23, r.—w.f. H	MissCotmore2nd	Berwick 1874	Mr. Hinckesman.

MISS ELLEN.

Red with white face, calved in the month of July, 1860.

Bred by the late Mr. Josiah Davies, Ivington, Leominster, the property of Mrs. Ann Davies, Ivington, Leominster; got by Young Cholstrey (1808), dam (Beauty) by Corner Cop (2481), g.d. (Young Hereford) by Truelove (2840), g.g.d. (Hereford) bred by Mr. Edwards, Brinsop.

1862, Nov. 15, r.—w.f. H	Rosebud	Conqueror	Mr. Davies.
1863, Sep. 24, r.—w.f. B	Steer	Chance 1908	do.
1864, Dec. 18, r.—w.f. H	Ellen 2nd	do.	do.

MISS GAY THE SECOND.

Red with white face, calved October 3, 1858.

Bred by the late Mr. Gwillim, Breinton, Hereford, the property of Mr. P. R. Jackson, Blackbrook, Skenfrith, Monmouth; got by Murphy (1331), dam (Miss Gay) by Gaylad (400), g.d. — by Berrington (435), g.g.d. — by Dewshall (358).

PRODUCE IN	NAME.	BY WHAT BULL.	BY WHOM BRED.
1862, July 1, r.—w.f. B	Defiance 2493	Carlisle 923	Mr. Jackson.
1863, Aug. 9, r.—w.f. B	Skenfrith 2781	do.	do.
1864, Dec. 2, r.—w.f. B	Steer	Young William	do.

MISS GREY.

Grey, calved in the year 1859.

Bred by Mr. Thomas Carter, Alcaston, Church Stretton, the property of Mr. A. R. Boughton Knight, Downton Castle, Ludlow; got by Bowdler (1516), dam — by a son of Walford (871).

1861, — r.—w.f.		Tiprey 1774	Mr. T. Carter.
1862, Sep. 21, g.—w.f. H	Gaylass	Lord Grey 2085	Mr. Knight.
1864, June 24, g.—w.f. H	Graceful	do.	do.

MISS HASTINGS THE SECOND.

Red with white face, calved July 13, 1862.

Bred by Mr. Thomas Roberts, Ivington Bury, Leominster, the property of Mr. John Baldwin, Luddington, Stratford-on-Avon; got by Sir Thomas (2228), dam (Lady Hastings), by Master Butterfly (1313), g.d. (Prima Donna) by King James (978), g.g.d. (Long Horns) by Andrew the Second (619), g.g.g.d. (Pigeon) by Prince by Dayhouse (299).

1864, Dec. 25, r.—w.f. B	Hesiod 2570	Battersea 1895	Mr. Badlwin.

Miss Hastings the Second was a winner of the First Prize in her class at the Worcester, Newcastle and Plymouth Meetings of the Royal Agricultural Society of England; First at the Bristol, and Second at the Hereford Meeting of the Bath and West of England Society; First at the meeting of the Herefordshire Agricultural Society, 1863; besides other local prizes.

COWS.

MISS JULIA.
Red with white face, calved in the year 1857.

Bred by Mr. Turner, Court of Noke, Leominster, the property of Mr. J. P. Apperley, Fownhope, Hereford; got by Sir David (349), dam — bred by Mr. J. Muscott, Westonbury.

PRODUCE IN	NAME.	BY WHAT BULL.	BY WHOM BRED.
1860, Dec. 5. r.—w.f. B	Bargain 2397	The Baron of Noke 1862	Mr. Apperley.
1861, Dec. 27, r.—w.f. H	Lark	Cornet 1934	do.
1863, Jan. 4, r.—w.f. B	Steer	do.	do.
1863, Dec. 29, r.—w.f. H	Tidy	do.	do.

MISS LEEDS.
Red with white face, calved July 12, 1861.

Bred by and the property of Mr. Morris, Therrow, Llyswen, Hay; got by Gay Boy (2010), dam (Miss Noble the Third) by Newton (344), g.d. (Miss Noble the Second) by Young Byron (832), g.g.d. (Miss Noble) by Noble (543), g.g.g.d. (Favourite) by Sovereign (404).

1864, — r.—w.f. H		Prince Imperial 2171	Mr. Morris.

MISS MORRIS.
Red with white face, of 1852, vols. iv. and v., pp. 154, 236.

Bred by and the property of Mr. Richard Shirley, Baucott, Munslow, Church Stretton; got by Northampton (600), dam (Old Miss Morris) by Albert (330), g.d. — by Chance (355).

1862, May 13, r.—w.f. H	Miss Hughes	Pilot 1036	Mr. Shirley.

MISS NOBLE.
Red with white face, of 1856, vols. iv. and v., pp. 155, 237.

Bred by Mr. W. S. Powell, Hinton Court, Hereford, the property of Mr. John Partridge, Bishop's Wood, Ross; got by Cardinal Wiseman (1168), dam (Noble) by Young Byron

COWS.

(832), g.d. Miss (Noble) by Noble (543), g.g.d. (Favourite) by Sovereign (404), g.g.g.d. (Damsel) by Young Wellington (505).

PRODUCE IN	NAME.	BY WHAT BULL.	BY WHOM BRED.
1863, Mar. 24, r.—w.f. H	Miss Wood 3rd	Geologist 2012	Mr. Partridge.
1864, May 30, r.—w.f. B	Garway 2nd 2537	Garway 2536	do.

MISS NOBLE.

Red with white face, of 1857, vol. v., p. 237.

Bred by Mr. W. S. Powell, Hinton Court, Hereford, the property of Mr. Wright, Halston Hall, Oswestry; got by Brecon (918), dam (Lady Noble) by Young Byron (832), g.d. (Miss Noble) by Noble (543), g.g.d. (Favourite) by Sovereign (404), g.g.g.d. (Damsel) by Young Wellington (505).

1859, July 27, r.—w.f. B	Halston 2024	Magnet 2nd 989	Mr. Wright.
1862, July 2, r.—w.f. B	Prince 2691	do.	do.
1863, July 24, r.—w.f. B	Duke	do.	do.
1864, July 14, r.—w.f. H	Baroness	Hero 2039	do.

MISS PAGE.

Red with white face, calved April 3, 1858.

Bred by and the property of Mr. Richard Shirley, Baucott, Church Stretton; got by Baucott (1507), dam (Old Miss Page) by Knockerell (1630), g.d. — by Holdgate (1615),

1861, May 13, r.—w.f. H	Wild Fire	Pilot 1036	Mr. Shirley.
1862, April 14, r.—w.f. B	Steer	do.	do.
1863, April 23, r.—w.f. B	Boniface 2429	do.	do.
1864, April 6, r.—w.f. H	Miss Miller	do.	do.

MISS PEGGY.

Red with white face, calved July 23, 1861.

Bred by and the property of Mr. Naylor, Leighton Hall, Montgomeryshire; got by Tom of Lincoln (1099), dam (Peggy) by Silvester (797), g.d. (Pleasant) by Henry the Second (966), g.g d. (Young Emma) bred by Mr. Maybery, Brecon.

1864, July 18, r.—w.f. H	Peony	Salisbury 2204	Mr. Naylor.

COWS.

MISS RUTH.
Red with white face, calved July 12, 1860.

Bred by and the property of Mr. Naylor, Leighton Hall, Montgomeryshire; got by Admiral (1481), dam (Rebecca) by Silvester (797), g.d. (Lady Elinor) by Big Ben (248), g.g.d. (Butterfly) by Prince (251), g.g.g.d. (Nell) by a son of Sir Andrew (183).

PRODUCE IN	NAME.	BY WHAT BULL.	BY WHOM BRED.
1863, June 29, r.—w.f. H	Miss Ruth 2nd	Salisbury 2204	Mr. Naylor.

MISS SALLY.
Red with white face, calved in the year 1860.

Bred by Mr. J. B. Green, Marlow, Shrewsbury, the property of Capt. Crawshay, Danypark, Crickhowell; got by Vanguard (1109), dam (Sally) by Beefy Ben (1869), g.d. — by Cholstrey (217), g.g.d. — by Zest of Oxford (2352), g.g.g.d. — by Discount (339).

1863, Dec. 17, r.—w.f. B	Dead	Zeal 2342	Mr. Green.
1864, — r.—w.f. B	.	Tom King 2830	Capt Crawshay.

MISS SEVERN.
Red with white face, calved August 15, 1861.

Bred by the late Lord Berwick, Cronkhill, Salop, the property of Mr. P. J. Kearney, Miltown House, Clonmellon, Ireland; got by Severn (1382), dam (Young Peggy) by Young Walford (1820), g.d. — by Emperor (221).

1864, Jan. 26. r.—w.f. B	Steer	Silver Stream 2214	Mr. Kearney.
1864, Dec. 24, r.—w.f. H	Silky	do.	do.

MISS STANTON.
Red with white face, calved in the month of July, 1860.

Bred by and the property of Mr. B. Rogers, The Grove, Pembridge; got by The Grove (1764), dam (Miss Stanton) by Severus (1062), g.d. (Miss Stanton) by Gaylad the Second) (1589), g.g.d. (Miss Stanton) by Portrait (372).

PRODUCE IN		NAME.	BY WHAT BULL.	BY WHOM BRED.
1863, July	r.—w.f. H	Lovely	Interest 2046	Mr. Rogers.
1864, Sep 2,	r.—w.f. B	Premier 2686	Grove 2nd 2556	do.

MISS WOOD.
Red with white face, calved August 20, 1860.

Bred by and the property of Mr. John Partridge, Bishop's Wood, Ross; got by The Jew (2266), dam (Miss Noble) by Cardinal Wiseman (1168), g.d. (Noble) by Young Byron (832), g.g.d. (Miss Noble) by Noble (543), g.g.g.d. (Favourite) by Sovereign (404), g.g.g.g.d. (Damsel) by Young Wellington (505).

1863, May 2,	r.—w.f. B	Steer	Noble Tom 2135	Mr. Partridge.
1864, May 26,	r.—w.f. H	Woodbine	Garway 2536	do.

MISS WOOD THE SECOND.
Red with white face, calved September 1, 1861.

Bred by and the property of Mr. John Partridge, Bishop's Wood, Ross; got by The Jew (2266), dam (Miss Noble) by Cardinal Wiseman (1168), g.d. (Noble) by Young Byron (832), g.g.d (Miss Noble) by Noble (543), g.g.g.d. (Favourite) by Sovereign (404), g.g.g.g.d. (Damsel) by Young Wellington (505).

1864, Oct. 18,	r.—w.f. H	Rebecca	Garway 2536	Mr. Partridge.

MOSSROSE.
Red with white face, calved in the year 1860.

Bred by Mr. Price, Tillington, Hereford, the property of Mr. Thomas Powell, Bage, Madley, Hereford; got by Courtier (1194), dam (Rosabelle) by Mameluke (1307), g.d. (Rosabelle) by Pope (527), g.g.d. (Old Silver) by Wellington (507), g.g.g.d.

(Beauty) by Sovereign (404), g.g.g.g.d. (Old Gentle) by Chance (355).

PRODUCE IN	NAME.	BY WHAT BULL.	BY WHOM BRED.
1862, Nov. 10, r.—w.f. H	Tulip	Zeno 1825	Mr. Powell.
1863, Oct. 5, r.—w.f. H	Rosina	do	do.
1864, Nov. 9, r.—w.f. B	Comus 2476	Interest 2046	do.

MOSSYFACE.
Red with white face, calved in the year 1859.

Bred by and the property of Mr. G. T. Forester, High Ercall, Wellington, Salop; got by Discord (1217), dam (Rosette) by Darling (1202), g.d. (Rosa) by Governor (464), g.g.d. (Moss Rose) by Chance (348), g.g.g.d — bred by the late Mr. D. Williams, by Whitenob (345), g.g.g.g.d. — by Young Wellington (505).

PRODUCE IN	NAME.	BY WHAT BULL.	BY WHOM BRED.
1862, Aug. 26, r.—w.f. H	Misletoe	Severn 1382	Mr. Forester.
1863, July 4, r.—w.f. H	Rosina	do.	do.
1864, July 30, r.—w.f. B	Steer	do.	do.

MOTTLE.
Red with mottled face, calved in the month of October, 1858.

Bred by and the property of Mr. John Burlton, Luntley Court, Leominster; got by Sampson (1061), dam (Old Mottle) by Red Ben (768), g.d. (Gaylass) by the Count (351), g.g.d. (Gaylass) by Young Goldfinder by Goldfinder (383).

PRODUCE IN	NAME.	BY WHAT BULL.	BY WHOM BRED.
1862, Aug. 28, r.—w.f. B	Steer	Rifleman 2189	Mr. Burlton.
1863, Oct. 22, r.—w.f. H	Mary Ann	do.	do.

MOTTLE.
Red with white face, calved May 22, 1861.

Bred by Mr. Palmer, Bollitree, Weston, Ross, the property of Mr. John Wigmore, Bickerton Court, Dymock; got by Forester (1238), dam (Old Mottle) bred by Mr. Palmer.

PRODUCE IN	NAME.	BY WHAT BULL.	BY WHOM BRED.
1864, June 4, r.—w.f. B		Speculator 2240	Mr. Wigmore.

Mottle was one of eight winners of the Second Prize in their class at the meeting of the Herefordshire Agricultural Society, 1863; and with Speculator and their offspring, a First Prize at Monmouth, 1864; she was also one of a pair, winners of a First Prize at Monmouth, 1864.

MOTTLE.

Red with mottled face, calved June 1, 1860.

Bred by and the property of Mr. J. R. Paramore, Dinedor Court, Hereford; got by The General (2817), dam (Peggy) by a bull bred by the late Mr. Turner, Noke Court, g.d. (Silk) bred by the late Mr. Davies, Alder's End Tarrington.

PRODUCE IN		NAME.	BY WHAT BULL.	BY WHOM BRED.
1862, June 7,	r.—w.f. H	Mottle 2nd	The General 2817	Mr. Paramore.
1863, Sep. 8,	r.—w.f. H	Rebecca	The Jew 2266	do.
1864, Aug. 6,	r.—w.f. H	Rosabelle	Portly 2165	do.

Mottle the Second was awarded a Prize in the Extra Stock class at the meeting of the Herefordshire Agricultural Society, 1864.

MULBERRY.

Red with white face, calved in the month of January, 1858.

Bred by Mr. H. E. Powell, Great Brampton, Madley, the property of Mr. J. H. Whitehouse, Ipsley Court, Redditch; got by Mussulman (1333), dam (Mayoress) by Prince (524), g d. (Marchioness) by Young Sovereign (379), g.g.d. (Miss Cotmore the Second) by Lottery the Second (408), g.g.g d. (Miss Cotmore) by Cotmore (376), g.g.g.g.d. — by Conqueror (412).

1861, Feb.	r.—w.f. H	Marigold	Dutiful 1978	Mr. Powell.
1862, March	r.—w.f. H	Mermaid	Mentor 2112	do.
1863, March	r.—w.f. H	Melody	do.	do.
1864, April 27,	r.—w.f. B	Monarch 2642	Vincent 2858	Mr. Whitehouse.

Melody, sold to Mr. E. Lewis, Breinton, Hereford.

MY LADY.

Red with white face, calved May 13, 1860.

Bred by and the property of Mr. R. Shirley, Baucott Munslow, Church Stretton; got by Pilot (1036), dam (Tugford) by Baucott (1507), g d. (Old Tugford) by Knockerell (1630), g,g.d. — by Dolluggan (759).

1863, April 15,	r.—w.f. B	Steer	Zoar 2355	Mr. Shirley.
1864, April 10,	r.—w.f. H	Lady Jane	do.	do.

MYRTLE.

Red with white face, calved in the month of November, 1859.

Bred by the late Mr. H. E. Powell, Great Brampton, Hereford, the property of Mr. B. Rogers, the Grove, Pembridge; got by Dutiful (1978), dam (Musk Rose) by Mameluke (1307), g.d. (Marchioness) by Young Sovereign (379), g.g.d. (Miss Cotmore the Second) by Lottery the Second (408), g.g.g.d. (Miss Cotmore) by Cotmore (376), g.g.g g.d. — by Conqueror (412).

PRODUCE IN	NAME.	BY WHAT BULL.	BY WHOM BRED.
1862, Oct. 27, r.—w.f. H	Moss Rose	Doctor 1964	Mr. Powell.
1864, March r.—w.f. B	Dead	Vincent 2858	do.
1864, Dec. 25, r.—w.f. B	Discount 2498	Grove 2nd 2556	Mr. Rogers.

Mossrose sold to Mr. G. Pye, Cublington, Madley.

NANCY.

Red with white face, of 1856, vol. v., p. 241.

Bred by Mr. J. Bourn, Mawley Town Farm, Cleobury Mortimer, the property of Mr. W. B. Peren, Compton, South Petherton, Somerset; got by Kyrewood (2062), dam (Nutty the Second) by the Duke (2265) g.d. (Nutty) by Silurian (1386). g.g.d. (Old Pink) by The Bishop (2260).

1863, Aug. 10, r.—w.f. H	Norma	Cardinal 1526	Mr. Bourn.
1864, Oct. 13, r.—w.f. H	Nutmeg	do.	do.

Nutmeg sold to Mr. V. Gosport, Tanylan, Holywell, North Wales.

NANNIE.

Red with white face, of 1857, vol. v, p. 242.

Bred by Lord Bateman, Shobdon Court, Leominster, the property of Mr. J. R. Paramore, Dinedor Court; got by Carlisle (923), dam (Nancy) by Sparrington (2239).

1862, r.—w.f. B	Steer	Garibaldi 2003	Mr. Garrold.
1863, Oct 1, r.—w.f. B	Steer	Jew 2266	Mr. Paramore.
1864, July r.—w.f. B		Portly 2165	do.

COWS.

NAOMI.

Red with white face, calved in the year 1860.

Bred by Capt. Power, Hill Court, Ross, the property of Mr. Wm. Jones, Hill of Eaton, Ross; got by The Jew (2266) dam — by Vanguard (1109), g.d. — bred by the late Visct. Hereford, Tregoyd, Hay.

PRODUCE IN	NAME.	BY WHAT BULL.	BY WHOM BRED.
1863, March 27, r.—w.f. H	Skylark	Geologist 2012	Mr. Jones.
1864, March 23, r.—w.f. B	Steer	do.	do.

NECKLACE.

Red with white face, calved February 17, 1862.

Bred by and the property of Mr. F. W. Stone, Moreton Lodge, Guelph, Canada West; got by Patriot (2150), dam (Nelly) by Carlisle (923), g.d. (Peeress) by Monarch (504), g.g.d. by St. Germans (227), g.g.g.d. — bred by the late Mr. Turner, Noke, Leominster.

1864, Nov. 2, r.—w.f. H	Necklace 2nd	Commodore 2473	Mr. Stone.

Necklace was a winner of the Third Prize in her class at the Toronto, Kingston, and Hamilton Meetings of the Canadian Agricultural Society.

NENA THE SECOND.

Red with white face, calved August 25, 1861.

Bred by and the property of Mr. Warren Evans, Llandowlais, Usk; got by Oakley (1673), dam (Nena) by Chieftain (930), g.d. (Dahlia) by Cholstrey (217), g.g.d. (Dahlia) by Monarch (219), g.g.g.d. (Dahlia) by Portrait (372), g.g.g.g.d. (Dahlia) by Sovereign (404).

1864, Jan. 18. r.—w.f. H	Lady	Monaughty 2117	Mr. Rea.

cows.

NEWTON.

Red with white face, calved July 30, 1861.

Bred by and the property of Mr. C. H. Hinckesman, The Poles, Ludlow; got by Jackanapes (1285), dam (Kinlet) by General (1251), g.d. (Beauty) by Paragon (1343).

PRODUCE IN	NAME.	BY WHAT BULL.	BY WHOM BRED.
1864, Aug. 1, r.—w.f. H	Newton 2nd	Salopian 2739	Mr. Hinckesman.

NELL.

Red with white face, calved in the month of September, 1860.

Bred by and the property of Mr. John Burlton, Luntley Court, Leominster; got by Havelock (1609), dam (Ellen) by Pembridge (721), g.d. — by Prince Dangerous (362), g.g.d. — by the Sheriff (356), g.g.g.d. — by Crabstock (303).

1863, Aug. 20, r.—w.f. H	Primrose	Rifleman 2189	Mr. Burlton.

NELL GWYNNE.

Red with white face, calved February 10, 1862.

Bred by and the property of Mr. R. Davey, M.P., Polsue House, Grampound; got by Penhallow (2154), dam (Patience) by Great Eastern (1598), g.d. (Cheerful) by Invincible (592), g.g.d. (Cherry) by Reform (508), g.g.g.d. (Rosebud) by Byron (440).

1864, May 20, r.—w.f. B		Castor 1900	Mr. Davey.

NELL GWYNNE.

Red with white face, calved March 5, 1862.

Bred by and the property of Mr. J. W. James, Mappowder Court, Blandford, Dorset.; got by Happy Land (2561), dam

(Broad) by Chance (2452), g.d. (Star) by Young Sovereign (2895).

PRODUCE IN	NAME.	BY WHAT BULL.	BY WHOM BRED.
1864, Sep. 8, r.—w.f. H	Bessie Gwynne	Bird's Eye 2420A	Mr. James.

Nell Gwynne was a winner of the Third Prize in her class at the Exeter Meeting of the Bath and West of England Society, and first as one of a pair at the meetings of the Sturminster, Sherborne, and Yeovil Agricultural Societies, 1863; also a first at Sherborne, 1864.

NELLY.

Red with white face, calved November 9, 1859.

Bred by and the property of Mr. J. Bourn, Mawley Town Farm, Cleobury Mortimer; got by Wigmore (1800), dam (Nutty the Second) by the Duke (2265), g.d. (Nutty) by Silurian (1386), g.g.d. (Old Pink) by the Bishop (2260).

1862, July 9, r.—w.f. H	Nell	Cardinal 1526	Mr. Bourn.
1863, July 7, r.—w.f. H	Nina	do.	do.

Nina, sold to Mr. V. Gosford, Tanylan, Holywell, North Wales.

NELLY.

Red with white face, calved in the month of March, 1856.

Bred by and the property of Mr. R. Keene, Pencraig, Caerleon, Monmouth; got by Prince Albert (2168), dam (Spot) by Young David (2325), g.d. (Old Spot) by Foxhall (2520).

1859, Jan.	r.—w.f. H	Sadlier	General Wyndham 1590	Mr. Keene.
1860, Feb.	r.—w.f. H	Dead		do.
1861, Jan.	r.—w.f. B	Steer	Odd Trick 1674	do.
1862, Jan.	r.—w.f. B	Steer	do.	do.
1863, Feb.	r.—w.f. B	Abernant 2369	Pencraig 2671	do.
1864, Feb.	r.—w.f. B	Steer	Cholstrey 2nd 1919	do.

NOBLE.

Red with white face, of 1858, vol. v., p. 243.

Bred by and the property of Mr. E. Wright, Halston Hall, Oswestry; got by Garrick (1248), dam (Lady Noble) by Young Byron (832), g.d. (Miss Noble) by Noble (543), g.g.d.

Favourite) by Sovereign (404), g.g.g.d. (Damsel) by Young Wellington (505).

PRODUCE IN		NAME.	BY WHAT BULL.	BY WHOM BRED.
1862, Oct. 9,	r.—w.f. B	Steer	Magnet 2nd 989	Mr. Wright.
1863, Sep. 1,	r.—w.f. H	Marchioness	Hero 2093	do.
1864, July 27,	r.—w.f. H	Countess	do.	do.

NOKE.
Red with white face, calved in the year 1858.

Bred by Mr. Turner, Court of Noke, Leominster; the property of Mr. J. P. Apperley, Fownhope, Hereford; got by Sir Roderick (1893), dam (Beauty) by Andrew the Second (619).

1861, Nov. 25,	r.—w.f. B	Steer	Cornet 1934	Mr. Apperley.
1862, Dec. 30,	r.—w.f. H	Miss Noke	do.	do.
1863, Dec. 15,	r.—w.f. B	Steer	do.	do.

NONPAREIL.
Red with white face, calved March 11, 1860.

Bred by and the property of Mr. Philip Turner, The Leen, Pembridge; got by Bertram (1513), dam (Exquisite) by Sir David (349), g.d. (Nell Gwynne) by the Knight (185), g.g.d. (Belle) by Sir Walter (352), g.g.g.d. (Myrtle) by Commerce (354), g.g.g.g.d (Sylph) by Old Court the Second (1341).

1862, July 28,	r.—w.f. B	Claudio 2463	Bolingbroke 1883	Mr. Turner.
1864, May 31,	r.—w.f. B	Steer	do.	do.

NONSUCH.
Red with white face, calved August 15, 1859.

Bred by Mr. James Rea, Monaughty, Knighton, the property of Mr. R. H. Capper, The Northgate, Ross; got by Wellington (1112), dam (Fairlass) by Chieftain (930), g.d. (Fairmaid) by Cholstrey (217), g.g.d. (Fairmaid) by Gallant (239), g.g.g.d. (Fairmaid) by Portrait (372).

1863, June 15,	r.—w.f. B	Sir Robert 2775	Sir Benjamin 1387	Mr. J. Rea.
1864, Sep. 29,	r.—w.f. B	Nonpariel 2654	Garibaldi 2008	Mr. Capper.

COWS.

NOSEGAY.
Red with white face, calved in the year 1861.

Bred by the Misses Abley, Norton, Presteign, the property of Mr. R. G. Price, M.P., Norton Manor, Presteign; got by Havelock (2563).

PRODUCE IN	NAME.	BY WHAT BULL.	BY WHOM BRED.
1864, Aug. 30, r.—w.f. B		Lord of the Manor 2622	Mr. R. G. Price.

NOSEGAY THE SECOND.
Red with white face, of 1858, vol. v., p. 245.

Bred by and the property of Mr. J. Taylor, Stretford Court, Leominster; got by St. Oswall (1378), dam (Nosegay) bred by the late Mr. J. Bowen, Monkland.

1862, Aug. 20, r.—w.f. B		Croft 937	Mr. Taylor.

NUT BUSH.
Red with white face, calved March 21, 1858.

Bred by and the property of Mr. Richard Shirley, Baucott, Munslow, Church Stretton; got by Baucott (1507), dam (Nutty) by Sir Henry (1730), g.d. (Big Nutty) by Byron (559), g.g.d. — by a Tully bull.

1861, Mar, 18, r.—w.f. B	Steer	Pilot 1036	Mr. Shirley.
1862, April 1, r.—w.f. H	Beech Nut	do.	do.
1863, April 18, r.—w.f. B	RattlingJack 2712	do.	do.
1864, June 13, r.—w.f. B	Shropshire 2753	do.	do.

NUTTY THE SECOND.
Red with white face, of 1851, vol. v., p. 245.

Bred by and the property of Mr. Bourn, Mawley Town Farm, Cleobury Mortimer; got by the Duke (2265), dam (Nutty) by Silurian (1386), g.d. (Old Pink) by The Bishop (2260).

1863, July 10, r.—w.f. H	Nora	Cardinal 1526	Mr. Bourn.

Nora, sold to Mr. V. Gosford, Tanylan, Holywell, North Wales.

NUTTY THE SECOND.
Red with white face, calved in the month of July 1859.

Bred by Mr. Richard Hickman, Bosbury, Ledbury, the property of Mr. James Gregg, Fencote Abbey, Leominster; got by Chester (1538), dam (Nutty) by Goldsmith (1258), g.d. (Damsel) by Young Governor, g.g.d. (Damsel) by Lot (364), g.g.g.d. (Daisy) by Discount (339).

PRODUCE IN	NAME.	BY WHAT BULL.	BY WHOM BRED.
1862, May 12, r.—w.f. H	Daisy	Fencote 1989	Mr. Gregg.
1863, May 29, r.—w.f. B	Steer	do.	do.
1864, July 14, r.—w.f. B	Steer	do.	do.

NYMPH.
Red with white face, of 1856, vol. v., p. 246.

Bred by Lord Bateman, Shobdon Court, the property of Mr. Wright, Halston Hall, Oswestry; got by Carlisle (923), dam (Little Beauty) by Andrew (619).

1863, Mar. 22, r.—w.f. H	Argenta	Magnet 2nd 989	Mr. Wright.
1864, July 26, r.—w.f. H	Alba	Prince 2691	do.

ORANGE THE FIRST.
Red with white face, of 1858, vol. v., p. 247.

Bred by and the property of Mr. John Partridge, Bishop's Wood, Ross; got by Uncle Tom (1108), dam (Young Original the Second) by Young Sir David (1137), g.d. (Young Original) by Half-sovereign (964), g.g.d. (Original) by Original the First (455).

1862, July 1, r.—w.f. H	Orange 2nd	Noble Tom 2135	Mr. Partridge.
1863, Oct. 1, r.—w.f. B	Dead	do.	do

PALE FACE.
Red with white face, calved in the month of July, 1861.

Bred by Mr. Rogers, Hereford, the property of Mr. S. Gilliland, Brook Hall, Londonderry, Ireland.

1864, May 20, r.—w.f. B	Jolly Miller 15th 2588	Jolly Miller 10th 2584	Mr. Gilliland.

COWS.

PANSY.
Red with white face, of 1857, vol. v., p. 247.

Bred by and the property of Mr. James Bourn, Mawley Town Farm, Cleobury Mortimer; got by Kyrewood (2062), dam (Pink the Third) by the Duke (2265), g.d. (Pink) by Silurian (1386), g g.d. (Old Pink) by The Bishop (2260).

PRODUCE IN	NAME.	BY WHAT BULL.	BY WHOM BRED.
1862, Aug. 15, r.—w.f. B	Steer	Cardinal 1526	Mr. Bourn.
1863, July 14, r.—w.f. B	Pentagon	do.	do.
1864, Sep. 16, r.—w.f. B		do.	do.

PATIENCE.
Red with white face, of 1855, vol. iv., p. 163.

Bred by and the property of Mr. Naylor, Leighton Hall, Montgomeryshire; got by Silvester (797), dam (Blowdy) by Big Ben (248), g.d. (Mottle) by Prince (251), g.g.d. (Beauty) by Claret (253), g.g.g.d. (Spot) by Trump (490).

1859, Aug. 25, r.—w.f. B	Steer	Leighton 1632	Mr. Naylor.
1860, July 16, r.—w.f. H	Dewdrop	Tom of Lincoln 1099	do.
1861, June 10, r.—w.f. H	Dead	do.	do.
1862, June 26, r.—w.f. B	Dead	Admiral 1481	do.
1863, Sep. 17, r.—w.f. B	Llywellyn 2609	Salisbury 2204	do.
1864, Sep. 1, r.—w.f. B	Waxy 2869	do.	do.

PATIENCE.
Red with white face, calved January 19, 1859.

Bred by Mr. J. Sobey, Penhallow, Grampound, the property of Mr. R. Davey, M.P., Polsue House, Grampound, Cornwall; got by Great Eastern (1598), dam (Cheerful) by Invincible (592), g.d. (Cherry) by Reform (508), g.g.d. (Rosebud) by Byron (440).

1862, Feb. 19, r.—w.f. H	Nell Gwynne	Penhallow 2154	Mr. Davey.
1863, Mar. 5, r.—w.f. H	Patty	do.	do.
1864, Apr. 17, r.—.w.f. {B/H}	Twins	Zippor 2354	do.

Patty, sold to Mr. Pollock, Galway, Ireland.

PATRONESS.

Red with white face, calved August 4, 1861.

Bred by and the property of Mr. Thomas Roberts, Ivington Bury, Leominster; got by Duke of Marlborough (1974), dam (Patroness) by King James (978), g.d. (Patroness the Second) by Fairboy (617), g.g.d. (Patroness) by Original (216), g.g.g.d. (Perry) by Woodman (255).

PRODUCE IN	NAME.	BY WHAT BULL	BY WHOM BRED.
1863, Sep. 8, r.—w.f. H	Lovely	Sir Thomas 2228	Mr. Roberts.
1864, Aug. 4, r.—w.f. B	Patron 2669	do.	do.

PAULINA.

Red with white face, calved May 28, 1861.

Bred by and the property of Mr. P. Turner, The Leen, Pembridge; got by Bertram (1513), dam (Dorcas) by Tom of Lincoln (1099), g.d. (Olive) by Andrew the Second (619), g.g.d. (Piety) by Cobden (668), g.g.g.d. (Beauty) by Sir Walter (352) g.g.g.g.d. (Gaylass) by Cupid (1950), — (Violet) by Lottery the Second (987a).

1864, Feb. 15, r.—w.f. B		Bolingbroke 1883	Mr. Turner.

PEACH.

Red with white face, calved December 30, 1860.

Bred by the late Lord Berwick, Cronkhill, Salop, the property of Mr. F. W. Stone, Moreton Lodge, Guelph, Canada West; got by Albert Edward (859), dam (Cherry the Thirteenth) by Walford (871), g.d. (Red Cherry) by Tom Thumb (243), g.g.d. (Cherry the Fifth) by Cholstrey (868), g.g.g.d. (Cherry the Fourth) by Green's Grey Bull (850a), — (Cherry the Third) by Chancellor (156), — (Cherry the Second) by Thickset (1769).

PRODUCE IN	NAME.	BY WHAT BULL.	BY WHOM BRED.
1863, Nov. 9, r.—w.f. B	Canadian Chief 2441	Sailor 2200	Mr. Stone.
1864, Oct. 12, r.—w.f. B	Wellington Hero 2870	do.	do.

Peach was a winner of the Second Prize in her class at the Toronto and Kingston meetings of the Canadian Agricultural Society, 1862 and 1863.

PEARL.
Red with white face, calved October 28, 1860.

Bred by and the property of Mr. W. Tudge, Adforton, Leintwardine, Ludlow; got by Carbonel (1525), dam (Pleasant) by The Doctor (1083), g.d. (Cherry) by Nelson (1021), g.g.d. (Cherry) by Turpin (300), g.g.g.d. (Cherry) by a Tully Bull.

1863, July 24, r.—w.f. H	Alexandra	Pilot 2156	Mr. Tudge.
1864, July 3, r.—w.f. H		do.	do.

PEERESS.
Red with white face, calved in the month of October, 1860.

Bred by the Rev. Archer Clive, Whitfield, Hereford, the property of Mr. R. H. Capper, The Northgate, Ross; got by Wellington (1112), dam (Purity) by Regent (891), g.d. (Duchess) by Gallant (239), g.g.d. — by Old Court (306).

1863, Oct. 4, r.—w.f. H	Peeress 2nd	Garibaldi 2008	Mr. Capper.
1864, Sept 21, r.—w.f. B	Young Royal 2892	do	do.

PEGGY.
Red with white face, calved August 24, 1859.

Bred by the late Mr. Sobey, Tencreek, Liskeard, the property of Mr. W. R. Grose, Penpont, Wadebridge, Cornwall; got by Big Ben (1875), dam (Countess) by Coningsby (718).

1862, Mar. 11, r.—w.f. H	Dolly	Penhallow 2154	Mr. Grose.
1863, Jan. 20, r.—w.f. B	Steer	Sir Hugh 2223	do.
1864, Jan. 10, r.—w.f. B		do.	do.

PEONY.
Red with white face, calved August 4, 1859.

Bred by Mr. J. Bourn, Mawley Town Farm, Cleobury Mortimer, the property of Mr. W. B. Peren, Compton, South Petherton; got by Wigmore (1800), dam (Pansy) by Kyrewood (2062), g.d. (Pink the Third) by The Duke (2265), g.g.d. (Pink) by Silurian (1386), g.g.g.d. (Old Pink) by The Bishop (2260).

PRODUCE IN	NAME.	BY WHAT BULL.	BY WHOM BRED.
1863, July 14, r.—w.f. H	Pet	Cardinal 1526	Mr. Bourn.
1864, July 21, r.—w.f. B	Bertie 2416	do.	do.

Pet, sold to Mr. V. Gosport, Tanylon, Holywell, North Wales.

PERFECTION.
Red with white face, calved in the month of April, 1859.

Bred by and the property of Mr. J. Prosser, Honeybourne Grounds, Worcester; got by Medalist (1009).

1863, Feb. r.—w.f. B	Speculation	The Jew 2266	Mr. Prosser.
1864, Mar. 2, r.—w.f. B	Picture	do.	do.
1864, Dec. 30, r.—w.f. H	Myrtle	Billingsly 2420	do.

PERFECTION.
Red with white face of 1854, vol. iv., p. 164.

Bred by and the property of the Rev. Archer Clive, Whitfield, Hereford; got by Trader (1101), dam (Pigeon) by Andrew the Second (619), g.d. (Old Pigeon), by Prince by Dayhouse (299).

1863, June 18, r.—w.f. B	Kingstone	Bertram 1513	Rev. A. Clive.
1864, July 9, r.—w.f. H	Peach	Chepstow 1916	do.

PHEASANT THE SECOND.
Red with white face, calved October 15, 1859.

Bred by the late Mr. J. Rea, Monaughty, Knighton, the property of Mr. Phillips, Abbey-cwm-hyr, got by Wellington (1112), dam (Pheasant) by Grenadier (961), g.d. (Laura) by Regent (891), g.g.d. (Laura) by Caractacus (659), g.g.g.d. (Lovely) by Albert (330).

PRODUCE IN	NAME.	BY WHAT BULL.	BY WHOM BRED.
1862, July 9, r.—w.f. H	Pheasant 3rd	Sir Benjamin 1387	Mr. J. Rea.

Pheasant the Third, sold to Mr. J. R. Paramore, Dinedor Court.

PICOTEE.
Grey with white face, calved July 19, 1857.

Bred by Lord Berwick, Cronkhill, Salop, the property of Mr. W. Newbery, Fernhill, Kenilworth; got by Albert Edward (859), dam (Finchback) by Walford (871), g.d. (Young Damsel) by Tom Thumb (243), g.g.d. (Damsel) by Young Trueboy (1475), g.g.g.d. (Prettymaid) by Cholstrey (868).

1862, Dec. 25, r.—w.f. H	Pert	Comus 2477	Mr. Newbery.
1863, Nov. 27, r.—w.f. H	Princess	do.	do.

PICTURE.
Red with white face, calved August 8, 1857.

Bred by the late Mr. J. Rea, Monaughty, Knighton, the property of Mr. John Burlton, Luntley Court, Leominster; got by Grenadier (961), dam (Purity) by Regent (891), g.d. (Duchess) by Barrister (658), g.g.d. (Duchess) by Gallant (239), g.g.g.d. — (Old Court (306).

1864, Feb. 8, r.—w.f. B	Sir John 2770	Sir Benjamin 1387	Mr. Thos. Rea.

PIGEON.
Red with white face, calved in the year 1856.

Bred by and the property of Mr. Henry Gibbons, Hampton Bishop, Hereford; got by The Admiral (1078), dam (Young Beauty) by Young Gaylad (1463).

1859, Sep. 18, r.—w.f. H		Woodman 2nd 1459	Mr. Gibbons.
1860, July 29, r.—w.f. H	Miss Chance	do.	do.
1861, Oct. 5, r.—w.f. B	Steer	Shamrock 2nd 2210	do.
1862, Sep. 18, r.—w.f. B	Steer	do.	do.
1863, Aug. 24, r.—w.f. H	Ringdove	do.	do.
1864, Aug. 15, r.—w.f. H	Fantail	do.	do.

PIGEON.

Red with white face, of 1853, vol. iv., p. 165.

Bred by and the property of Mr. Naylor, Leighton Hall, Montgomeryshire; got by Uriconium (598), dam (Toby Pigeon) by Big Ben (248), g.d. (Duchess) by Tobias (487), g.g.d. (Duchess) by Sovereign (404), g.g.g.d. — bred by the late Mr. Turner, Noke.

PRODUCE IN	NAME.	BY WHAT BULL.	BY WHOM BRED.
1859, July 6, r.—w.f. B	Steer	Thickset 1768	Mr. Naylor.
1860, Oct. 23, r.—w.f. B	Steer	Admiral 1481	do.
1861, July 26, r.—w.f. B	Steer	do.	do.
1862, July 27, r.—w.f. B	Dead	Volunteer 2299	do.
1863, May 25, r.—w.f. B	Sambo 2740	Salisbury 2204	do.
1864, June 13, r.—w.f. B		do.	do.

PIGEON.

Red with white face, of 1853, vol. iii., p. 198.

Bred by Mr. William Perry, St. Oswald's, Cholstrey, Leominster, the property of Mr. William Stallard, Brockhampton, Ross; got by Newton (1023), dam (Miss Hewer) by Noble Boy (751), g.d. (Chadnor) by Turpin (365), g.g.d. — bred by Mr. John Hewer.

1858, Sep. 20, r.—w.f. B	Hatfield 2030	Monkland 3rd 1013	Mr. Perry.
1859, July 20, r.—w.f. B	Lord Nelson 2088	Noble Boy 1337	do.
1860, Sep. 2, r.—w.f. H	Young Pigeon	Van Tromp 1440	do.
1862, Jan. 18, r.—w.f. B	Sir Henry	Lord Wellington 2094	do. do.
1862, Dec. 22, r.—w.f. H	Pigeon 2nd	Cowarne 1942	do.
1863, Dec. 5, r.—w.f. H	Pigeon 3rd	Wallend 2864	do.

Pigeon the Third, sold to Mr. Wm. Stallard, Brockhampton, Ross.

PIGEON.

Red with white face, calved in the month of October, 1859.

Bred by the late Mr. Vaughan, Lawton, Leominster, the property of Messrs. T. and W. Vaughan, Lawton, Leominster;

got by Peaceful (2670), dam — by Lawton (980), g.d. — by Emperor (373).

PRODUCE IN		NAME.	BY WHAT BULL.	BY WHOM BRED.
1862, Jan.	r.—w.f. B	Steer	Caractacus 659	Mr. Vaughan.
1862, Dec.	r.—w.f. B		do.	do.
1863, Oct. 23,	r.—w.f. H	Pigeon 2nd	do.	Messrs. Vaughan.
1864, Sep. 30,	r.—w.f. H	Pigeon 3rd	Plunder 2nd 2681	do.

PIGEON.

Red with white face, calved in the year 1855.

Bred by and the property of Mr. R. Hickman, Bosbury, Ledbury; got by Defence (1207), dam (Pigeon) by Governor (464), g.d. — by Royal Prince (528), g.g.d. — by Discount (339).

1858, Sep. 20,	r.—w.f. H	Pink	Goldsmith 1258	Mr. Hickman.
1859, Oct. 30,	r.—w.f. H	Primrose	Chester 1538	do.
1860, Oct. 5,	r.—w.f. H	Princess	do.	do.
1861, Sep. 20,	r.—w.f. H	Perfection	Havelock 2033	do.

PIGEON.

Red with white face, of 1852, vol. iv., p. 166.

Bred by the Rev. T. Kevil Davies, Croft Castle, the property of the Rev. Archer Clive, Whitfield, Hereford; got by Andrew the Second (619), dam (Pigeon) by Prince by Dayhouse (299).

1862, Nov. 8,	r.—w.f. H	Promise	Ballarat 1858	Rev. A. Clive.
1863, Sep. 21,	r.—w.f. H	Pink	Bertram 1513	do.
1864, Sep. 8,	r.—w.f. B	Steer	Plato 2160	do.

PIGEON THE THIRD.

Red with white face, calved in the year 1861.

Bred by Mr. Vaughan, Cholstrey, Leominster, the property of Mr. James Gregg, Fencote Abbey, Leominster; got by Pirate by Plunder (1038), dam (Pigeon the Second) by Plunder (1038), g.d. — by Emperor (373).

1864, June	r.—w.f. H	Pigeon 4th	Young Salisbury 2893	Mr. Gregg.

COWS.

PINK.
Red with white face, calved in the month of October, 1858.

Bred by Mr. Matty, Grove, Dewchurch, Hereford, the property of Mr. J. R. Paramore, Dinedor Court, Hereford; got by Edgar (946), dam (Pink) by Young Gaylad (1463).

PRODUCE IN		NAME.	BY WHAT BULL.	BY WHOM BRED.
1662,	r.—w.f. B	Steer	Son of General 1251	Mr. Matty.
1863, Oct. 10,	r.—w.f. H	Rosette	The Jew 2266	Mr. Paramore.

PINK THE SECOND.
Red with white face, calved August 18, 1859.

Bred by Mr. R. Hickman, Bosbury, Ledbury, the property of Mr. J. Gregg, Fencote Abbey, Leominster; got by Chester (1538), dam (Pink) by Goldsmith (1258), g.d. (Pink) by Governor (464), g.g.d. (Pigeon) by Discount (339).

1862, May 24,	r.—w.f. H	Pink 3rd	Fencote 1989	Mr. Gregg.
1863,		Dead	do.	do.
1864,	r.—w.f. B	Steer	do.	do.

PLEASANT.
Red with white face, calved November 1, 1856.

Bred by and the property of Mr. Wm. Tudge, Adforton, Leintwardine, Ludlow; got by The Doctor (1083), dam (Cherry) by Nelson (1021), g.d. (Cherry) by Turpin (300), g.g.d. (Cherry) by a Tully Bull, g.g.g.d. (Cherry) by Crabstock (303),

1859, Dec. 22,	r.—w.f. B	Steer	Carbonel 1525	Mr. Tudge.
1860, Oct 28,	r.—w.f. H	Pearl	do.	do.
1861, Sep. 1,	r.—w.f. H	Ruby	The Grove 1764	do.
1862, Aug. 10,	r.—w.f. H	Patty	Sir Colin 2216	do.
1863, July 14,	r.—w.f. B	Steer	Harold 2029	do.

The Steer of 1859 was a winner of the First Prize in his class at Abingdon, and Third at the meeting of the Smithfield Club, 1862.

cows.

PLUM.
Red with white face, calved November 21, 1858.

Bred by and the property of Mr. Thos. Powell, Bage Madley, Hereford; got by Green Gage (1266), dam (Pigeon) by Son of Madley (1301), g.d. (Blossom), by a son of Young Sovereign (379), g.g.d. (Blossom) bred by the late Mr. Fluck, Didley.

PRODUCE IN		NAME.	BY WHAT BULL.	BY WHOM BRED.
1861, Nov.	r.—w.f. H	Cherry	Son of Delight 1564	Mr. Powell.
1862, July 30,	r.—w.f. B	Credenhill	Harliquin 2031	do.
1863, Aug. 19,	r.—w.f. B	Paragon 2665	Portly 2165	do.
1864, Sep. 14,	r.—w.f. H	Beautiful	Interest 2046	do.

PLUM.
Red with white face, of 1853, vol. v., p. 253.

Bred by Viscount Hereford, Tregoyd, Hay, the property of Major-General the Hon. A. N. Hood, Cumberland Lodge, Windsor; got by Phantom (1035), dam (Plum) by Severn (245).

1863, Mar. 31, r.—w.f. H	Penelope	Ajax 1843	Gen. Hon. A. N. Hood.

PLUM THE THIRD.
Red with white face, calved November 15, 1858.

Bred by the late Mr. Thos. Rea, Westonbury, Leominster, the property of Mr. Wiseman, Dorset; got by Sir Benjamin (1387), dam (Plum) by Regent (891), g.d. (Curly) by Barrister (658), g.g.d (Curly) by Old Court (306).

1861, Sep. 26, r.—w.f. B	Dead	Lord Nelson 2088	Mr. T. Rea.
1862, July, 8, r.—w.f. H	Plum 4th	Whitfield	do.
1863, Aug. 5, r.—w.f. H	Plum 5th	Artful 2391	do.

Plum the Fourth, sold to Mr. J. Lumsden, Auchry House, Aberdeenshire.
Plum the Fifth, sold to Mr. J. D. Allen, Pyt House, Tylsbury, Wilts.

POLLY.
Red with white face, calved July 25, 1861.

Bred by and the property of Mr. Naylor, Leighton Hall, Montgomeryshire; got by Admiral (1481) dam (Mary Ann) by Tom of Lincoln (1099), g.d. (Mary) by Silvester (797), g.g.d. (Lady Elinor) by Big Ben (248), g.g.g.d. (Butterfly) by Prince (251), g.g.g.g.d. — by a son of Sir Andrew (183).

PRODUCE IN	NAME.	BY WHAT BULL.	BY WHOM BRED.
1864, July 12, r.—w.f. B		Salisbury 2204	Mr. Naylor.

POLYANTHUS.
Red with white face, of 1858, vol. v., p. 254.

Bred by the late Lord Berwick, Cronkhill, Salop, the property of the Hon. Rev. Noel Hill, Berrington, Salop; got by Albert Edward (859), dam (Primrose) by Walford (871), g.d. Young Nutty) by Tom Thumb (243), g.g.d. (Nutty the Third) by the Count (351), g.g.g.d. (Nutty the Second) by Young Trueboy (1475), g.g.g.g.d. (Nutty) by Cholstrey (860), — (Old Betsy) bred by the late Mr. Knight, Downton Castle.

1862, July 20, r.—w.f. B	Steer	Conqueror 1929	Hon. and Rev. H. Noel Hill.
1863, July 19, r.—w.f. H	Peach	Van Tromp 2291	do.
1864, July 15, r.—w.f. { B H	Cronkhill 2487 Petunia 2nd	Couqueror 1929	do.

PRETTYLASS.
Red with white face, calved June 22, 1860.

Bred by Mr. Duckham, Baysham Court, Ross, the property of Mr. R. Davey, M.P., Polsue House, Grampound, Cornwall; got by Franklyn (1576), dam (Prettymaid) by Colossus (591), g.d. (Beauty) by Young Royal (1136), g.g.d. (Prettymaid) by Mercury (1317), g.g.g.d. bred by Mr. D. Pearce, Stretton Court, Hereford.

1863, Dec. 21, r.—w.f. H	Zillah	Zippor 2354	Mr. Davey.

Prettylass was a winner of the Second Prize in her class at the meeting of the Royal Cornwall Agricultural Society, 1863.

COWS.

PRETTYMAID.

Red with white face, calved October 18, 1855.

Bred by and the property of Mr. Henry Gibbons, Hampton Bishop, Hereford; got by The Admiral (1078), dam (Moap) by Zephyr (1826).

PRODUCE IN		NAME.	BY WHAT BULL.	BY WHOM BRED.
1858, Oct. 22,	r.—w.f. B	Steer	Defence 2nd 1208	Mr. Gibbons.
1859, Feb. 12,	r.—w.f. H		Woodman 2nd 1459	do.
1860, Feb. 1,	r.—w.f. B	Marcle 2102	do.	do.
1861, Jan. 5,	r.—w.f. H	Cherry	Shamrock 2nd 2210	do.
1862, Dec. 22,	r.—w.f. H	Countess	do.	do.
1863, Nov. 25,	r.—w.f. H	Prettylass	do.	do.
1864, Sep. 30,	r.—w.f. H	Pleasant	do.	do.

Countess with her sire and dam were winners of the Second Prize in their class at the Worcester Meeting of the Royal Agricultural Society of England.

PRETTYMAID.

Red with white face, calved in the year 1859.

Bred by Mr. J. B. Green, Marlow, Shrewsbury, the property of Captain Crawshay, Danypark, Crickhowel; got by Beefy Ben (1869), dam — by Cholstrey (217), g.d. — by Zest of Oxford (2352), g.g.d. — by Discount (339).

1862,	r.—w.f. B	Steer	Zealot 2344	Mr. J. B. Green.
1863, July 24,	r.—w.f. H	Prettylass	do.	do.
1864,	r.—w.f. B		Zeal 2342	Capt. Crawshay

PRETTYMAID.

Red with white face, calved in the month of January, 1860.

Bred by and the property of Mr. B. Rogers, The Grove, Pembridge; got by The Grove (1764), dam (Prettymaid the Second) by Young Royal (1470), g.d. (Prettymaid) by Prince (251), g.g.d. (Curly) by Charity the Second (1535), g.g.g.d. (Curly) by Portrait (372).

1863, July	r.—w.f. B	Steer	Interest 2046	Mr. Rogers.
1864, May 10,	r.—w.f. H	Prettymaid 2nd	North Star 2138	do.

PRETTYMAID.
Red with white face.

Bred by Mr. John Hewer, Vern House, Marden, the property of Mr. J. Sparksman, Little Marcle, Ledbury; got by General (1251), dam (Red Rose) by Chance (355), g.d. Rosebud) by Young Wellington (507), g.g.d. (Old Red Rose) by Waxy (403), g.g.g.d. (Prettymaid) by Old Wellington (505), g.g.g.g.d. — by Silver (540).

PRODUCE IN	NAME.	BY WHAT BULL.	BY WHOM BRED.
1854, Mar. 6, r.—w.f. B	Mars 1312	Marden 1309	Mr. Hewer.
1855, Jan. 20. r.—w.f. H		Brecon 918	Mr. W. S. Powell.
1856, Dec. 3, r.—w.f. H	Hinton	Silurian 1386	do.
1857, Dec. 6, r.—w.f. H	Damsel	do.	Mr. Hewer.
1858, Nov. 15, r.—w.f. H	Bessie	Deceiver 1206	do.
1859, Dec. 30, r.—w.f. B		Mameluke 1307	do.
1860, Dec. 28, r.—w.f. H	Pigeon	Abdel Kader 1837	do.
1861, Dec. 31, r.—w.f. B	Stockwell 2244	Mameluke 1307	Mr. Sparksman.
1863, Feb. 28, r.—w.f. H	Young Redrose	Cyrus 1199	do.
1864, Feb. 29, r.—w.f. B	Coronet 2nd 2482	Coronet 1936	do.

Hinton, sold to Mr. Yeomans, Stretton Court, Hereford.

PRETTYMAID.
Red with white face, calved July 5, 1862.

Bred by and the property of Mr. T. Olver, Penhallow, Grampound, Cornwall; got by Conservative (1931), dam (Patience) by Colossus (591), g.d. (Cheerful) by Invincible (592), g.g.d. (Cherry) by Reform (508), g.g.g.d. (Rosebud) by Old Byron (440).

1864, Oct. 27, r.—w.f. H	Pearl	Zippor 2354	Mr. Olver.

PRETTYMAID.
Red with white face, calved in the year 1859.

Bred by Messrs. F. and C. Bodenham, Hereford, the property of Mr. John Rogers, Letchmoor, Presteign; got by Priam

(1039), dam — by Young Dewshall (1125), g.d. — bred by the late Sir R. Price, Foxley.

PRODUCE IN		NAME.	BY WHAT BULL.	BY WHOM BRED.
1862, July,	r.—w.f. H	Prettymaid 2nd	Matchless 2110	Mr. Rogers.
1863, Aug.	r.—w.f. H	Prettymaid 3rd	do.	do.
1864, Nov. 6,	r.—w.f. B	Young Matchless 2890	do.	do.

PRETTYMAID.
Red with white face, calved January 12, 1856.

Bred by and the property of Mr. James Bourn, Mawley Town Farm, Cleobury Mortimer; got by Kyrewood (2062), dam (Pink the Third) by the Duke (2265), g.d. (Pink) by Silurian (1386), g.g.d. (Old Pink) by The Bishop (2260).

1862, June 25, r.—w.f. H	Patience	Cardinal 1526	Mr. Bourn.
1863, June 27, r.—w.f. H		do.	do.

PRETTYMAID.
Red with white face, calved September 29, 1861.

Bred by Mr. Thos. Davies, Burlton Court, Hereford, the property of Lieut.-Colonel Feilden, Dulas Court, Hereford; got by Zouave (2359), dam (Handsome) by Newton (1022), g.d. — by Monarch (219), g.g.d. — by Gallant (239).

1864, Aug. 12, r.—w.f. B	Gift 2nd 2543	Gift 1254	Lt.-Col. Feilden.

PRETTYMAID THE SECOND.
Red with white face, of 1852, vol. v., p. 255.

Bred by and the property of Mr. B. Rogers, The Grove, Pembridge, Leominster; got by Young Royal (1470), dam (Prettymaid) by Prince (251), g.d. (Curly the Fourth) by Charity the Second (1535), g.g.d. (Curly the Third) by Portrait (372), g.g.g.d. (Curly the Second) by Sovereign the Second (1739).

1862, Oct.	r.—w.f. H	Dead	The Grove 1764	Mr. Rogers.
1863, July	r.—w.f. H	Daisy	Bolingbroke 1883	do.

Daisy, sold to Mr. Gilliland, Brook Hall, Londonderry, Ireland.

PRETTYMAID THE SECOND.
Red with white face, of 1855, vol. v., p. 257.

Bred by and the property of Mr. Thomas Edwards, Wintercott, Leominster; got by Croft (937), dam (Prettymaid) by Coningsby the Second (1552), g.d. (Beauty) by Big Ben (248).

PRODUCE IN	NAME.	BY WHAT BULL.	BY WHOM BRED.
1863, Mar. 16, r.—w.f. H	Barmaid	Royal George 2197	Mr. Edwards.
1864, Sep. 19, r.—w.f. H	Dairymaid	Adforton 1839	do.

PRIMA DONNA.
Red with white face, of 1856, vol. v., p. 258.

Bred by and the property of Mr. T. Roberts, Ivington Bury, Leominster; got by King James (978), dam (Long Horns) by Andrew the Second (619), g.d. (Pigeon) by Dayhouse (299).

1862. Aug. 10, r.—w.f. H	Prima Donna 2nd	Sir Thomas 2228	Mr. Roberts.
1863, Aug. 3, r.—w.f. B	Prime Minister 2689	do.	do.
1864, July 26, r.—w.f. H	Prima Donna 3rd	do.	do.

PRIMA DONNA THE SECOND.
Red with white face, calved August 10, 1862.

Bred by and the property of Mr. Thomas Roberts, Ivington Bury, Leominster; got by Sir Thomas (2228), dam (Prima Donna) by King James (978), g.d. (Long Horns) by Andrew the Second (619), g.g.d. (Pigeon) by Dayhouse (299).

1864, Dec. 29, r.—w.f. H	Young Prima Donna 2nd	Adforton 1839	Mr. Roberts.

PRIMROSE.
Red with white face, calved in the month of March, 1860.

Bred by the late Mr. James Elliott, Wormhill, the property of Mr. Thomas Wheeler, Wormhill, Eaton Bishop, Hereford; got by Medallist (1009), dam (Daisy) by Governor (464).

1861, Dec. 1, r.—w.f. B	Steer	Son of Medallist	Mr. Wheeler.
1863, Aug. 30, r.—w.f. H	Primrose 2nd	Mentor 2112	do.
1864, Aug. 30, r.—w.f. H	Primrose 3rd	Washington 2868	do.

PRIMROSE.
Red with white face, calved September 15, 1860.

Bred by and the property of Mr. Philip Turner, The Leen, Pembridge; got by Bertram (1513), dam (Woodbine) by Silurian (1064), g.d. (Fatima) by Andrew the Second (619), g.g.d. (Jessamine) by Cobden (668), g.g.g.d. (Dainty) by Commerce (354).

PRODUCE IN	NAME.	BY WHAT BULL.	BY WHOM BRED.
1864, May, 20, r.—w.f. H	Primrose 2nd	Bolingbroke 1883	Mr. Turner.

PRIMROSE.
Red with white face, of 1853, vols. iv. and v., pp. 172, 260.

Bred by the late Mr. James Rea, Monaughty, Knighton, the property of Lord Wenlock, Wenlock, Salop; got by Glendower (898), dam (Primrose) by Cholstrey (217), g.d. (Primrose) by Gallant (239).

1862, Oct. 1, r.—w.f. B	Steer	Lord Nelson 2088	Mr. T. Rea.
1864, Mar. 9, r.—w.f. B	Westonbury	Artful 2391	do.

PRIMROSE THE SECOND.
Red with white face, calved December 14, 1860.

Bred by and the property of Mr. W. Lane, Compton Casey, Andoversford; got by Casey (1527), dam (Primrose) by Hospodar (1621), g.d. (Tulip) by Whittington (1797).

1863, Dec. 28, r.—w.f. H	Primrose 3rd	Hardy 2027	Mr. Lane.

PRIMROSE THE SECOND.
Red with white face, calved November 27, 1860.

Bred by Mr. T. Rea, Westonbury, Leominster, the property of Mr. J. H. Whitehouse, Ipsley Court, Redditch; got by Sir Benjamin (1387), dam (Primrose) by Glendower (898), g.d.

(Primrose) by Cholstrey (217), g.g.d. (Primrose) by Gallant (239).

PRODUCE IN	NAME.	BY WHAT BULL.	BY WHOM BRED.
1863, June 17, r.—w.f. H	Primrose 3rd	Chieftain 930	Mr. Rea.
1864, Nov. 19, r.—w.f. B	Promise 2705	Zeno 1825	Mr. Whitehouse.

Primrose the Third, sold to Mr. Cadle, Longcroft, Westbury-on-Severn.

PRINCESS.
Red with white face, calved in the year 1857.

Bred by Mr. Turner, Court of Noke, the property of Mr. J. P. Apperley, Fownhope, Hereford; got by Severus (1062), dam (Nancy Belle) by Sir David (349).

1860, Oct. r.—w.f. B	Steer	Baron of Noke 1862	Mr. Turner.
1861, Nov. 4, r.—w.f. B	Steer	Cornet 1934	Mr. Apperley.
1862, Oct. 25, r.—w.f. B	Steer	do.	do.
1863, Oct. 1, r.—w.f. B	Steer	do.	do.
1864, Oct. 31, r.—w.f. H	Nancy	Capt. Perry 2444	do.

PRINCESS.
Red with white face, calved November 2, 1859.

Bred by Mr. Edward Price, Pembridge, Leominster, the property of the Duke of Bedford, Woburn Park, Beds.; got by Goldfinder the Second (959), dam (Princess) by Magnet the Second (989), g.d. (Victoria) by Sir David (349), g.g d. (Blowdy) by Hope (439), g.g.g.d. (Beauty) by Sovereign (404).

1862, July 12, r.—w.f. H	Princess Alice	Carbonel 1525	Duke of Bedford.
1863, July 17, r.—w.f. B	Steer	do.	do.

PRINCESS.
Red with white face.

Bred by Mr. Goode, Felton, Bromyard, the property of Mr. W. H. Apperley, Withington, Hereford; got by Berrington (435), dam (Trojan) by Young Sir David (1137), g.d. — by a bull bred by the late Mr. Jeffries, The Grove.

1862, August, r.—w.f. H		Young Monk 2332	Mr. Apperley.
1863, r.—w.f. B	Steer	do.	do.
1864, Sep. r.—w.f. H	Princess 2nd	do.	do.

COWS.

PRINCESS.
Red with white face of 1857, vol. v., p. 261.

Bred by and the property of Mr. Thos. Morris, Therrow, Llyswen, Hay; got by Aberhonddu (903), dam (Miss Noble the Third) by Newton (344), g.d. (Miss Noble the Second) by Young Byron (832), g.g.d. (Miss Noble) by Noble (543), g.g.g.d. (Favourite) by Sovereign (404), g.g.g.g.d. (Damsel) by Young Wellington (505).

PRODUCE IN		NAME.	BY WHAT BULL.	BY WHOM BRED.
1862,	r.—w.f. B	Steer	Druid 1220	Mr. Morris.
1863, June,	r.—w.f. B	Steer	Guardsman	do.
1864, July,	r.—w.f. H	Princess of Wales	Prince Imperial 2171	do.

PRINCESS.
Red with white face, calved in the month of November, 1859.

Bred by Mr. Olver, Penhallow, Grampound, the property of Mr. R. Davey, M.P., Polsue House, Grampound, Cornwall; got by Goldfinder the Second (959), dam (Queen) by Forester (398), g.d. (Blowdy) by Hope (439), g.g.d. (Beauty) by Sovereign (404).

1862, April 17, r.—w.f H	Peeress	Conservative 1931	Mr. Davey.
1863, Mar. 21, r.—w.f. H	Prudence	Penhallow 2154	do.
1864, April 12, r.—w f. H	Baroness	Zippor 2354	do.

PRINCESS.
Red with white face, calved November 13, 1856.

Bred by Mr. W. Tudge, Adforton, Leintwardine, Salop, the property of Mr. H. Bettridge, Hanney, Wantage; got by The Doctor (1083), dam (Spot) by Turpin (300), g.d. (Cherry) by a Tully bull, g.g.d. (Cherry) by Crabstock (303).

1859, Nov. 2, r.—w.f. H	Pansy	Sir David Bendigo	Mr. Tudge.
1860, Dec. 28, r.—w.f. B	Steer	Carbonel 1525	do.
1861, Oct. 28, r.—w.f. H	Spot	The Grove 1764	do.
1862, Oct. 8, r.—w.f. H	Phillis	Sir Colin 2216	do.

PRINCESS.

Red with white face, calved September 28, 1861.

Bred by and the property of Mr. H. Gibbons, Hampton Bishop, Hereford; got by Shamrock the Second (2210), dam (Tulip) by The Admiral (1078), g.d. (Old Tulip) by Young Gaylad (1463).

PRODUCE IN	NAME.	BY WHAT BULL.	BY WHOM BRED.
1864, June 11, r.—w.f. B		Trumpeter 2282	Mr. Gibbons.

PRINCESS ALEXANDRA.

Red with white face, calved August 30, 1861.

Bred by Sir H. J. Brydges, Bart, Boultibrook, Presteign, the property of Mr. R. Green Price, M.P., Norton Manor, Presteign; got by Sir Colin (2216), dam (Miss Coxall) by Coxall (1196), g.d. — by Grateful (1260), g.g.d. — by Confidence (367).

1864, Aug. 15, r.—w.f. B		Lord of the Manor 2622	Mr. Price.

PRINCESS ALICE.

Red with white face, calved in the month of February 1861.

Bred by and the property of Mr. Keene, Pencraig, Caerleon, Monmouth; got by General Wyndham (1590), dam (Nancy) by Prince Albert (2168), g.d. (Gaylass) by Young David (2325), g.g.d. (Old Gaylass) by Foxhall (2520).

1863,	r.—w.f. H	Dead	Dolward 1966	Mr. Keene.
1864,	r.—w.f. H	Countess	Cholstrey 2nd 1919	do.

PRINCESS ROYAL.

Red with white face, calved in the year 1840.

Bred by and the property of Mr. W. H. Oatley, Wroxeter, Salop;

got by Emigrant (1980), dam a (Sovereign cow) by Sovereign (404).

PRODUCE IN		NAME.	BY WHAT BULL.	BY WHOM BRED.
1843, Oct. 23,	r.—w.f. B	Wroxeter 386	Cotmore 376	Mr. Oatley.
1844, Oct. 25,	r.—w.f. B	Steer	Camel 384	do.
1845, Nov. 2,	r.—w.f. B	Norton	Speculation 387	do.
1846, Sep. 23,	r.—w.f. H	Rowena	Herald 389	do.
1847, Sep. 5,	r.—w.f. H	Fausta	Bryony 599	do.
1848, Sep. 1,	r.—w.f. H	Dead	do.	do.
1849, May 20,	r.—w.f. H	Long Horns	Surprise 779	do.

PRINCESS ROYAL.

Red with white face, calved in the year 1858.

Bred by Mr. Sheriff, Coxall, Ludlow, the property of Mr. R. H. Capper, The Northgate, Ross; got by Coxall (1196), dam — by Brilliant (1518), g.d. — by Confidence (367), g.g.d. — by Emperor (221).

1863, Aug. 24,	r.—w.f. H	Princess Charlotte	Sir Colin 2216	Mr. Sheriff.
1864, Aug. 14,	r.—w.f. B	Prince Royal 2699	Garibaldi 2008	Mr. Capper.

PRIZE DAISY.

Red with white face, calved January 4, 1860.

Bred by and the property of Mr. Thos. Roberts, Ivington Bury, Leominster; got by Sir Benjamin (1387), dam (Prize Flower) by Arthur Napoleon (910), g.d. (Long Horns) by Andrew the Second (619), g.g.d. (Pigeon) by Dayhouse (299).

1862,	r.—w.f. H	Prize Daisy 2nd	Sir Thomas 2228	Mr. Roberts.
1863,	r.—w.f. B	Dead	Lord Warwick 2093	do.
1864,	r.—w.f. B		Sir Thomas 2228	do.

PROMISE.

Red with white face, calved in the year 1856.

Bred by and the property of Mr. G. T. Forester, High Ercall, Wellington, Salop; got by Darling (1202), dam (Query) by Wonder (420), g.d. (Hopeless) by Hope (439), g.g.d. (Johanna) by Lottery the Second (408), g.g.g.d. — bred by Mr. J. Hewer.

PRODUCE IN	NAME.	BY WHAT BULL.	BY WHOM BRED.
1859, Oct. r.—w.f. H	Prophetess	Discord 1217	Mr. Forester.
1860, Oct. r.—w.f. H	Dead	do.	do.
1861, Dec. 27, r.—w.f. B		do.	do.
1862, Dec. 13. r.—w.f. H	Fairy	Severn 1382	do.
1863, Dec. 31, r.—w.f. H	Sibyl	do.	do.
1864, Nov. 12, r.—w.f. B		do.	do.

PROSERPINE.
Red with white face, calved in the year 1859.

Bred by and the property of Mr. G. T. Forester, High Ercall, Wellington, Salop; got by Discord (1217), dam (Ceres) by Darling (1202), g.d. (Sicklehorn) by Wonder (420), g.g.d. (Twilight) by Byron (380), g.g.g.d. (Number Thirty) by Confidence (367), g.g.g.g.d. (Lady Chance) by Chance (355), — bred by by the late Mr. Jeffries, The Grove, by Sovereign (404).

1862, Aug. 18, r.—w.f. B	Steer	Severn	Mr. Forester.
1863, July 17, r.—w.f. H	Cyanè	do.	do.
1864, June 10, r.—w.f. H	Arethusa	do.	do.

PRUDENCE.
Red with white face, calved May 18, 1861.

Bred by Mr. J. Bourn, Mawley Town Farm, Cleobury Mortimer, the property of Mr. V. Gosport, Tanylan, Holywell, North Wales; got by Cardinal (1526), dam (Prettymaid) by Kyrewood (2062), g.d. (Pink the Third) by The Duke (2265), g.g.d. (Pink) by Silurian (1386), g.g.g.d. (Old Pink) by The Bishop (2260).

1864, July 10, r.—w.f. B	Colenso 2466	Sir Colin 1389	Mr. Bourn.

PRUDENCE.
Red with white face, calved June 28, 1861.

Bred by and the property of Mr. T. Olver, Penhallow, Grampound; got by Volunteer (2300), dam (Patience) by Colossus

(591), g.d. (Cheerful) by The Invincible (592), g.g.d. (Cherry) by Reform (508), g.g.g.d. (Rosebud) by Old Byron (440).

PRODUCE IN	NAME.	BY WHAT BULL.	BY WHOM BRED.
1864, Sep. 6, r.—w.f. H	Primrose	Zippor 2354	Mr. Olver.

PRUDENCE.
Red with white face, calved January 12, 1860.

Bred by and the property of Mr. P. Turner, The Leen, Pembridge; got by Felix (953), dam (Charity) by Andrew the Second (619), g.d. (Novice) by The Duke (2265), g.g d. (The Nun) by Sir Walter (352), g.g.g.d. (Silvia) by Commerce (354).

PRODUCE IN	NAME.	BY WHAT BULL.	BY WHOM BRED.
1862, Dec. 8. r.—w.f. H	Darling	Bolingbroke 1883	Mr. Turner.

PRUDENCE.
Red with white face, of 1856, vol. iv., p. 176.

Bred by and the property of Mr. Naylor, Leighton Hall, Welshpool; got by Silvester (797), dam (Blowdy) by Big Ben (248), g.d. (Mottle) by Prince (251), g g.d (Beauty) by Claret (253), g.g.g.d. (Spot) by Trump (490).

PRODUCE IN	NAME.	BY WHAT BULL.	BY WHOM BRED.
1859, Dec. 9, r.—w.f. H	Dead	Tom of Lincoln 1099	Mr. Naylor.
1860, Oct. 22, r.—w.f. H	Prudence 2nd	do.	do.
1861, Dec. 16, r.—w.f. B	Steer	do.	do.
1863, Feb. 4, r.—w.f. H	Prudence 3rd	Salisbury 2204	do.
1864, Jan. 19, r.—w.f. B	Commerce 2470	do.	do.

PRUDENCE THE SECOND.
Red with white face, calved July 28, 1859.

Bred by the late Mr. J. Rea, Monaughty, Knighton, the property of Mr. J. Farr, Pontrilas, Hereford; got by Balaclava (1505), dam (Prudence) by Nelson (1021), g.d. (Lady) by Brampton (917), g.g.d. (Lady) by Monarch (219), g.g.g.d. — by Regulator (360), g.g.g.g.d. — by Crabstock (303).

PRODUCE IN	NAME.	BY WHAT BULL.	BY WHOM BRED.
1861, Nov. 25, r.—w.f. H	Prudence 3rd	Sir Benjamin 1387	Mr. Rea.
1862, r.—w.f. B	Steer	do.	do.

Prudence the Third, sold to Mr. Baldwin, Luddington, Stratford-on-Avon.

COWS.

PRUNE.

Red with white face, calved May 4, 1861.

Bred by the Rev. A. Clive, Whitfield, Hereford, the property of Mr. J. R. Paramore, Dinedor Court, Hereford; got by General (1251), dam (Prudence) by Silurian (1064), g.d. (Promise) by Sir Walter (352), g.g.d. (Promise) by Commerce (354), g.g.g.d. Mottle) by Old Court the Second (1341), g.g.g.g.d. (Lively) by Lottery the Second (987).

PRODUCE IN	NAME.	BY WHAT BULL.	BY WHOM BRED.
1863, Sep. 19, r.—w.f. B	Steer	The Jew 2266	Mr. Paramore.
1864, July 29, r.—w.f. H	Prudence	Grateful 1596	do.

PURITY.

Red with white face, calved in the year 1858.

Bred by Mr. E. Price, Court House, Pembridge, the property of Mr. H. Haywood, Blakemore, Hereford; got by Goldfinder the Second (959), dam — by Magnet (823), g.d. — by Forester (398).

1861, r.—w.f. H			Mr. Price.
1862, r.—w.f. H			do.
1864, Jany 3, r.—w.f. H	Prizeflower	Zeno 1825	Mr. Haywood.

PURITY THE SECOND.

Red with white face, calved May 6, 1859.

Bred by the late Mr. J. Rea, Monaughty, Knighton, the property of Mr. Thomas Cadle, Longcroft, Westbury-on-Severn; got by Balaclava (1505), dam (Purity) by Regent (891), g d. (Duchess) by Barrister (658), g.g.d. (Duchess) by Gallant (239), g.g.g.d. — by Old Court (306).

1863, Feb. 16, r.—w.f. H	Pussy	Sir Benjamin 1387	Mr. Rea.
1864, Jan. 24, r.—w.f. H	Polly	do.	Mr. Cadle.

Pussy, sold to Mr. T. Cadle.

PYAT.

Red with white face, calved February 10, 1856.

Bred by and the property of Mr. H. Gibbons, Hampton Bishop, Hereford; got by the Admiral (1078), dam (Beauty) by Young Gaylad (1463), g.d. (Old Beauty) by Zephyr (1826).

PRODUCE IN	NAME.	BY WHAT BULL.	BY WHOM BRED.
1859, Sep. 6, r.—w.f. H	Pyat 2nd	Woodman 2nd 1459	Mr. Gibbons.
1860, July 24, r.—w.f. H		do.	do.
1861, Oct. 24, r.—w.f. H	Duchess	Shamrock 2nd 2210	do.
1862, Oct. 28, r.—w.f. B	Steer	do.	do.
1863, Sep. 11, r.—w.f. B	Gladstone 2546	do.	do.
1864, Nov. 12, r.—w.f. H		do.	do.

QUEEN'S GILLIFLOWER

Red with white face, calved July 19, 1856.

Bred by Mr. J. Rea, Monaughty, Knighton, the property of Mr. W. Stallard, Brockhampton, Ross; got by The Doctor (1083), dam (Dame's Violet) by Young Conrad (2322), g.d. (Dahlia) by Caractacus (659), g.g.d. (Dahlia) by Old Court (306).

1860, July 3, r.—w.f. H	Wallflower	Wellington 1112	Mr. Rea.
1861, July 28, r.—w.f. B	Dead	Shamrock 2nd 2210	Mr. Stallard.
1862, Aug. 21, r.—w.f. B	Seignior 2745	Chieftain 2nd 1917	do.
1863, Oct. 29. r.—w.f. B	Soothsayer 2785	do.	do.

Wallflower, sold to Mr. Stallard, Brockhampton. Ross.

QUEEN OF BEAUTY.

Red with white face, calved in the year 1858.

Bred by the Misses Abley, Norton, Presteign, the property of Mr. R. G. Price, M.P., Norton Manor, Presteign; got by Trump (2842).

1864, Aug. 19, r.—w.f. B	Norton	Lord of the Manor 2622	Mr. Price.

QUICKSET.
Red with white face, calved June 3, 1861.

Bred by and the property of Mr. P. Turner, The Leen, Pembridge; got by Logic (2079), dam (Hawthorn) by Felix (953), g.d. (Maythorn) by Sir David (349), g.g.d. (Princess) by Andrew the Second (619), g.g.g.d. (Brenda) by Viscount (816), g.g.g.g.d. (Rarity) by Cupid (1950).

PRODUCE IN	NAME.	BY WHAT BULL.	BY WHOM BRED.
1864, Feb. 15, r.-w.f. H	Eglantine	Bolingbroke 1883	Mr. Turner.

RARITY.
Red with white face, of 1856, vol. v., p. 264.

Bred by and the property of Mr. G. T. Forester, High Ercall, Wellington, Salop; got by Darling (1202), dam (Result) by Governor (464), g.d. (Maid o' the Mill) by Hope (439), g.g.d. (Moreton) by Royal (331), g.g.g.d. — bred by Mr. Yeomans, Moreton, Hereford.

1862, Aug. 15, r.—w.f. H	Gaylass	Severn 1382	Mr. Forester.
1863, July 27, r.—w.f. B	Roden 2727	do.	do.
1864, June 19, r.—w.f. B	Steer	do.	do.

REBECCA.
Red with white face, of 1854, vol. iv., p. 178.

Bred by and the property of Mr. Naylor, Leighton Hall, Montgomeryshire; got by Silvester (797), dam (Lady Elinor) by Big Ben (248), g.d. (Butterfly) by Prince (251), g.g.d. (Nell) by a Son of Sir Andrew (183).

1859, July 2, r.—w.f. H	Dead	Admiral 1481	Mr. Naylor.
1860, July 12, r.—w.f. H	Ruth	do.	do.
1861, Aug. 7, r.—w.f. B	Steer	do.	do.
1862, July 26, r.—w.f. B	Dead	Salisbury 2204	do.
1863, July 22, r.—w.f. B	Steer	do.	do.
1864, July 1, r.—w.f. B	Steer	Gladstone 2547	do.

COWS.

RECOVERY.
Red with white face, calved in the month of October, 1859.

Bred by and the property of Mr. G. T. Forester, High Ercall, Wellington, Salop; got by Discord (1217), dam (Relapse) by Darling (1202), g.d. (Result) by Governor (464), g.g.d. (Maid o' the Mill) by Hope (439), g.g.g.d. (Moreton) by Royal (331).

PRODUCE IN		NAME.	BY WHAT BULL.	BY WHOM BRED.
1862, Oct. 1,	r.—w.f. H	Spot	Severn 1382	Mr. Forester.
1863, Aug. 2,	r.—w.f. H	Resemblance	do.	do.
1864, July 9,	r.—w.f. H	Recompense	do.	do.

RED CAP.
Red with white face, of 1858, vol. v., p. 265.

Bred by and the property of Mr. T. Morris, Therrow, Llyswen, Hay; got by Telegraph (1404), dam (Blossom) by Newton (344), g.d. (Silver) by White Nob (345), g.g.d. (Beauty) by Counsellor (422), g.g.g.d. (Lovely) by Charity the Second (516).

1863,		r.—w.f. B	Steer	Prince Imperial 2171	Mr. Morris.
1864, Nov.		r.—w.f. B	Rufus 2735	do.	do.

RED ROSE.
Red with white face, calved November 1, 1861.

Bred by Mr. John Smith, Sevenhampton, the property of Mr. John Barton, Coln, Fairford; got by St. Michael (1718), dam (Gaylass)) by Magnum Bonum (1303), g.d. (Gaily) by Garrick (1248), g.g.d. (Old Gaily) g.g.g.d. (Countess the Third) by Sir Charles (1388), g.g.g.g.d. (Countess the Second) by Sovereign (404), — (Old Prettymaid) by Old Wellington (507), — (Old Primrose) by Silver (540).

1864, July,	r.—w.f. B		Young Cardinal Wiseman 2882	Mr. Barton.

COWS.

RED ROSE.
Red with white face, calved in the year 1861.

Bred by the Misses Abley, Norton, Presteign, the property of Mr. R. G. Price, M.P., Norton Manor, Presteign; got by Havelock (2563).

PRODUCE IN	NAME.	BY WHAT BULL.	BY WHOM BRED.
1864 July 28, r.—w.f. B		Lord of the Manor 2622	Mr. Price.

RED ROSE THE SECOND.
Red with white face, calved in the year 1860.

Bred by Mr. William Berrow, The Green, Allensmoor, Hereford, the property of Mr. Samuel Gilliland, Brook Hall, Londonderry, Ireland; got by Napoleon (1018), dam — by Widgeon (1799).

1864, Jan. 12, r.—w.f. H	Red Rose 4th	JollyMiller 11th 2585	Mr. Gilliland.

Red Rose the Second was a winner of First Prizes at the Londonderry and the North West Agricultural Societies Meetings in the years 1861, 1862, and 1863, and First at the Londonderry, and Second at the North West Meetings, 1864; also First Prizes at the Belfast and Sligo Meetings of the Royal Agricultural Society, Ireland, 1861 and 1864.

RED ROSE THE SECOND.
Red with white face, calved September 18, 1859.

Bred by Mr. R. Hickman, Bosbury, Ledbury, the property of Mr. J. Gregg, Fencote Abbey, Leominster; got by Chester (1538), dam (Red Rose) by Goldsmith (1258), g.d. (Rosa) by Discount (339), g.g.d. (Rosebud) by Chance (348).

1862, July 9, r.—w.f. H	Redrose 3rd	Fencote 1989	Mr. Gregg.
1864, May 12, r.—w.f. H	Redrose 4th	do.	do.

RED ROSE THE THIRD.
Red with white face, calved May 16, 1860.

Bred by and the property of Mr. Samuel Gilliland, Brook Hall,

Londonderry; got by Jolly Miller the Fifth (2583), dam (Rose Bud) by Jolly Miller the Third (2581), g.d. (Red Rose).

PRODUCE IN	NAME.	BY WHAT BULL.	BY WHOM BRED.
1863, Feb. 27, r.—w.f. B	Dead	Jolly Miller 4th 2582	Mr. Gilliland.
1864, Feb. 26, r.—w.f. B	Rose Bud 2nd	Jolly Miller 10th 2584	do.

REDWING.

Red with white face, calved October 12, 1859.

Bred by and the property of Mr. P. Turner, The Leen, Pembridge; got by Felix (953), dam (Duchess) by Silurian (1064), g.d. (Princess) by Andrew the Second (619), g.g.d. (Brenda) by Viscount (816), g.g.g.d. (Rarity) by Cupid (1950).

1863, Feb. 14, r.—w.f. B	Steer	Bolingbroke 1883	Mr. Turner.
1864, April 20, r.—w.f. H	Kathleen	do.	do.

RINGDOVE.

Red with white face, calved August 27, 1859.

Bred by Mr. J. Sobey, Penhallow, Grampound, the property of Mr. R. Davey, M.P., Polsue House, Grampound, Cornwall; got by Great Eastern (1598), dam (Young Pigeon) by a son of Confidence (367).

1862, Feb. 13, r.—w.f. H	Mabel	Penhallow 2154	Mr. Davey.
1863, Feb. 3, r.—w.f. B	Steer	do.	do.
1863, Dec. 26, r.—w.f. B	Steer	Zippor 2354	do.

RINGDOVE.

Red with white face, calved in the year 1861.

Bred by and the property of Mr. R. Green Price, M.P., Norton Manor, Presteign; got by Stanage (1742).

1864, June 11, r.—w.f. H	Turtle Dove	Lord of the Manor 2622	Mr. Price.

COWS.

RINGLET.
Red with white face, calved December 4, 1860.

Bred by and the property of Mr. T. Olver, Penhallow, Grampound, Cornwall; got by Earl Derby (1979), dam (Ringdove) by Young Walford (1820).

PRODUCE IN	NAME.	BY WHAT BULL.	BY WHOM BRED.
1863, June 12, r.—w.f. B	Steer	Conservative 1931	Mr. Olver.
1864, May 27, r.—w.f. H	Rosebud	Zippor 2354	do.

Ringdove was a winner of the First Prize in her class at the Truro Meeting of the Royal Cornwall Agricultural Society, 1863.

ROSA.
Red with white face, calved September 20, 1862.

Bred by and the property of Mr. T. Olver, Penhallow, Grampound, Cornwall; got by Conservative (1931), dam (Ringdove) by Young Walford (1820).

1864, Nov. 20, r.—w.f. H	Ruby	Zippor 2354	Mr. Olver.

ROSA.
Red with white face, of 1857, vol. v., p. 269.

Bred by and the property of Mr. T. Duckham, Baysham Court, Ross; got by Colossus (591), dam (Rose) by Mercury (1317), g.d. (Prettymaid) bred by Mr. Pearce, Stretton Court, Hereford.

1862, May 11, r.—w.f. H	Rosette	Caster 1900	Mr. Duckham.
1863, June 2, r.—w.f. B	Steer	Victory 2296	do.
1864, July 14, r.—w.f. B	Steer	Cato 1902	do.

ROSA.
Red with white face, calved February 23, 1859.

Bred by Mr. T. Olver, Penhallow, Grampound, Cornwall, the property of Mr. Thomas Golding, Callington, Cornwall; got by Great Eastern (1598), dam (Lady) by The Earl (1761), g.d, — bred by Earl St. Germans.

1864, Sep. 30, r.—w.f. B	Pleasant 2679	Zippor 2354	Mr. Olver.

COWS.

ROSA.
Red with white face, calved in the year 1861.

Bred by the Rev. Archer Clive, Whitfield, Hereford, the property of Lieut.-Colonel Feilden, Dulas Court, Hereford; got by Kilpeck (1626), dam (Rebecca) by Quicksilver (353), g.d. (Rebecca) by Royalty (1374).

PRODUCE IN	NAME.	BY WHAT BULL.	BY WHOM BRED.
1863, Aug. 7, r.—w.f. H	Violet	Gift 1254	Col. Feilden.
1864, Aug. 1, r.—w.f. B		do.	do.

ROSA.
Red with white face, calved December 30, 1861.

Bred by Mr. J. Hollings, the property of Mr. J. A. Hollings, Hillend, Hereford; got by St. Clement (2201), dam (Rose the Sixth) by Noke (1338), g.d. (Rose the Fifth) by Voltigeur (1445), g.g.d. (Rose the Fourth) by Byron (380), g.g.g.d. (Rose the Third) by Herald (2037), g.g.g.g.d. (Rose the Second) by Cornet (1933), — (Rose the First) by Young Waterloo (2341).

1864, Nov. 3, r.—w.f. H	Chieftain's Rosa	Chieftain 2nd 1917	Mr. J.A. Hollings.

ROSA.
Red with white face, calved in the year 1860.

Bred by Mr. Sheriff, Coxall, the property of Mr. R. Harcourt Capper, The Northgate, Ross; got by Coxall (1196), dam — by Brilliant (1518), g.d. — by Young Emperor (1811), g.g.d. — by Royal (331).

1863, Aug. 2, r.—w.f. B	Steer	Sir Colin 2216	Mr. Sheriff.
1864, Aug. 9, r.—w.f. H	Rosette	Garibaldi 2008	Mr. Capper.

ROSABELLE.
Red with white face, of 1856, vol. v., p. 270.

Bred by Mr. Duckham, Baysham Court, Ross, the property of Mr. R. Davey, M.P., Polsue House, Grampound; got by Sellack

Grove (1722), dam (Rose) by Mercury (1317), g d. (Prettymaid) bred by Mr. Pearce, Stretton Court, Hereford.

PRODUCE IN	NAME.	BY WHAT BULL.	BY WHOM BRED.
1863, Jan. 22, r.—w.f. H	Isabelle	Penhallow 2154	Mr. Davey.

Isabelle, sold to Mr. Pollock, Galway, Ireland.

ROSABELLE.
Red with white face, calved in the year 1856.

Bred by Mr. John Davies, Tillington, Hereford, the property of Mr. Thomas Powell, Bage, Madley, Hereford; got by Mameluke (1307), dam (Old Rosabelle) by Pope (527), g.d. (Old Silver) by Old Wellington (507), g.g.d. (Beauty) by Sovereign (404), g.g.g.d. (Old Gentle) by Chance (355).

1860, r.—w.f. H	Mossrose	Courtier 1194	Mr. Price.
1861, r.—w.f. B	Portly 2165	Zeno 1825	Mr. Powell.
1862, Oct. 7, r.—w.f. B	Victor 2855	do.	do.
1863, Sep. 1, r.—w.f. H	Redrose	Troubadour 1780	do.
1864, Aug. 3, r.—w-f. { B	Stockwell 2792	Interest 2046	do.
{ H	Rosa		

Rosa, sold to Mr. Paramore, Dinedor Court, Hereford.

ROSE.
Red with white face, calved in the month of February, 1859.

Bred by and the property of Mr. W. Lane, Compton Casey, Andoversford; got by Tyro (1786), dam (Rose) by Hospodar (1621).

1862, Feb. 8, r.—w.f. B	Steer		Mr. Lane.
1863, Feb. 13, r.—w.f. H	Rosebud	Hardy 2027	do.

ROSE.
Red with white face, of 1853, vols. iv. and v., pp. 181, 270.

Bred by and the property of Mr. Thomas Edwards, Wintercott, Leominster; got by Stretford (1749), dam (Beauty) by Big Ben (248).

1862, Feb. 18, r.—w.f. H	Dead		Mr. Edwards.
1863, Feb. 17. r.—w.f. H	Rose 2nd	Royal George 2197	do.

ROSE.
Red with white face, calved May 10, 1861.

Bred by and the property of Mr. Philip Turner, The Leen, Pembridge; got by Logic (2079), dam (Daisy) by Felix (953), g.d. (Rosebud) by Andrew the Second (619), g.g.d. (Daisy) by Marmion (763), g.g.d. (Desdemona) by Sir Walter (352), g.g.g.g.d. (Daisy) by Old Court the Second (1341).

PRODUCE IN	NAME.	BY WHAT BULL.	BY WHOM BRED.
1864, Feb. 9, r.—w.f. H	Geraldine	Bolingbroke 1883	Mr. Turner.

ROSE THE SECOND.
Red with white face, of 1857, vol. v., p. 273.

Bred by Mr. B. Rogers, The Grove, Pembridge, the property of Mr. Ford; got by Gaylad the Second (1589), dam (Rose) by Young Royal (1470), g.d. (Primrose) by Prince (251).

1862, July, r.—w.f. H	Primrose 3rd	Bolingbroke 1883	Mr. Rogers.
1863, July, r.—w.f. B	Steer	Interest 2046	do.
1864, May 21, r.—w.f. B	Dead	Bolingbroke 1883	do.

ROSE THE SIXTH.
Red with white face, of 1855, vol. v., p. 274.

Bred by Mr. Hollings, Hillend, Hereford, the property of Mr. J. A. Hollings, Hillend, Hereford; got by Noke (1338), dam (Rose the Fifth) by Voltigeur (1445), g.d. (Rose the Fourth) by Byron (380), g.g.d. (Rose the Third) by Herald (2037), g.g.g.d. (Rose the Second) by Cornet (1933), g.g.g.g.d (Rose the First) by Young Waterloo (2341).

1863, Jan. 7, r.—w.f. H	Rosabelle	St. Clement 2201	Mr. Hollings.

ROSE BLOSSOM (A TWIN).
Red with mottled face, calved Sepember 22, 1861.

Bred by and the property of Captain Peploe, Garnstone,

Weobley; got by The Twin (1420), dam (Red Rose) by Musician (725) g.d. (Lovely) by Tyro (692), g.g.d. (Larkspur) by Victory (2297), g.g.g.d. (Larkspur) by Semplon (58).

PRODUCE IN	NAME.	BY WHAT BULL.	BY WHOM BRED.
1864, Sep. 11, r.—m.f. H	Rose of Garnstone	Leo 2070	Capt. Peploe.

ROSEBUD.

Red with white face, calved April 28, 1861.

Bred by and the property of Mr. T. Morris, Therrow, Llyswen, Hay; got by (Druid) (1220), dam (Sophia) by Prior (1359), g.d. (Fatrumps) by Enterprise (948), g.g.d. (Lovely) by Charity the Second (516), g.g.g.d. (Lovely) by White Nob (345).

1663, July, r.—w.f. H	Rose Leaf	Prince Imperial 2171	Mr. Morris,
1864, Sept., r.—w.f. H	Rose Bud 2nd	do.	do.

ROSEBUD.

Red with white face, calved June 17, 1859.

bred by and the property of Mr. Duckham, Baysham Court, Ross; got by Sambo (1719), dam (Rose) by Mercury (1317), g.d. ((Prettymaid) bred by Mr. Pearce, Stretton Court, Hereford.

1863, Mar. 18, r.—w.f. B	Steer	Garibaldi 2003	Mr. Duckham.
1864, April 8, r.—w.f. H	Rosaline	Cato 1902	do.

ROSEBUD.

Red with white face, of 1852, vol. v., p. 272.

Bred by the late Viscount Hereford, Tregoyd, Hay, the property of Mr. John Monkhouse, The Stow, Hereford.; got by Phantom (1035), dam bred by Mr. Trouncer, by a son of Goldfinder (383).

1862, Aug. 11, r.—w.f. H	Rosalind	Chieftain 930	Mr. Monkhouse.
1864, Jan. 16, r.—w.f. B	Steer	do.	do.

ROSEBUD.

Red with white face.

Bred by Mr. John Rogers, The Stocken, Presteign, the property of Mr. R. H. Ridler, Gattertop, Leominster; got by The Count (1760), dam — by Young Royal (1469), g.d. — by Young Albert, g.g.d. — by Portrait (372), g.g.g.d. — by Sovereign (404).

PRODUCE IN	NAME.	BY WHAT BULL.	BY WHOM BRED.
1862, June, r.—w.f. H	Rebe	Sir Benjamin 1387	Mr. J. Rogers.
1863, April 5, r.—w.f. B	Master Benjamin 2636	do.	Mr. Ridler.
1864, April 2, r.—w.f. H	Rose	Defiance 1957	do.

ROSEBUD.

Red with white face, calved April 13, 1861.

Bred by Mr. John Hewer, Vern House, Marden, the property of Mr. John Palmer, Hampton-on-the-Hill; got by General (1251), dam (Caroline) by Cardinal Wiseman (1168), g.d. (Hampton Lass) by Mark (424), g.g.d. (Miss Hampton) by Garrick (1248), g.g.g.d. (Lady Hampton) by Reform (508).

1864, Jan. 4, r.—w.f. H	Dead	Sir Edmund Lyons 2219	Mr. Palmer.
1864, Nov. 24, r.—w.f. B	Oliver Twist	My Lord 2647	do.

ROSEBUD.

Red with white face, calved in the year 1861.

Bred by Mr. Sheriff, Coxall, Ludlow, the property of Mr. R. H. Capper, The Northgate, Ross; got by Sir Colin (2216), dam (Belle) by Coxall (1196).

1864, Aug. 10, r.—w.f. H	Rosabelle	Garibaldi 2008	Mr. Capper.

ROSEBUD THE SECOND.

Red with white face, calved January 22, 1861.

Bred by and the property of Mr. Thomas Edwards, Wintercott, Leominster; got by Sir Newton (1731), dam (Rosebud) by Croft

COWS.

(937), g.d. (Daisy) by Stretford (1749), g.g.d. (Dainty) by Coningsby the Second (1552).

PRODUCE IN	NAME.	BY WHAT BULL.	BY WHOM BRED.
1863, Aug. 2, r.—w.f. B	Adforton 2nd 2371	Adforton 1839	Mr. Edwards.
1864, Sep. 28, r.—w.f. B	Steer	do.	do.

ROSEBUD (A TWIN).
Red with mottled face, calved September 22, 1861.

Bred by and the property of Capt. Peploe, Garnstone, Weobley; got by the Twin (1420), dam (Red Rose) by Musician (725), g.d. (Lovely) by Tyro (692) g.g.d. (Larkspur) by Victory (2297), g.g.g.d. (Larkspur) by Semplon (58).

1864, Aug. 15, r.—m.f. H	Moss Rose	Leo 2070	Capt. Peploe

ROSE OF MAPPOWDER.
Red with white face, calved March 1, 1862.

Bred by and the property of Mr. J. W. James, Mappowder Court, Blandford, Dorset; got by Happy Land (2561), dam (Picture) by Chance (2452), g.d. (Old Picture) by Young Sovereign (2895).

1863, Sep. 14, r.—w.f. H	Miriam	Bird's Eye 2420A	Mr. James.

Rose of Mappowder was a winner of the First Prize in her class at the Exeter Meeting of the Bath and West of England Society, and as one of a pair at the meetings of the Sturminster and Yeovil Agricultural Societies, 1863, and at Sherborn, 1864.

ROSE OF THE VALLEY.
Red with white face, calved October 1, 1859.

Bred by Mr. J. Rea, Monaughty, Knighton, the property of Mr. R. H. Capper, The Northgate, Ross; got by Wellington (1112), dam (Lily of the Valley) by Chieftain (930), g.d. (Lily) by Confidence (367), g.g.d. (Lily) by Old Court (306).

1863, June 23, r.—w.f. H	Lady of the Valley	Zenith 2350	Mr. Rea.

cows.

ROSE OF WESTON.

Red with white face, calved November 2, 1861.

Bred by Mr. John Hollings, the property of Mr. J. A. Hollings, Hillend, Hereford; got by Saint Clement (2200), dam (Rose the Fifth) by Voltigeur (1445), g.d. (Rose the Fourth) by Byron (380), g.g.d. (Rose the Third) by Herald (2037), g.g.g.d. (Rose the Second) by Cornet (1933), g.g.g.g.d. (Rose the First) by Young Waterloo (2341).

PRODUCE IN	NAME.	BY WHAT BULL.	BY WHOM BRED.
1864, Feb. 1, r.—w.f. B	Chieftain 4th 2458	Chieftain 2nd 1917	Mr. J. A. Hollings.

ROSETTE.

Red with white face, calved in the year 1862.

Bred by Mr. Sheriff, Coxall, Ludlow, the property of Mr. R. H. Capper, The Northgate, Ross; got by Sir Colin (2216), dam (Trinket) by Brilliant (1518).

1864, Aug. 1, r.—w.f. B		Florence 1991	Mr. Capper.

ROSINA.

Red with white face, of 1858, vol. v., p. 274.

Bred by the late Lord Berwick, Cronkhill, Salop, the property of The Hon. and Rev. Noel Hill, Berrington, Salop; got by Attingham (911), dam (Phillis) by Albert Edward (859), g.d. (Wood Pigeon) by The Count (351), g.g.d. (Pigeon) by Young Trueboy (1475), g.g.g.d. (Pigeon) by Ashley Moor White Bull (870), g.g.g.g.d. (Damsel) by Cholstrey (868), — (Old Damsel) by Coleman's Bull 1547), — (Old Daisy) by Chancellor (156).

1862, Nov. 20, r.—w.f. B	Crink	Conqueror 1929	Hon. & Rev. Noel Hill.
1863, Nov. 12, r.—w.f. H	Lady Victoria	Van Tromp 2291	do.

cows.

RUBY.

Red with white face, calved in the year 1860.

Bred by Mr. Thos. Sheriff, Coxall, Ludlow, the property of Capt. Crawshay, Danypark, Crickhowell; got by Coxall (1196), dam — by Brilliant (1518), g.d. — by Young Royal (1470), g.g.d. — by Walford (871).

PRODUCE IN		NAME.	BY WHAT BULL.	BY WHOM BRED.
1862,	r.—w.f. H	Lady Clyde	Sir Colin 2216	Mr. Sheriff.
1863, July 25,	r.—w.f. B	Steer	do.	do.
1864,	r.—w.f. {B H}	Twins	Zeal 2342	Capt. Crawshay.

RUTH.

Red with white face, calved in the year 1861.

Bred by Captain Power, Hill Court, Ross, the property of Mr. W. Jones, Hill of Eaton, Ross; got by The Jew (2266), dam — by Uncle Tom (1108), g.d. — bred by the late Mr. Phillipps Bryngwyn.

1864, Nov. 4, r.—w.f. H	Pink	Chieftain 3rd 2457	Mr. W. Jones,

RUTH.

Red with white face, of 1858, vol. v., p. 276.

Bred by the late Mr. J. Rea, Monaughty, Knighton, the property of Mr. J. Farr, Pontrilas, Hereford; got by Vanguard (1109), dam (Fairmaid the Fourth) by Chieftain (930), g.d. (Fairmaid) by Cholstrey (217), g.g.d. (Fairmaid) by Gallant (239), g.g.g.d. (Fairmaid) by Portrait (372).

1863, July 11, r.—w.f. B		Sir Benjamin 1387	Mr. Farr.
1864, Aug. 17, r.—w.f. B		Salford 2738	Mr. Rea.

RUTH.

Red with white face, calved November 1858.

Bred by the late Lord Berwick, the property of Mr. J. O. G.

Pollock, Mountainstown, Navan, Ireland; got by Attingham (911), dam (Rebecca) by Governor (464), g.d. (Old Prettymaid) by Young Sovereign (1472), g.g.d. — by White Nob (345), g.g.g.d. — by Young Wellington (505).

PRODUCE IN		NAME.	BY WHAT BULL.	BY WHOM BRED.
1863, July,	r.—w.f. H	Rose	Master Willie 2637	Mr. Pollock.
1864, July	r.—w.f. B	Roymon	Reindeer 2717	do.

SALLY.

Red with white face.

Bred by and the property of Mr. Richard Shirley, Baucott, Munslow, Church Stretton; got by Prime Minister (1696), dam (Giantess) by Marlow (2104), dam (Tasty) by Knockerell (1630), g.g.d. — by the Count (2263).

1863, April 14, r.—w.f. H	Our Sall	Pilot 1036	Mr. Shirley
1864, June 15, r.—w.f. H	Sarah	do.	do.

SHINY THE THIRD.

Red with white face, calved February 5, 1858.

Bred by and the property of Mr. Wright, Halston Hall, Oswestry; got by Magnet the Second (989), dam (Shiny) by Gratitude (1261), g.d. (Young Silver).

1862, Dec. 14, r.—w.f. H	Silky	Whittington 2313	Mr. Wright.
1863, Nov. 23, r.—w.f. B	Major	Hero 2039	do.
1864, Oct. 7, r.—w.f. H	Silk	do.	do.

SHUT.

Red with white face, calved August 10, 1857.

Bred by Lord Bateman, Shobdon Court, Leominster, the property of the representatives of the late Mr. C. Bulmer, Holmer, Hereford; got by Carlisle (923), dam (Strapper) by Monarch (504).

PRODUCE IN	NAME.	BY WHAT BULL.	BY WHOM BRED.
1860, Nov. 18, r.—w.f. H	Superb	Golden Horn 2015	Mr. Bulmer.
1861, Dec. 19, r.—w.f. B	Lord Bateman 2612	Lord Hereford 2617	do.
1862, Dec. 4, r.—w.f. H	Miss Hanbury	Garrick, jun., 2532	do.
1863, Oct. 16, r.—w.f. B	Steer	do.	do.

Shut was one of four winners of the First Prize in their class at the meeting of the Leominster Agricultural Society, 1859.

SILK.

Red with white face, calved in the year 1856.

Bred by Mr. William Berrow, The Green, Allensmoor, Hereford, the property of Mr. Samuel Gilliland, Brook Hall, Londonderry; got by Widgeon (1799), dam (Original) by Original the First (455), g.d. (Withrington).

1859, Dec. 3, r.—w.f. B	Duke of Wellington 2508	Widgeon 1799	Mr. Gilliland.
1861, Jan. 4, r.—w.f. B	Jolly Miller 10th 2584	Jolly Miller 4th 2582	do.
1862, May, 6, r.—w.f. B	Jolly Miller 12th	Duke of Wellington 2508	do.
1863, Apr. 10. r.—w.f. B	do.	Jolly Miller 11th 2585	do.
1864, May 4, r.—w.f. H	Silk the Second	Jolly Miller 13th 2586	do.

Silk was a winner of First Prizes at the meetings of the Londonderry and North West Agricultural Societies, 1861, 1862, and 1863; also a First Prize at the Belfast Meeting of the Royal Agricultural Society of Ireland, 1861.

SILK.

Red with white face, calved in the year 1859.

Bred by Mr. H. Gibbons, Hampton Bishop, Hereford, the property of Mr. J. R. Paramore, Dinedor Court, Hereford; got by Admiral (1078), dam (Young Beauty) by Young Gaylad (1463), g.d. (Old Beauty) by Zephyr (1826), g.g.d. — by a son of Dewshall (358).

1861, Sep. 24, r.—w.f. H	Cowslip	Shamrock 2nd 2210	Mr. Gibbons.
1863, Oct. 6, r.—w.f. B	Dead	The Jew 2266	Mr. Paramore.
1864, Aug. 20, r.—w.f. H	Silky	Grateful 1596	do.

COWS.

SILK.

Red with white face, calved March 29, 1859.

Bred by and the property of Mr. Henry Gibbons, Hampton Bishop, Hereford; got by Hampton (1272), dam (Young Beauty) by The Admiral (1078), g.d. (Beauty) by Young Gaylad (1463).

PRODUCE IN	NAME.	BY WHAT BULL.	BY WHOM BRED.
1862, May 29, r.—w.f. H	Pleasant	Shamrock 2nd 2210	Mr. Gibbons.
1863, June 21, r.—w.f. H	Satin	do.	do.
1864, June 6, r.—w.f. B	Steer	do.	do.

SILKY.

Red with white face, calved in the month of December, 1860.

Bred by the late Lord Berwick, Cronkhill, Salop, the property of the Hon. and Rev. Noel Hill, Berrington, Salop; got by Severn (1382), dam (Fatrumps) by Albert Edward (859).

1863, June, r.—w.f. B		Van Tromp 2291	Hon. & Rev. Noel Hill.

SILKY.

Red with white face, of 1852, vol. iv., p. 184.

Bred by the Rev. Kevil Davies, Croft Castle, the property of the Rev. Archer Clive, Whitfield; got by Andrew the Second (619), dam (Cowslip) by The Baron.

1863, Sep. 14, r.—w.f. H	Silver	Bertram 1513	Rev. A. Clive.
1864, Oct. 5, r.—w.f. H	Snowdrop	Plato 2160	do.

SILKY.

Red with white face, calved in the year 1860.

Bred by the late Mr. Williams, Chapel Clun, Salop, the property of Mr. J. Rogers, Letchmoor, Presteign; got by Jerry (976).

1863, r.—w.f. B	Steer	Plato 2161	Mr. Williams.
1864, June r.—w.f. H	Young Silky	do	Mr. Rogers.

SILKY *alias* LILY.

Red with white face, calved September 6, 1861.

Bred by and the property of Mr. R. Davey, M.P., Polsue House, Grampound, Cornwall; got by Penhallow (2154), dam (Fairmaid) by Big Ben (1875), g.d. (White Rose) bred by the late Earl St. Germans.

PRODUCE IN	NAME.	BY WHAT BULL.	BY WHOM BRED.
1864, June 14, r.—w.f. B		Castor 1900	Mr. Davey.

SILKY.

Red with white face, of 1859, vol. v., p. 278.

Bred by and the property of Mr. B. Hawkins, Orleton, Ludlow; got by Merry Andrew (1011), dam (Silver) by Northampton (600), g.d. (Silver) by Young Chance (449).

1862, June 1, r.—w.f. H	Sal	The Grove 1764	Mr. Hawkins.
1863, Apr. 15, r.—w.f. H	Silk	do.	do.

SILKY MOTTLE FACE.

Red with mottled face, of 1857, vol. v, p. 279.

Bred by and the property of Mr. R. Shirley, Baucott, Church Stretton, Salop; got by Marlow (2104), dam (Mottle Silky) by Knockerell (1630), g.d. (Silky) by Dollgan (759), g.g.d. (Tidy) by The Count (2263).

1862, May 19, r.—m.f. H	Curly Silky	Pilot 1036	Mr. Shirley.
1863, Mar. 27, r.—w.f. B	Glossy 2551	do.	do.
1864, June 4, r.—w.f. B	Star of England 2791	do.	do.

SILVER.

Red with white face, calved April 14, 1859.

Bred by and the property of Mr. T. Duckham, Baysham Court, Ross; got by Colossus (591), dam (Sylph) by Pope (527), g.d.

(Eywood) by Cotmore the Second (1191), g.g.d. — bred by the late Earl of Oxford.

PRODUCE IN	NAME.	BY WHAT BULL.	BY WHOM BRED.
1862, Mar. 14, r.—w.f. H	Sylvia	Castor 1900	Mr. Duckham.
1863, Feb. 7, r.—w.f. B	Steer	do.	do.
1863, Dec. 17, r.—w.f. B	Dead	Victory 2296	do.
1864, Nov. 26, r.—w.f. B	Francisco 2523	Franky 1243	do.

SILVER.
Red with white face, of 1855, vol. v., p. 283.

Bred by and the property of Mr. Thomas Edwards, Wintercott, Leominster; got by Croft (937), dam (Rosa) by Coningsby the Second (1552).

1862, Sep. 2, r.—w.f. B	Dead	Sir Newton 1731	Mr. Edwards.
1863, Sep. 8, r.—w.f. H	Silvery	Adforton 1839	do.

SILVER.
Red with white face, calved August 18, 1857.

Bred by the late Mr. J. Rea, Monaughty, Knighton, the property of Mr. P. R. Jackson, Blackbrook, Skenfrith, Monmouth; got by Treasurer (1105), dam (Glow Worm) by Young Conrad (2322), g.d. (Blossom) by Caractacus (659), g.g.d. (Blossom) by Old Court (306).

1862, Aug. 22, r.—w.f. H	Silver 2nd	Sir Benjamin 1387	Mr. Rea,
1864, Nov. 3, r.—w.f. B	Earl of Monmouth 2511	Florence 1991	Mr. Jackson.

Silver the Second, sold to Mr. J. R. Paramore, Dinedor Court, Hereford.

SILVER.
Red with white face, calved in the year 1860.

Bred by Mr. B. Rogers, The Grove, Pembridge, the property of Mr. Thos. Rogers, Coxall, Ludlow; got by Claret (1921), dam (Silver) by Emperor the Second (1572), g.d. — by Young Royal (1470), g.g.d. — by Gaylad the Second (1589), g.g.g.d. — by Sovereign the Second (1739).

1864, Sep. 9, r.—w.f. B	Speculation 2789	Grove 2nd 1764	Mr. Rogers.

SILVER.
Red with white face, of 1852, vol. v., p. 281.

Bred by Mr. Thos. Hawkins, Sugwas Court, Hereford, the property of Mr. Benjamin Hawkins, Orleton, Ludlow; got by Northampton (600), dam (Silver) by Young Chance (449).

PRODUCE IN	NAME.	BY WHAT BULL.	BY WHOM BRED.
1862, May 8, r.—w.f. H	Sarah	The Grove 1764	Mr. Hawkins
1863, Apr. 27, r.—w.f. H	Sally	do.	do.

SILVER.
Red with white face, of 1849, vols. iv. and v., pp. 185, 282.

The property of Mr. T. Elsmere, Berrington, Salop; got by Emperor (221).

1862, Nov. 20, r.—w.f. B	Alderman 2383	Franky 1243	Mr. Elsmere.
1863, Nov. 11, r.—w.f. H	Alma	Van Tromp 2291	do.
1864, Dec. 27, r.—w.f. B	Andrew 2389	Albert 2380	do.

SILVER.
Red with white face of 1853, vol. v., p. 284.

Bred by Mr. John Taylor, Stretford Court, the property of Mr. Jas. Taylor, Stretford Court, Leominster; got by King John (830), dam (Silver) bred by the late Mr. Bowen, Monkland.

1862, July 28, r.—w.f. B	Steer	Croft 937	Mr. Jas. Taylor.

This Steer was one of three winners of the First Prize in their class at the meeting of the Leominster Agricultural Society, 1863, and a Second Prize as one of four at Ludlow, 1864.

SILVER.
Red with white face, calved in the year 1861.

Bred by and the property of Mr. R. G. Price, M.P., Norton Manor, Presteign; got by Havelock (2563).

1864, Apr. 5, r.—w.f. B		Lord of the Manor 2622	Mr. Price.

SILVER THE SECOND.
Red with white face, calved March 30, 1861.

Bred by and the property of Mr. T. Duckham, Baysham Court,

COWS.

Ross; got by Colonist (1925), dam (Carlisle) by Albert Edward (859), g.d. (Silver) by Emperor (221).

PRODUCE IN	NAME.	BY WHAT BULL.	BY WHOM BRED.
1864, Apr. 22, r.—w.f. H	Florence	Cato 1902	Mr. Duckham.

SILVER THE SECOND.

Red with white face, calved in the month of August, 1860.

Bred by and the property of Mr. J. Jones, Llwyn-y-Gaer, Raglan; got by Chancellor (1172), dam (Silver) by Patron the Second (1678), g.d. (Silver) by Killough (1625), g.g.d. (Silver) bred by Mr. P. Morgan, Abbeydore.

1863, Dec. 30. r.—w.f. B		Bold David 1881	Mr. Jones.

Silver the Second was one of eight, winners of a prize at the meeting of the Monmouthshire Agricultural Society, 1862.

SILVER THE SECOND.

Grey, calved in the month of September, 1860.

Bred by and the property of Mr. Thomas Smith, Bodenham, Dymock; got by Abdel Kader (1837), dam (Beauty) by Orlando (2143), g.d. (Silver).

1863,	r.—w.f. B	King Charles 2593	Mr. Smith.
1864,	r.—w.f. B	do.	do.

Silver was one of a pair, winners of the First Prize in their class at the meeting of the Gloucestershire Agricultural Society, 1845.

SILVER THE FIFTH.

Red with white face, of 1858, vol. v., p. 284.

Bred by and the property of Mr. Jas. Taylor, Stretford Court, Leominster; got by St. Oswall (1378), dam (Silver), bred by the late Mr. Bowen, Monkland.

1862, July 25, r.—w.f. B	Steer	Croft 937	Mr. Taylor.
1863, Aug. 10, r.—w.f. B	Steer	Trustful 2845	do.
1864, July 19, r.—w.f. H	Silver 10th	do.	do.

COWS.

SILVER THE FIFTH.
Red with white face, calved in the month of October, 1858.

Bred by Mr. B. Rogers, The Grove, Pembridge, the property of Mr. J. Wigmore, Weston, Ross; got by Severus (1062), dam (Silver the Fourth) by Young Royal (1470), g.d. (Silver the Third) by Gaylad the Second (1589), g.g.d. (Silver the Second) by Portrait (372), g.g.g.d. (Silver) by Sovereign the Second (1739).

PRODUCE IN		NAME.	BY WHAT BULL	BY WHOM BRED.
1861, Aug.	r.—w.f. B	Steer	The Grove 1764	Mr. Rogers.
1862, Aug. 10,	r.—w.f. B	Grove 2nd 2556	Bolingbroke 1883	do.
1863, Dec.	r.—w.f. B	Comus 2475	do.	do.
1864, Dec. 25,	r.—w.f. B		Monkland 3rd 1013	Mr. Wigmore.

SILVER THE SIXTH.
Red with white face, calved in the month of August, 1859.

Bred by Mr. B. Rogers, The Grove, Pembridge, the property of Mr. Thomas Rogers, Coxall, Ludlow; got by Severus (1062), dam (Silver the Fourth) by Young Royal (1470), g.d. (Silver the Third) by Gaylad the Second (1589), g.g.d. (Silver the Second) by Portrait (372).

1862, Aug.,	r.—w.f. B		Bolingbroke 1883	Mr. B. Rogers.
1863, Dec.,	r.—w.f. H	Silver 7th	do.	do.

SKYLARK.
Red with white face, calved January 19, 1859.

Bred by and the property of Mr. Naylor, Leighton Hall, Montgomeryshire; got by Tom of Lincoln (1099), dam (Cress) by Silvester (797), g.d. (Lily) by Young Persian, g.g.d. (Greystock) by Young Charity, g.g.g.d. — by a bull bred by the late Mr. Tully, g.g.g.g.d. — by Blood Royal bull.

1862, July 26,	r.—w.f. H	Dead	Volunteer 2299	Mr. Naylor.
1863, June 16,	r.—w.f. B	Cromwell 2485	Salisbury 2204	do.

COWS.

SNOW.
Red with white face, calved September 1, 1861.

Bred by and the property of Mr. J. P. Apperley, Fownhope, Hereford; got by Coroner (1555), dam (Snowdrop) by Wonder (1458), g.d. (Snowberry) by Wonder (420), g.g.d. (Snowball) by Young Kingsland (1464), g.g.g.d. (Miss Seeward) by Hector (181).

PRODUCE IN		NAME.	BY WHAT BULL.	BY WHOM BRED.
1864, June 3,	r.—w.f. B		Capt. Perry 2444	Mr. Apperley.

SNOWDROP.
Red with white face, calved in the year 1853.

Bred by the late Mr. J. Davies, Ivington, the property of Mrs. Ann Davies, Ivington, Leominster; got by Sutton (1752).

		Name	By what bull	By whom bred
1856, Jan.,	r.—w.f. B	Steer	Aymestry 1504	Mr. Davies.
1856, Dec.,	r.—w.f. H	Strawberry	Newton 1668	do.
1857, Nov.,	r.—w.f. H	Dainty	do.	do.
1858, Oct. 9,	r.—w.f. B	Goliah	Giant 1411	do.
1859, Oct. 16,	r.—w.f. H	Miss Perry	Monk	do.
1860, Oct. 25,	r.—w.f. H	Miss Bury	Master Butterfly 1313	do.
1861, Oct. 1,	r.—w.f. B	Chance 1908	do.	do.

SNOWDROP.
White, calved in the month of April, 1856.

Bred by Mr. T. Carter, Alcaston, Church Stretton, the property of Mr. A. R. Boughton Knight, Downton Castle, Ludlow; got by Orleton (901), dam — bred by Mr. T. Roberts, Ivington Bury.

		Name	By what bull	By whom bred
1858,	g.—w.f. H	Stately	Young Walford 1820	Mr. Carter.
1861, Sep. 14,	white H	Dewdrop	Tiprey 1774	Mr. Knight.
1862, Nov. 7,	g.—w.f. H	Lady Jane Grey	Lord Grey 2085	do.
1863, Oct. 16,	g.—w.f. H	Greyling	do.	do.
1864, Aug. 19,	white B		do.	do.

Stately, sold to Mr. Knight.

Lady Jane Grey was a winner of the Second Prize in her class at the Worcester Meeting of the Royal Agricultural Society of England, and Third at the Bristol Meeting of the Bath and West of England Society.

Greyling was a winner of the Second Prize in her class at the Newcastle Meeting of the Royal Agricultural Society of England.

COWS.

SNOWDROP.

Red with white face, calved September 12, 1860.

Bred by and the property of Mr. John Monkhouse, The Stow, Hereford; got by Chieftain (930). dam (Primrose) by Cantab (717), g.d. (Violet) by Sir Andrew (183), g.g.d. (Violet) by Westonbury (187), g.g.g.d. (Little Violet) by Whitney (105).

PRODUCE IN		NAME.	BY WHAT BULL.	BY WHOM BRED.
1863, June 20, r.—w.f .H		Crocus	Sir Richard 1734	Mr. Monkhouse.
1864, July 14, r.—w.f.	{B/B}	Twins	Llowes 2608	do.

SNOWDROP.

Red with white face, of vols. iii. and v., pp. 214, 286.

Bred by Mr. H. Chamberlain, Desford, Leicestershire, the property of Mr. J. P. Apperley, Fownhope, Hereford; got by Wonder (1458), dam (Snowberry) by Wonder (420), g.d. (Snowball) by Young Kingsland (1464), g.g.d. (Miss Seeward) by Hector (181).

1862, Oct. 3, r.—w.f. B	Abbot 2367	Coroner 1555	Mr. Apperley.
1864, Jan. 20, r.—w.f. B	Hermit 2568	Volunteer 2861	do.

SOVEREIGN.

Red with white face, calved in the year 1859.

Bred by and the property of Mr. J. Rogers, Letchmoor, Presteign; got by Mr. Green's bull, dam — by a bull bred by the late Mr. Moor, Norton.

1862, Aug.,	r.—w.f. H	Sovereign 2nd	Matchless 2110	Mr. Rogers.
1863, July,	r.—w.f. H	Sovereign 3rd	do.	do.
1864, June,	r.—w.f. H	Sovereign 4th	do.	do.

cows.

SOVEREIGN COW.

Red with white face, calved in the year 1820.

Bred by the late Mr. Jeffries, The Church House, Lycnshall, Kington, the property of Mr. W. H. Oatley, Wroxeter, Salop; got by Sovereign (404).

PRODUCE IN		NAME.	BY WHAT BULL.	BY WHOM BRED.
1839,	r.—w.f. B		Westonbury 187	Mr. Jeffries.
1840,	r.—w.f. H	Princess Royal	Emigrant 1980	Mr. Oatley.

SPANGLE.

Grey, calved in the year 1861.

Bred by the Misses Abley, Norton, Presteign, the property of Mr. R. Green Price, M.P., Norton Manor, Presteign; got by Havelock (2563).

1864, July 14, r.—w.f. B		Lord of the Manor 2622	Mr. Price.

SPANGLE THE SECOND.

Red with white face, calved September 16, 1860.

Bred by the late Mr. J. Rea, Monaughty, Knighton, the property of Mr. J. Baldwin, Luddington, Stratford-on-Avon; got by Wellington (1112), dam (Spangle) by Chieftain (930), g.d. (Young Venus) by Cholstrey (217), g.g.d. (Venus) by Albert) (330), g.g.g.d. (Countess) by Old Court (306).

1863, May 2,	r.—w.f. B	Yeoman 2880	Agriculturist 1842	Mr. J. Rea.
1864, Apr. 17,	r.—w.f. B	Sir Frank 2762	Sir Richard 1734	Mr. T. Rea.

Spangle was one of a pair, winners of the First Prize in their class at the Worcester Meeting of the Royal Agricultural Society of England; she was also the winner of a First Prize at their Newcastle Meeting.

COWS.

SPARK.

Red with white face, calved in the month of January, 1860.

Bred by and the property of Mr. B. Rogers, The Grove, Pembridge; got by The Grove (1764), dam (Fairmaid the Second) by Madley (1301), g.d. (Fairmaid) by Gaylad the Second (1589), g.g.d. (Fairmaid) by Prince (251).

PRODUCE IN		NAME.	BY WHAT BULL.	BY WHOM BRED.
1863, June,	r.—w.f. H	Spark 2nd	Interest 2046	Mr. Rogers.
1864, May 1,	r.—w.f. H	Spark 3rd	North Star 2138	do.

SPARK.

Red with white face, calved July 15, 1860.

Bred by and the property of Mr. Naylor, Leighton Hall, Montgomeryshire; got by Thickset (1768), dam (Star) by Uriconium (598), g.d. (Beauty) by Hampden (1603), g.g.d. (Venus) by Big Ben (248), g.g.g.d. (Twist) by Prince (251), g.g.g.g.d. (Gipsy) by Trump (490).

1863, July 2,	r.—w.f. B		Salisbury 2204	Mr. Naylor.

SPECK.

Red with white face, calved April 22, 1861.

Bred by and the property of Mr. John Wigmore, Bickerton Court, Dymock; got by Forester (1238), dam (Mottle) by Dolphin (2500), g.d. — bred by Mr. Jones, The Hollow, Dinedor.

1863, Sep.	r.—w.f. B	Steer	Speculator 2240	Mr. Wigmore.
1864, Sep.	r.—w.f. H	Spot	do.	do.

Speck was one of a pair, winners of the First Prize in their class at the meeting of the Monmouthshire Agricultural Society, 1864; she was also a winner of a Third Prize at the Cheltenham meeting of the Gloucestershire Society, 1864.

COWS.

SPLENDOUR.

Red with white face, calved July 21, 1859.

Bred by and the property of Mr. Philip Turner, The Leen, Pembridge, Leominster; got by Sorcerer (1737), dam (Sal) by Sir David (349), g.d. (Gaudy) by Defiance (1209), g.g.d. (Beauty) by Old Court (306), g.g.g.d. — bred by Mr. Child, Wigmore Grange.

PRODUCE IN	NAME.	BY WHAT BULL.	BY WHOM BRED.
1862, Mar. 15, r.—w.f. H	Spangle	Bolingbroke 1883	Mr. Turner.
1863, Feb. 27, r.—w.f. H	Fairy	do.	do.
1864, May 16, r.—w.f. H	Dahlia	do.	do.

SPONSARD.

Red with white face, calved in the year 1861.

Bred by Mr. Sheriff, Coxall, Ludlow, the property of Mr. R. H. Capper, The Northgate, Ross; got by Sir Colin (2216), dam (Stately) by Brilliant (1518).

1864, Oct. 25, r.—w.f. H	Patty	Garibaldi 2008	Mr. Capper.

SPOT.

Red with white face, calved September 1858.

Bred by and the property of Mr. J. H. Arkwright, Hampton Court, Leominster; got by Riff Raff (1052), dam — by Quicksilver the Second, g.d. — by Jupiter (1289), g.g.d. — by Reliance (278).

1861, Sep., r.—w.f. H	Violet	Sheriff	Mr. Arkwright.
1862, Sep., r.—w.f. B	Steer	do.	do.
1863, Oct. 15, r.—w.f. B	Agitator 2377	Dan O'Connel 1952	do.
1864, Oct., r.—w.f. B	Alligator 2386	do.	do.

Violet was one of a pair winners of the Second Prize in their class at the meeting of the Herefordshire Agricultural Society, 1862, and First at their meeting, 1863; she was also the winner of a Third Prize at Birmingham, and a Second at Smithfield, 1864.

COWS.

SPOT.

Red with white face, calved in the year 1861.

Bred by Mr. Thomas Burlton, The Vern, Bodenham, Hereford, the property of Mr. W. H. Apperley, Withington, Hereford; got by The Doctor (1083).

PRODUCE IN	NAME.	BY WHAT BULL	BY WHOM BRED.
1864, Sep. 19, r.—w.f. H	Spot 2nd	Sir Harry 2222	Mr. Burlton.

Spot the Second, sold to Mr. W. H. Apperley.

SPOT.

Red with white face, calved December 10, 1855.

Bred by and the property of Mr. H. Gibbons, Hampton Bishop, Hereford; got by The Admiral (1078), dam (Spot) by Young Gaylad (1463).

1858, Oct. 8, r.—w.f. H	Lovely	Defence 2nd 1208	Mr. Gibbons.
1859, Dec. 13, r.—w.f. H	Spot 2nd	Woodman 2nd 1459	do.
1860, Nov. 18, r.—w.f. B	Steer	do.	do.
1861, Sep. 12, r.—w.f. H	Spot 3rd	Shamrock 2nd 2210	do.
1862, Sep. 2, r.—w.f. B	Glendower 2550	do.	do.
1863, Sep. 20, r.—w.f. B	Steer	do.	do.
1864, Aug. 30, r.—w.f. H	Spot 4th	do.	do.

SPOT.

Red with white face, calved in the year 1857.

Bred by Mr. Turner, Court of Noke, Leominster, the property of Mr. J. P. Apperley, Fownhope, Hereford; got by Sir David (349), dam (Woodhouse) by Confidence (367).

1860, Sep., r.—w.f. H	Spot	Baron of Noke 1862	Mr. Turner.
1861, Sep. 2, r.—w.f. H	Woodhouse	Cornet 1934	Mr. Apperley.
1862, Nov. 7, r.—w.f. B	Steer	do.	do.

SPOT.

Red with white face, calved September 1860.

Bred by Mr. Turner, Court of Noke, Leominster, the property of Mr. J. P. Apperley, Fownhope, Hereford; got by Baron of

Noke (1862), dam (Spot) by Sir David (349), g.d. (Woodhouse) by Confidence (367).

PRODUCE IN	NAME.	BY WHAT BULL.	BY WHOM BRED.
1863. Oct. 12, r.—w.f. H	Star	Cornet 1934	Mr. Apperley.

SPOT THE SECOND.
Red with white face, calved July 10, 1860.

Bred by Mr. Rea, Monaughty, Knighton, the property of Mr. Warren Evans, Llandowlas, Usk; got by Sir Benjamin (1387), dam (Spot) by Gratitude (1261), g d, (Spot) by Conrad (1183).

PRODUCE IN	NAME.	BY WHAT BULL.	BY WHOM BRED.
1864, June 17, r.—w.f. B		Monaughty 2117	Mr Evans.

STAR.
Red with white face, calved in the month of December, 1860.

Bred by Mr. Stedman, Bedstone Hall, Salop, the property of Lord Wenlock, Bourton Grange, Wenlock; got by Kinlet (1293), dam (Perfection) by Bedstone (2411), g.d. — by Perfection (538), g.g.d. (Violet) by Dinedor (395), g.g.g.d. — by Trojan (542), g.g.g.g.d. — by a son of Waterloo.

PRODUCE IN	NAME.	BY WHAT BULL.	BY WHOM BRED.
1863, Oct. 6, r.—w.f. B	Firebrand 2516	Dreadnought 1973	Lord Wenlock.

STAR.
Red with white face, of 1854, vol. iv., p. 190.

Bred by and the property of Mr. Naylor, Leighton Hall, Montgomeryshire; got by Uriconium (598), dam (Beauty) by Hampden (1603), g.d. (Venus) by Big Ben (248), g.g.d. (Twist) by Prince (251), g.g.g.d. (Gipsy) by Trump (490).

PRODUCE IN	NAME.	BY WHAT BULL.	BY WHOM BRED.
1859, July 6, r.—w.f. B	Steer	Thickset 1768	Mr. Naylor.
1860, July 15, r.—w.f. H	Spark	do.	do.
1861, July 22, r.—w.f. B	Steer	Admiral 1481	do.

STATELY.

Red with white face, calved in the year 1858.

Bred by Mr. Sheriff, Coxall, Ludlow, the property of Captain Crawshay, Danypark, Crickhowell; got by Brilliant (1518), dam — by Young Emperor (1811), g.d. — by Grateful (1260), g.g.d. — by Old Court (306).

PRODUCE IN	NAME.	BY WHAT BULL.	BY WHOM BRED.
1861, r.—w.f. H	Stately 2nd	Sir Colin 2216	Mr. Sheriff.
1862, r.—w.f. H	Stately 3rd	Defender 1956	do.
1863, Nov. 22, r.—w.f. B	Steer	Sir Colin 2216	do.

STATELY.

Red with white face, calved in the year 1861.

Bred by Mr. Sheriff, Coxall, Ludlow, the property of Mr. R. H. Capper, The Northgate, Ross; got by Sir Colin (2216), dam (Lily of the Valley) by Brilliant (1518).

1864, Dec. 27, r.—w.f. H	Dignity	Orphan 2662	Mr. Capper.

STATELY.

Grey with white face, calved in the year 1858.

Bred by Mr. Thomas Carter, Alcaston, Church Stretton, the property of Mr. A. R. Boughton Knight, Downton Castle, Ludlow; got by Young Walford (1820), dam (Snowdrop) by Orleton (901), g.d. — bred by Mr. Roberts of Ivington Bury.

1862, Aug. 30, r.—w.f. B	Steer	Lord Grey 2085	Mr. Knight.
1863, Oct. 26, g.—w.f. B	Downton Pippin	Garibaldi 2003	do.
1864, Sep. 30, g.—w.f. H	Strawberry	do.	do.

STATELY THE SECOND.

Red with white face, calved March 16, 1860.

Bred by and the property of Mr. H. R. Evans, Swanstone Court, Leominster; got by Rambler (1046), dam (Stately) by Swanstone (1072), g.d. (Juno) by Emperor (373), g g.d.

(Countess) by Coningsby (718), g g.g.d. (Lovely) by Young Trueboy (1475), g g.g.g.d. (Lovely) by Ashby Moor White Bull (870). — (Old Damsel) by Coleman's Bull (1547).

PRODUCE IN	NAME.	BY WHAT BULL.	BY WHOM BRED.
1863, July 23, r.—w.f. H	Worcester Lass	Chatham 1914	Mr. Evans.
1864, July 14, r.—w.f. H	Lofty	do.	do.

Stately the Second was one of a pair winners of the Second Prize in their class at the Worcester Meeting of the Royal Agricultural Society of England.

STATELY THE THIRD.

Red with white face, calved October 4, 1860.

Bred by and the property of Mr. John Wigmore, Bickerton Court, Dymock; got by Forester (1238), dam (Stately) by Dolphin (2500), g.d. — bred by Mr. Jones, The Hollow, Dinedor.

1863. July 3, r.—w.f. H	Boss	Speculator 2240	Mr. Wigmore.

Stately the Third was a winner of the First Prize in her class, together with the extra prize of £20 given for the best breeding animal exhibited at the Tredegar Meeting, 1863; and with Speculator and Boss, the First Prize in their class at the Meetings of the Gloucestershire and the Monmouthshire Agricultural Societies, 1863, and Third at Hereford the same year; also First at Cheltenham and Second at Hereford, 1864. She was also one of a pair winners of First Prizes at Monmouth and Ross, 1863.

STATELY THE FOURTH.

Red with white face, calved in the month of August, 1859.

Bred by Mr. Rogers, The Grove, Pembridge, the property of Mr. John Wigmore, Bickerton Court, Dymock; got by The Grove (1764), dam (Stately) by Young Royal (1470), g.d. (Stately) by Gaylad the Second (1589), g.g.d. — by Prince (251).

1862, Aug., r.—w.f. H	Stately 5th	Interest 2046	Mr. Rogers
1863, June, r.—w.f. H	Stately 6th	do.	do.
1864, June, r.—w.f. H	Stately 7th	North Star 2138	do.

COWS.

STELLA.

Red with white face, calved January 30, 1859.

Bred by and the property of Mr. Naylor, Leighton Hall, Montgomeryshire; got by Tom of Lincoln (1099), dam (Young Beauty) by Uriconium (598), g.d. (Beauty) by Hampden (1603), g.g.d. (Venus) by Big Ben (248), g.g.g.d. (Twist) by Prince (251), g.g.g.g.d. (Gipsy) by Trump (490).

PRODUCE IN	NAME.	BY WHAT BULL.	BY WHOM BRED.
1862, July 3, r.—w.f. B	Rifleman 2724	Volunteer 2299	Mr. Naylor.
1863, July 19, r.—w.f. B	Steer	Blondin 1880	do.
1864, July 15, r.—w.f. H	Stella 2nd	Salisbury 2204	do.

STOKE.

Red with white face, calved July 12, 1860.

Bred by and the property of Mr. Naylor, Leighton Hall, Montgomeryshire; got by Admiral (1481), dam (Young Stately) by Venison the Second (1442), g.d. (Stately) by Dinedor (395).

1863, June 26, r.—w.f. B	Steer	Salisbury 2204	Mr. Naylor.
1864, June 23, r.—w.f. H	Stoke 2nd	do.	do.

STRAPPER.

Red with white face, calved December 1, 1858.

Bred by and the property of Mr. H. Gibbons, Hampton Bishop, Hereford; got by Defence the Second (1208), dam (Strapper) by The Admiral (1078), g.d. (Old Strapper) by Young Gaylad (1463).

1861, Aug. 18, r.—w.f. B	Steer	Shamrock 2nd 2210	Mr. Gibbons.
1862, June 26, r.—w.f. H	Marigold	do.	do.
1863, July 31, r.—w.f. H	Tulip	do.	do.
1864, Sep. 15, r.—w.f. B	Gambler 2528	do.	do.

STRAWBERRY.

Red with white face, calved in the year 1856.

Bred by Lord Berwick, Cronkhill, Salop, the property of Mr. W. R. Grose, Penpont, Wadebridge, Cornwall; got by Attingham (911), dam (Young Oak Apple) by Tom Thumb (243), g.d. (Oak Apple) by Commerce (354), g.g.d. (Strawberry), bred by the late Mr. Jeffries, The Grove, Leominster.

PRODUCE IN		NAME.	BY WHAT BULL.	BY WHOM BRED.
1860, Oct. 20, r.—w.f.	H	Lady	Conservative 1931	Mr. Sobey.
1861, Oct. 6, r.—w.f.	B	Steer	Penhallow 2154	Mr. Grose.
1862, Sep. 24, r.—w.f.	B	Steer	Sir Hugh 2223	do.
1863, Sep. 20, r.—w.f.	B	Steer	do.	do.
1864, Oct. 18, r.—w.f.	H	Phœbe	do.	do.

STRAWBERRY.

Red with white face, calved in the month of August, 1858.

Bred by Mr. B. Rogers, The Grove, Pembridge, the property of Colonel Feilding, Dulas Court, Hereford; got by Severus (1062), dam (Strawberry) by Young Royal (1470), g.d. (Miss Grove the Second) by Gaylad the Second (1589), g.g.d. (Miss Grove) by Defiance (1209), g.g.g.d bred by the late Mr. Jeffries, The Grove.

1861, Aug., r.—w.f.	B	Dead	The Grove 1764	Mr. Rogers.
1862, Aug., r.—w.f.	H	Strawberry 2nd	Bolingbroke 1883	do.
1863, July, r.—w.f.	B	Steer	Interest 2046	do.
1864, Aug. 20, r.—w.f.	B	Mars 2635	North Star 2138	do.

STRAWBERRY.

Red with white face, calved October 5, 1859.

Bred by and the property of Mr. H. Gibbons, Hampton Bishop, Hereford; got by Woodman the Second (1459), dam (Strawberry) by The Admiral (1078), g.d. (Moap) by Zephyr (1826).

1862, June 21, r.—w.f.	H	Lucy	Shamrock 2nd 2210	Mr. Gibbons.
1863, Aug. 21, r.—w.f.	H	Lucy 2nd	do.	do.
1864, Sep. 17, r.—w.f.	H	Lucy 3rd	do.	do.

COWS.

STRAWBERRY THE THIRD.

Red with white face, of 1856, vol. v., p. 294.

Bred by Mr. John Taylor, Stretford Court, the property of Mr. James Taylor, Stretford Court, Leominster; got by King James (978), dam (Strawberry) by Andrew (1495), g d. — by Cotmore (376), g.g.d. — by Eyton (557).

PRODUCE IN	NAME.	BY WHAT BULL.	BY WHOM BRED.
1862, July 29, r.—w.f. B	Steer	Croft 937	Mr. Jas. Taylor.
1863, July 21, r.—w.f. B	Steer	Garibaldi 2004	do.
1864, July 10, r.—w.f. H	Strawberry 7th	Trustful 2845	do.

The Steer of 1862 was one of four, winners of the First Prize in their class at the meeting of the Leominster Agricultural Society, 1863, and a Second Prize at Ludlow, 1864. The Steer of 1863 was one of four winners of a First Prize at Ludlow, 1864, and one of a pair winners of First Prizes at Leominster and Hereford the same year.

STRAWBERRY THE FIFTH.

Red with white face, calved March 24, 1861.

Bred by and the property of Mr. James Taylor, Stretford Court, Leominster; got by Croft (937), dam (Strawberry) by Andrew (1495), g d. — by Cotmore (376), g.g.d. — by Eyton (557).

1863, Sep. 14, r.—w.f. B	Steer	Trustful 2845	Mr. Taylor.
1864, July 25, r.—w.f. H		do.	do.

Strawberry the Fifth, with six others, were winners of the First Prize in their class at the meeting of the Herefordshire Agricultural Society, 1864; the steer of 1863 was one of four, winners of a First Prize at Ludlow, 1864; and one of a pair winners of a Second Prize at Hereford the same year.

SULTANA.

Red with white face, calved Sepember 20, 1857.

Bred by the late Mr. James Rea, Monaughty, Knighton, the property of Mr. P. J. Kearney, Miltown House, Clonmellon, Ireland; got by Grenadier (961), dam (Gaylass) by Canute

(890), g.d. (Gaylass) by Confidence (367), g.g.d. — by Charity (375).

PRODUCE IN	NAME.	BY WHAT BULL.	BY WHOM BRED.
1863, June 29, r.—w.f. B	Sir Cupiss Ball 2761	Sir Benjamin 1387	Mr. Rea.
1864, May 31, r.—w.f. H	The Belle	Sir Richard 1734	Mr. Kearney.

Sultana was the winner of a prize at the Clonmel Meeting of the Royal Agricultural Society of Ireland.

SULTANA.
Red with white face.

Bred by Mr. Price, Court House, Pembridge, the property of Mr. J. O. G. Pollock, Mountains Town, Navan, Ireland; got by Goldfinder the Second (959), dam (Queen) by Forester (398), g.d. (Blowdy) by Hope (439), g.g.d. (Beauty) by Sovereign (404).

1861, Aug. 25, r.—w.f. B	Rifleman	Salisbury 2204	Mr. Price.
1862, Sep. 2, r.—w.f. H	Sultana 2nd	do.	do.
1863, July 29, r.—w.f. H	Sultana 3rd	Earl Derby 2nd 2510	do.
1864, Sep. 12, r.—w.f. B	Steer	do.	Mr. Pollock.

SULTANA.
Red with white face, calved March 3, 1860.

Bred by Mr. John Hewer, Vern House, Marden, Hereford, the property of Mr. J. Baldwin, Luddington, Stratford-on-Avon; got by Mameluke (1307), dam (Caroline) by Cardinal Wiseman (1168), g.d. (Hampton Lass) by Mark (424), g.g.d. (Miss Hampton) by Garrick (1248), g.g.g.d. (Lady Hampton) by Reform (508).

1862, June 3, r.—w.f. B	Steer	Byron 3rd	Mr. Baldwin.
1863, Aug. 25, r.—w.f. B	Steer	Battersea 1865	do.
1863, Aug. 25, r.—w.f. H	Lady Battersea	do.	do.
1864, Aug. 30, r.—w.f. H	Rose of Battersea	do.	do.

SUNBEAM.
Red with white face, calved in the year 1859.

Bred by Mr. J. Sheriff, Burrington, Ludlow, the property of

COWS.

Mr. W. Stallard, Brockhampton, Ross; got by Brilliant (1518) dam — by Confidence (367), g.d. — by Byron (440).

PRODUCE IN	NAME.	BY WHAT BULL.	BY WHOM BRED.
1861, June 26, r.—w.f. B	Steer	Sir Colin 2216	Mr. Stallard.
1863, Apr. 17, r.—w.f. B	Steer	Shamrock 2nd 2210	do.
1864, June 9, r.—w.f. H	Sunbeam 2nd	Chieftain 3rd 2457	do.

SUNFLOWER.
Red with white face, calved September 4, 1857.

Bred by the late Mr. J. Rea, Monaughty, Knighton, the property of Major-General the Hon. A. N. Hood, Cumberland Lodge, Windsor; got by Grenadier (961), dam (Fatty) by Regent (891), g.d. (Fatty) by Patron (888), g.g.g.d. (Fatty) by Old Court (306).

1862, July 21, r.—w.f. H	Sunbeam	Sir Benjamin 1387	Mr. Rea.
1863, Dec. 10, r.—w.f. H	Jeannette	do.	Hon. A. N. Hood
1864, Nov. 17, r.—w.f. H	Lady Mary Hood	Ajax 1843	do.

Sunbeam, sold to Mr. Paramore, Dinedor Court, Hereford.

SUNFLOWER.
Red with white face, calved in the year 1858.

Bred by and the property of Mr. R. Green Price, M.P., Norton Manor, Presteign; got by David the Third, bred by the late Mr. Rogers, Stocken

1864, Sep. 25, r.—w.f. B		Lord of the Manor 2622	Mr. Price.

SUPERB.
Red with white face, calved November 18, 1860.

Bred by and the property of Mr. C. Bulmer, Holmer, Hereford; got by Golden Horn (2015), dam (Shut) by Carlisle (923), g.d. (Strapper) by Monarch (504), g.g.d. — bred by Lord Bateman.

1863, Jan. 3, r.—w.f. H	Nel Gwynne	Garrick Junior 2532	Mr. Bulmer.
1864, Dec. 1, r.—w.f. H	Lady Garrick	do.	do.

COWS.

SUPERB.
Red with white face, of 1856, vol. v., p. 295.

Bred by Lord Bateman, Shobdon Court, Leominster, the property of Mr. E. Wright, Halston Hall, Oswestry; got by Carlisle (923), dam (Strapper) by Monarch (504), g.d. — by Andrew the Second (619).

PRODUCE IN	NAME.	BY WHAT BULL.	BY WHOM BRED.
1862, Nov. 9, r.—w.f. H	Superior	Magnet 2nd 989	Mr. Wright.
1863, Dec. 1, r.—w.f. H	Sprightly	do.	do.
1864, Nov. 3, r.—w.f. H	Odora	Prince 2691	do.

SWEETBRIAR THE SECOND.
Red with white face, calved August 9, 1859.

Bred by Mr. J. Rea, Monaughty, Knighton, the property of Major-General the Hon. A. N. Hood, Cumberland Lodge, Windsor; got by Wellington (1112), dam (Sweetbriar) by Grenadier (961), g.d. (Storrel).

1863, June 13, r.—w.f. H	Agnes	Sir Benjamin 1387	Mr. Rea.

SWEETHEART.
Red with white face, calved December 21, 1860.

Bred by the late Lord Berwick, Cronkhill, Salop, the property of Mr. F. W. Stone, Moreton Lodge, Guelph, Canada West; got by Albert Edward (859), dam (Whiteheart) by Attingham (911), g.d. (Wonder Cherry) by Windsor (420), g.g.d. Cherry the Fifth) by Cholstry (868), g.g.g.d. (Cherry the Fourth) by Green's Grey Bull (850A), — (Cherry the Third) by Chancellor (156), — (Cherry the Second) by Thickset (1769).

1864, Feb. 24, r.—w.f. H	Heartsease	Patriot 2150	Mr. Stone.

Sweetheart was a winner of the First Prize in her class at the Toronto and Kingston Meetings of the Canadian Agricultural Society, 1862 and 1863.

COWS.

SYLPH.
Red with white face, calved August 12, 1860.

Bred by the late Mr. James Rea, Monaughty, Knighton, the property of Mr. P. J. Kearney, Miltown House, Clonmellon, Ireland; got by Sir Benjamin (1387), dam (Sylph), bred by Mr. Stephens, Sheephouse, Hay.

PRODUCE IN	NAME.	BY WHAT BULL.	BY WHOM BRED.
1863, June 5, r.—w.f. H	Bountiful	Baron of Boultibroke 1861	Mr. Rea.
1864, Sep. 4, r.—w.f. B	Steer	Silverstream 2214	Mr. P. J. Kearney

Bountiful, sold to Mr P. J. Kearney, Miltown House, Clonmellon.

SYLPH.
Red with white face, of 1851, vol. v., p. 296.

Bred by Mr. J. Turner, Court of Noke, Pembridge, the property of Mr. J. A. Hollings, Hillend, Hereford; got by The Knight (185), dam — by Monarch (504).

1863, Mar. 21, r.—w.f. H	Sylph 3rd	St. Clement 2201	Mr. J. A. Hollings
1864, Aug. 3, r.—w.f. B		Chieftain 2nd 1917	do.

SYLPH.
Red with white face, of 1853, vols. iii. and v., pp. 221, 296.

Bred by and the property of Mr. T. Duckham, Baysham Court, Ross; got by Pope (527), dam (Eywood) by Cotmore the Second (1191), g.d. — bred by the late Earl of Oxford.

1862, May 16, r.—w.f. H	Sylph 2nd	Castor 1900	Mr. Duckham.

SYLPH THE SECOND.
Red with white face, calved January 8, 1861.

Bred by and the property of Mr. W. Lane, Compton Casey, Cheltenham; got by Casey (1899), dam (Sylph) by Compton (1551), g.d. (Tulip) by Whittington (1797).

1863, Dec. 16, r.—w.f. H	Sylph 3rd	Hardy 2027	Mr. Lane.

COWS.

SYLPH THE SECOND.

Red with white face, calved in the month of October, 1854.

Bred by Mr. J. Hollings, Hillend, the property of Mr. J. A. Hollings, Hillend, Hereford; got by Voltigeur (1445), dam (Sylph) by The Knight (185), g.d. — by Monarch (504).

PRODUCE IN		NAME.	BY WHAT BULL.	BY WHOM BRED.
1857, Oct.,	r.—w.f. B	Steer	Noke 1338	Mr. Hollings.
1858, Sep.,	r.—w.f. H	Dead	do.	do.
1860, May,	r.—w.f. B	Steer	Woodman 1460	do.
1861, Feb. 3,	r.—w.f. B	Steer	do.	do.
1862, Jan. 18,	r.—w.f. B	Steer	St. Clement 2201	do.
1863, Jan. 5,	r.—w.f. B	Steer	Chieftain 2nd 1917	Mr. J. A. Hollings
1864, Jan. 11,	r.—w.f. H	Sylph 4th	do.	do.
1864, Dec. 22,	r.—w.f. B	Steer	St. Clement 2201	do.

SYLPH THE THIRD.

Red with white face, calved August 26, 1861.

Bred by the late Mr. T. Rea, Westonbury, Leominster, the property of Mr. J. D. Allen, Pyt House, Tisbury, Wilts; got by Lord Nelson (2088), dam (Sylph the Second) by Sir Benjamin (1387), g.d. — bred by Mr. Stephens, Sheephouse, Hay.

1864, Feb. 18,	r.—w.f. H	Sylph 4th	Iris 2047	Mr. Allen.

SYLVA.

Red with white face, calved in the year 1859.

Bred by and the property of Mr. T. Morris, Therrow, Llyswen, Hay; got by Telegraph (1404), dam (Brunette) by Newton (344), g.d. (Brunette) by Enterprise (948), g.g.d. (Cloudy) by White Nob (345), g.g.g.d. (Silver) by Charity the Second (516).

1862,	r.—w.f. H	Dead	Druid 1220	Mr. Morris.
1863,	r.—w.f. B	Steer	Prince Imperial 2171	do.
1864,	r.—w.f. B		do.	do.

SYLVIA.

Red with white face, calved May 7, 1860.

Bred by Mr. Duckham, Baysham Court, Ross, the property of

Mr. R. Davey, M.P., Polsue House, Grampound, Cornwall; got by Franklyn (1576), dam (Sylva) by Colossus (591), g.d. (Sylph) by Pope (527), g.g.d. (Eywood) by Cotmore the Second (1191), g.g.g.d. —, bred by the late Earl of Oxford.

PRODUCE IN	NAME.	BY WHAT BULL.	BY WHOM BRED.
1863, Nov. 28, r.—w.f. B	Santiago	Castor 1900	Mr. Davey.

SYMMETRY.
Red with white face, calved May, 1858.

Bred by Mr. Thomas Hawkins, Sugwas Court, Hereford, the property of Mr. Benjamin Hawkins, Orleton, Ludlow; got by Merry Andrew (1011), dam (Silver) by Northampton (600), g.d. (Silver) by Chance (449).

1861, May 6, r.—w.f. B	Steer	Young Chancellor 1124	Mr. Hawkins
1862, April 15, r.—w.f. B	Gainful 2526	The Grove 1764	do.
1863, April 28, r.—w.f. B	Gamester 2530	do.	do.
1864, April 8, r.—w.f. H	Symmetry 2nd	Noble Boy 1337	do.

TAFFY.
Red with white face, calved February 14, 1861.

Bred by and the property of Mr. W. Lane, Compton Casey, Cheltenham; got by Casey (1899), dam (Taffy) by Tyro (1786).

1863, Oct. 27. r.—w.f. H	Lassie	Hardy 2027	Mr. Lane.

THE DOVE.
Red with white face, calved July 10, 1861.

Bred by and the property of Mr. E. Wright, Halston Hall, Oswestry, Shropshire; got by Halston (2024), dam (Winsome) by Carlisle (923), g.d. (Young Lady) by Andrew the Second (619).

1864, Jan. 4, r.—w.f. H	Pink	Hero 2039	Mr. Wright.
1864, Dec. 1, r.—w.f. H	Plum	do.	do.

COWS.

THEORA.
Red with white face, calved October 16, 1859.

Bred by and the property of Mr. J. M. Read, Elkstone, Cheltenham; got by Sebastopol (1381), dam (Cherry the Seventh) by Hotspur (855), g.d. (Cherry the Fifth) by Cholstrey (868), g.g.d. (Cherry the Fourth) by Green's Grey Bull (850A), g.g.g.d (Cherry the Third) by Chancellor (156), g.g.g.g.d. (Cherry the Second) by Thickset (1769), — (Cherry) bred by the late Mr. Knight, Downton Castle.

PRODUCE IN		NAME.	BY WHAT BULL.	BY WHOM BRED.
1862, Dec. 5,	r.—w.f. B	The Anchorite 2809	Ariconium 1498	Mr. Read.
1864, Jan. 9,	r.—w.f. H	Alethe	Caliban 1163	do.
1864, Dec. 1,	r.—w.f. B	Orcus 2659	Colesborne 2467	do.

THINGEHILL.
Red with white face, of 1856, vol. v., p. 298.

Bred by and the property of Mr. T. Powell, Castle Froome, Bromyard; got by Sir David Thingehill (1066), dam (Young Miss Chance the Fourth) by Young Sir David (1137), g.d. (Miss Chance) by Confidence (367), g.g.d. (Miss Chance) by Chance (355), g.g.g.d. (Old Miss Chance) bred by the late Mr. T. Jeffries, The Grove.

1862, Dec. 29,	r.—w.f. B	Steer	Promise 2175	Mr. Powell.
1863, Sep. 21,	r.—w.f. H	Pleasant	Croft 937	do.
1864, Aug. 19,	r.—w.f. B	Steer	Promise 2175	do.

TOPSY.
Red with white face, calved in the month of March, 1861.

Bred by and the property of Mr. John Barton, Coln, Fairford; got by Cardinal Wiseman (1168), dam (Beauty) by Guinea (963), g.d. (Handsome) bred by Mr. P. Matthews.

1864, Feb.,	r.—w.f. H	Butterfly	Coleshill 1923	Mr. Barton.

TRINKET.

Red with white face, calved September 25, 1860.

Bred by the late Lord Berwick, Cronkhill, Salop, the property of the Hon. and Rev. Noel Hill, Berrington, Salop; got by Albert Edward (859), dam (Finchback) by Walford (871), g.d. (Young Damsel) by Tom Thumb (243), g.g.d. (Damsel) by Young Trueboy (1475), g.g.g.d. (Prettymaid) by Cholstrey (868), g.g.g.g.d. (Damsel) by Coleman's bull (1547), — (Old Daisy) by Chancellor (156), — (Cherry the Second) by Thickset (1769).

PRODUCE IN	NAME	BY WHAT BULL.	BY WHOM BRED.
1863, Nov. 4, r.—w.f. H	Jewel	Van Tromp 2291	Hon. & Rev. Noel Hill.
1864, Sep. 30, r.—w.f. H	Cornelian	Conqueror 1929	do.

TULIP.

Red with white face, calved January 3, 1859.

Bred by and the property of Mr. Philip Turner, The Leen, Pembridge; got by Felix (953), dam (Moss Rose) by Silurian (1064), g.d. (Ada) by Andrew the Second (619), g.g.d. (Johanna) by Albert Edward (754), g.g.g.d. (Lady) by Sir Walter (352), g.g.g.g.d. (Peeress) by Viscount (816), — (Peeress) by Lottery the Second (987A).

1860, May 22, r.—w.f. B	Steer	Logic 2079	Mr. Turner.
1862, May 12, r.—w.f. B	Demetrius 2494	Bolingbroke 1883	do.
1863, Apr. 28, r.—w.f. H	Dead	do.	do.
1864, Mar. 18, r.—w.f. H	Flirt	do.	do.

TULIP.

Red with white face, of 1858, vol. v., p. 300.

Bred by and the property of Mr. John Monkhouse, The Stow, Hereford; got by Formidable (1240), dam (Dahlia) by Madoc (899), g.d. (Dahlia) by Guy Fawkes (581), g.g.d. (Daffodil) by a son of Charity (375), g.g.g.d. (Tulip) by Sir Andrew (183).

1862, Oct. 16, r.—w.f. H	Picture	Chieftain 930	Mr. Monkhouse.
1863, Sep. 9, r.—w.f. H	Pencil	Llowess 2608	do.
1864, Sep. 21, r.—w.f. H	Sunflower	Chieftain 930	do.

COWS.

TULIP.

Red with white face, calved March 21, 1858.

Bred by and the property of Mr. John Barton, Coln, Fairford; got by Guinea (963), dam — bred by Mr. Peter Matthews.

PRODUCE IN		NAME.	BY WHAT BULL.	BY WHOM BRED.
1861, Feb. 2,	r.—w.f. B	Dead	Young Cardinal Wiseman 2882	Mr. Barton.
1862, Apr. 3,	r.—w.f. H	Nancy	do.	do.
1863, Mar. 6,	r.—w.f. H	Noble	do.	do.

TULIP.

Red with white face, calved in the month of February 1859.

Bred by the late Mr. Keene, Pencraig, the property of Mr. Rees Keene, Pencraig, Caerleon; got by General Wyndham (1590), dam (Browny) by Prince Albert (2168), g.d. (Browny) by Young David (349), g.g.d. (Brown) by Nonpareil (696).

1862, Mar.,	r.—w.f. H	Gentle	Odd Trick 1674	Mr. Keene.
1863, Mar.,	r.—w.f. B	Dead	Dolward 1966	do.
1864, Mar.,	r.—w.f. H	Belle	Cholstrey 2nd 1919	do.

TULIP.

Red with white face, calved in the year 1861.

Bred by Mr. Thomas Burlton, The Vern, Bodenham, Hereford, the property of Mr. W. H. Apperley, Withington, Hereford; got by The Doctor (1083).

1864, Sep. 20, r.—w.f. { H | Lady H | Rose }		Sir Harry 2222	Mr. Burlton.

Lady and Rose sold to Mr. W. H. Apperley.

TURTLE DOVE.

Red with white face, calved October 29, 1861.

Bred by and the property of Mr. Naylor, Leighton Hall, Montgomeryshire; got by Admiral (1481), dam (Venus) by Big Ben (248), g.d. (Twist) by Prince (251), g.g.d. (Gipsy) by Trump

(490), g.g.g.d. (Venus) by Tobias (487), g.g.g.g.d. (Duchess) by Young Cupid (259).

PRODUCE IN	NAME.	BY WHAT BULL.	BY WHOM BRED.
1864, Aug. 17, r.—w.f. B		Gladstone 2547	Mr. Naylor.

VANITY.

Red with white face, of 1855, vol. v., p. 300.

Bred by and the property of Mr. John Monkhouse, The Stow, Hereford; got by Madoc (899), dam (Lofty) by Phantom (1035), g.d. (Stately) by Sir Andrew (182), g.g.d. (Stately) by A Son of Sovereign (404).

PRODUCE IN	NAME.	BY WHAT BULL.	BY WHOM BRED.
1862, Sep. 13, r.—w.f. H	Gaudy	Chieftain 930	Mr. Monkhouse.
1863, Aug. 14, r.—w.f. B	Vainhope 2853	do.	do.
1864, Aug. 6, r.—w.f. B	Dandy 2489	do.	do.

VANQUISH.

Red with white face, calved July 23, 1862.

Bred by and the property of Mr. F. W. Stone, Moreton Lodge, Guelph, Canada West; got by Patriot (2150), dam (Verbena) by Carlisle (923), g.d. (Flower) by Radnor, g.g.d. (Old Fancy) bred by the late Mr. Galliers, Shobdon.

PRODUCE IN	NAME.	BY WHAT BULL.	BY WHOM BRED.
1864, Dec. 21, r.—w.f. B	Sir Thomas 2777	Sailor 2200	Mr. Stone.

VENUS.

Red with white face, of 1846, vol. iv., p. 198.

Bred by Mr. Thomas Yeld, The Broome, Pembridge, the property of Mr. Naylor, Leighton Hall, Montgomeryshire; got by Big Ben (248), dam (Twist) by Prince (251), g.d. (Gipsy) by Trump (490), g.g.d. (Venus) by Tobias (487), g.g.g.d. (Duchess) by Young Cupid (259).

PRODUCE IN	NAME.	BY WHAT BULL.	BY WHOM BRED.
1859, Aug. 2, r.—w.f. B	Garibaldi 2005	Admiral 1481	Mr. Naylor.
1860, Aug. 13, r.—w.f. H	Venus 2nd	do.	do.
1861, Oct. 29, r.—w.f. H	Turtle Dove	do.	do.

VENUS.

Red with white face, calved in the year 1859.

Bred by Mr. J. Sheriff, Burrington, Ludlow, the property of Mr. W. Stallard, Brockhampton, Ross; got by Brilliant (1518), dam — by Young Emperor (1811), g.d. — by Confidence (367).

PRODUCE IN	NAME.	BY WHAT BULL.	BY WHOM BRED.
1861, July 24, r.—w.f. H	Venus 2nd	Sir Colin 2216	Mr. Stallard.
1862, May 25, r.—w.f. H	Venus 3rd	St. Clement 2201	do.
1863, July 2, r.—w.f. B	Steer	Chieftain 2nd 1917	do.
1864, June 6. r.—w.f. H	Venus 4th	do.	do.

VENUS.

Red with white face, of 1860, vol. v., p. 316.

Bred by and the property of Mr. John Partridge, Bishop's Wood, Ross; got by Noble Tom (2135), dam (Grove the Fourth) by Young Sir David (1137), g.d. (Grove the Second) by Favourite (952), g.g.d. (Grove) by Confidence (367), g.g.g.d. — bred by the late Mr. Jeffries, The Grove.

1864, May 21, r.—w.f. H	Miss Grove	Garway 2536	Mr. Partridge.

VENUS.

Red with white face, calved November 13, 1859.

Bred by and the property of Mr. T. Olver, Penballow, Grampound, Cornwall; got by Great Eastern (1598), dam (Venus) by Duke of Cornwall (1569).

1864, Jan. 12, r.—w.f. { H / H	Victress / Virtue	} Zippor 2354	Mr. Olver.
1864, Dec. 10, r.—w.f. H	Vanity	do.	do.

VENUS.

Red with white face.

Bred by Mr. E. Price, Court House, Pembridge, Leominster, the property of Mr. J. O. G. Pollock, Mountainstown, Navan, Ireland; got by Goldfinder the Second (959), dam — by Magnet

(823), g.d. (Queen) by Forester (398), g.g.d. (Blowdy) by Hope (439), g.g.g.d. (Beauty) by Sovereign (404), g.g.g.g.d. (Speculation) by Crabstock (303).

PRODUCE IN		NAME.	BY WHAT BULL.	BY WHOM BRED.
1862, July 10, r.—w.f. B		Dead	Salisbury 2204	Mr. Price
1863, Aug. 6, r.—w.f. B		Steer	Earl Derby 2nd 2510	do.
1864, Oct. 5, r.—w.f.	{B / H}	Twins	do.	Mr. Pollock.

VENUS THE SECOND.
Red with white face, calved August 13, 1860.

Bred by and the property of Mr. Naylor, Leighton Hall, Montgomeryshire; got by Admiral (1481), dam (Venus) by Big Ben (248), g.d. (Twist) by Prince (251), g.g.d. (Gipsy) by Trump (490), g.g.g.d. (Venus) by Tobias (487), g.g.g.g.d. (Duchess) by Young Cupid (259).

1863, June 19, r.—w.f. H	Venus 3rd	Salisbury 2204	Mr. Naylor.
1864, July 1, r.—w.f. H	Venus 4th	do.	do.

VENUS THE THIRD.
Red with white face, calved April 15, 1859.

Bred by the late Mr. Josiah Davies, Ivington, the property of Mrs. Ann Davies, Ivington. Leominster; got by Claret (1542), dam (Curly) by Aymestry (1504), g.d. — bred by Mr. Boughton, Clifton-on-Teme.

1861, Nov., r.—w.f. H	Venus 4th	Master Butterfly 1313	Mr. Davies.
1862, Sep. 6, r.—w.f. B	Steer	Conqueror.	do.
1863, July 24, r.—w.f. H	Venus 5th	do.	do.
1864, July 19, r.—w.f. H	Victoria	Chance 1908	do.

VENUS THE FOURTH.
Red with white face, calved in the month of November 1861.

Bred by the late Mr. Josiah Davies, Ivington, the property of Mrs. Ann Davies, Ivington, Leominster; got by Master Butterfly (1313), dam (Venus the Third) by Claret (1542), g.d. (Curly) by Aymestrey (1504), g.g.d. — bred by Mr. Boughton, Clifton-on-Teme.

PRODUCE IN	NAME.	BY WHAT BULL.	BY WHOM BRED.
1863, Nov. 24, r.—w.f. B	Steer	Chance 1908	Mr. Davies.
1864, Oct. 21, r.—w.f. H	Stately	do.	do.

VENUS THE FIFTH.
Red with white face, of 1848, vol. iv., p. 200.

Bred by Mr. James Rea, Monaughty, the property of Mr. James Taylor, Stretford Court, Leominster; got by Albert (330), dam (Winifred) by Monaughty (220), g.d. (Venus the Fourth) by Duke (304), g.g d. (Venus the Third) by Regulator (360), g.g.g.d. (Venus the Second) by Noble (238), g.g.g.g.d. (Venus) by Crabstock (303).

1862, Oct. 12. r.—w.f H	Venus 7th	Unity 2287	Mr. Taylor.

VENUS THE SIXTH.
Red with white face, calved July 29, 1860.

Bred by the late Mr. Rea, Monaughty, the property of Mr. J. Baldwin, Luddington, Stratford-on-Avon; got by Wellington (1112), dam (Venus the Fifth) by Albert (330), g.d. (Winifred) by Monaughty (220), g.g.d. (Venus the Fourth) by Duke (304), g g.g.d. (Venus the Third) by Regulator (360).

1863, Aug. 29, r.—w.f. B	Steer	Sir George	Mr. Baldwin.
1864, July 18, r.—w.f. { B / H	Twins	Battersea 1865	do.

VERBENA.
Red with white face, calved in the month of July, 1856.

Bred by the late Lord Berwick, Cronkhill, the property of Mr. J. O. G. Pollock, Mountainstown, Navan, Ireland; got by Attingham (911), dam (Young Rebecca) by Young Hope (343), g.d. (Rebecca) by Governor (464).

1862, Sep., r.—w.f. H	Vinca	Sir Robert 2227	Mr. Pollock.

VICTORIA.

Red with white face, of 1856, vol. v., p. 305.

Bred by H.R.H. the Prince Consort, Windsor Castle, the property of Major-General the Hon. A. N. Hood, Cumberland Lodge, Windsor; got by Brecon (918), dam (Gwenllian) by Young Dewshall (1125), g.d (Gwenllian), g.g.d. (Young Princess), g.g.g.d. (Princess) by Sovereign (404).

PRODUCE IN	NAME.	BY WHAT BULL.	BY WHOM BRED.
1862, Jan. 17, r.—w.f. B	Steer	Windsor 1456	Gen. the Hon. A. N. Hood.
1863, Jan. 12, r.—w.f. H	Princess of Wales	Maximus 1650	do.
1863, Nov. 30, r.—w.f. H	Irene	Ajax 1843	do.
1864, Nov. 4, r.—w.f. B	Prince Arthur 2693	Garibaldi 2007	do.

VICTORIA.

Red with white face, calved in the year 1854.

Bred by Mr. Thomas Sheriff, Coxall, the property of Mr. Thomas Rogers, Coxall, Ludlow; got by Young Royal (1470), dam — Dinedor (395), g.d. — by Emperor (221), g.g.d. — by Confidence (367).

1863, July 25, r.—w.f. B	Coxall	Sir Colin 2216	Mr. T. Sheriff.
1864, July 20, r.—w.f. H	Maid of Coxall	North Star 2138	Mr. Rogers.

VICTORINE.

Red with white face, calved August 29, 1859.

Bred by the late Lord Berwick, Cronkhill, Salop, the property of the Hon. and Rev. Noel Hill, Berrington, Salop; got by Sir David (349), dam (Young Vic) by Wonder (420).

1862, Oct. 1, r.—w.f. H	Victoria	Van Tromp 2291	Hon. and Rev. Noel Hill.
1863, Sep. 22, r.—w.f. { B / H	Victor 2856 / Victress	do.	do.
1864, Nov. 15, r.—w.f. B	Valiant 2854	Albert 2380	do.

VICTORY.

Red with white face, calved January 10, 1860.

Bred by and the property of Mr. John Monkhouse, The Stow, Hereford; got by Chieftain (930), dam (Victoria) by Young Hope (343), g.d. — by Governor (464), g.g.d. — by Chance (348).

PRODUCE IN	NAME.	BY WHAT BULL.	BY WHOM BRED.
1863, Sep. 1, r.—w.f. H	Victress	Llowess 2608	Mr. Monkhouse.
1864, July 17, r.—w.f. H	Heroine	do.	do.

VIOLET.

Red with white face, calved July 18, 1860.

Bred by Mr. J. Rea, Monaughty, the property of Mr. Stallard, Brockhampton, Ross; got by Wellington (1112), dam (Sunbeam) by Grenadier (961), g.d. (Lustre) by Patron (888), g.g.d. (Lustre) by Cholstrey (217), g.g.g.d. — by Old Court (306).

1863, May 16, r.—w.f. H	Hare Bell	Shamrock 2nd 2210	Mr. Stallard.
1864, May 16, r.—w.f. H	Blue Bell	Chieftain 2nd 1917	do.

VIOLET.

Red with white face, calved in the year 1855.

Bred by Mr. Mason, Yatton, Aymestrey, the property of Mr. T. Rogers, Coxall, Ludlow; got by Malcolm (1305).

1863, Dec. 18, r.—w.f. H	Violet 2nd	Plunder 1038	Mr. Rogers.
1864, Dec. 5, r.—w.f. B	High Sheriff 2571	The Grove 1764	do.

VIOLET.

Red with white face, calved September 23, 1859.

Bred by and the property of Mr. J. Monkhouse, The Stow, Whitney, Hereford; got by Madoc (899), dam (Primrose) by Cantab (717), g.d. (Violet) by Sir Andrew (183), g.g.d. (Violet) by Westonbury (187), g.g.g.d. (Little Violet) by Whitney (105).

1862,	Dead	Chieftain 930	Mr. Monkhouse.
1863, Feb. 28, r.—w.f. H	Isabel	do.	do.
1864, July 4, r.—w.f. H	Josephine	do.	do.

VIOLET.
Red with white face, calved in the year 1858.

Bred by the late Mr. Thomas Longmore, Bucton, Salop, the property of Mr. P. R. Jackson, Blackbrook, Skenfrith, Monmouth; got by Young Sir David (1818), dam (Rosebud) by Young Walford (1820).

PRODUCE IN	NAME.	BY WHAT BULL.	BY WHOM BRED.
1862, Apr. 1, r.—w.f. H	Verbena	Carlisle 923	Mr. Jackson.
1863, Aug. 15, r.—w.f. B	Llandilo 2605½	do.	do.
1864, Oct. 15. r.—w.f. B	Skenfrith 2781	Defiance 2493	do.

VIOLET THE SECOND.
Red with white face, calved September 16, 1859.

Bred by and the property of Mr. Naylor, Leighton Hall, Montgomeryshire; got by Admiral (1481), dam (Violet) by Uriconium (598), g.d. (Wren) by Big Ben (248), g.g.d. (Tidy) bred by Mr. Rea, Monaughty, g.g.g.d. — by Oldcourt (306).

1862, July 31, r.—w.f. H	Dead	Salisbury 2204	Mr. Naylor.
1863, June 10, r.—w.f. H	Violet 3rd	do.	do.

VIRGIN.
Red with white face, calved July 15, 1860.

Bred by and the property of Mr. J. Monkhouse, The Stow, Hereford; got by Chieftain (930), dam (Damsel) by Formidable (1240), g.d. (Maiden) by Madoc (899).

1863, Sep. 4, r.—w.f. H	Vestal	Llowess 2608	Mr. Monkhouse.

WALLFLOWER.
Red with white face, calved July 3, 1860.

Bred by Mr. J. Rea, Monaughty, the property of Mr. W. Stallard, Brockhampton, Ross; got by Wellington (1112), dam (Queen's Gilliflower) by The Doctor (1083), g.d. (Dame's Violet) by Young Conrad (2322), g.g.d. (Dahlia) by Caractacus (659), g.g.g.d. (Dahlia) by Old Court (306).

1863, May 6. r.—w.f. H	Walnut	Shamrock 2nd 2210	Mr. Stallard.
1864, Mar. 30, r,—w.f. H	Woodbine	Chieftain 2nd 1917	do.

COWS.

WAXY THE SECOND.

Red with white face, calved in the month of October, 1860.

Bred by Mr. B. Rogers, The Grove, Pembridge, the property of Mr. Thomas Rogers, Coxall, Ludlow; got by The Grove (1764), dam (Waxy) by Royalty (1374).

PRODUCE IN		NAME.	BY WHAT BULL.	BY WHOM BRED.
1863, Dec.,	r.—w.f. H	Waxy 3rd	Bolingbroke 1883	Mr. B. Rogers.
1864, Sep. 17,	r.—w.f. B	Steer	Grove 2nd 2556	do.

WELCOME.

Red with white face, calved in the month of August, 1860.

Bred by and the property of Mr. J. H. Arkwright, Hampton Court, Leominster; got by Mortimer (1013), dam (Welcome) by Confidence the Second), g.d. (Cherry) by Sir David (349), g.g.d. (Lovely) by Prince Dangerous (362), g.g.g.d. (Luck's-all the Younger) by The Sheriff (356).

1863, Aug.,	r.—w.f. H	Lovely	Sir Oliver 2nd 1733	Mr. Arkwright.
1864, Aug.,	r.—w.f. H	Graceful	Dan O'Connell 1952	do.

WELLBRED.

Red with white face, calved October 15, 1855.

Bred by Mr. E. Price, Court House, Pembridge, Leominster, the property of Mr. J. O. G. Pollock, Mountainstown, Navan, Ireland; got by Magnet (823), dam (Symmetry) by Sir David (349), g.d. (Curly) by Prince Dangerous (362), g.g.d. (Countess) by The Sheriff (356), g.g.g.d. (Tidy) by Forester (398), g.g.g.g.d. (Silk) by Crabstock (303).

1859, Sep. 17,	r.—w.f. B	Garibaldi	Goldfinder 2nd 959	Mr. Price.
1860, Aug. 10,	r.—w.f. H	Wellbred 2nd	do.	do.
1861, July 8,	r.—w.f. H	Wellbred 3rd	Salisbury 2204	do.
1862, Sep. 5,	r.—w.f. H	Wellbred 4th	do.	do.
1863, Aug. 1,	r.—w.f. B	Steer	Earl Derby 2nd 2510	do.
1864,	r.—w.f. B	Walloon	do.	do.

WHITE ROSE.

Red with white face, calved in the month of November, 1857.

Bred by Mr. J. Hollings, Hillend, the property of Mr. J. A. Hollings, Hillend, Hereford; got by Noke (1338), dam (Rose the Fourth) by Byron (380), g.d. (Rose the Third) by Herald (2037), g.g.d. (Rose the Second) by Cornet (1933), g.g.g d. (Rose the First) by Young Waterloo (2341).

PRODUCE IN		NAME.	BY WHAT BULL	BY WHOM BRED.
1860, Oct.	r.—w.f. H	Silver Rose	Woodman 1460	Mr. J. Hollings.
1861, Nov. 3,	r.—w.f. B	Steer	St. Clement 2201	do.
1862, Nov. 9,	r.—w.f. B	Steer	do.	do.
1863, Nov. 25,	r.—w.f. B	Steer	do.	do.
1864, Dec 27,	r.—w.f. H	Clements Rose	do.	do.

WHITE STOCKING.

Red with white face, calved in the month of October, 1859.

Bred by and the property of Mr. G. T. Forester, High Ercall, Wellington, Salop; got by Discord (1217), dam (Rarity) by Darling (1202), g.d. (Result) by Governor (464), g.g.d. (Maid o'-the-Mill) by Hope (439), g.g.g.d. (Moreton) by Royal (331).

1862, July 31,	r.—w.f. H	Peerless	Severn 1382	Mr. Forester.
1863, Sep. 4,	r.—w.f. B	Dead	do.	do.
1864, July 24,	r.—w.f. B	Steer	do.	do.

WILD ROSE.

Red with white face, calved in the month of December, 1858.

Bred by Mr. J. Hollings, Hillend, the property of Mr. J. A. Hollings, Hillend, Hereford; got by Noke (1338), dam (Rose the Fourth) by Byron (380), g.d. (Rose the Third) by Herald (2037), g.g.d. (Rose the Second) by Cornet (1933), g.g.g.d. (Rose) by Young Waterloo (2341).

1861, Sep. 2,	r.—w.f. B	Steer	St. Clement 2201	Mr. J. Hollings.
1862, Dec. 13,	r.—w.f. H	Clementine	do.	do.
1863, Nov. 17,	r.—w.f. H	Wild Rose 2nd	do.	do.
1864, Dec. 1,	r.—w.f. B	Steer	do.	do.

COWS.

WILD ROSE.
Red with white face, calved December 8, 1860.

Bred by Lord Berwick, Cronkhill, Salop, the property of Mr. W. Stallard, Brockhampton, Ross; got by Albert Edward (859), dam (Rosebud) by Walford (871), g.d. (Big Damsel) by The Count (351), g.g.d. (Prettymaid) by Cholstrey (868), g.g.g.d. (Old Damsel) by Coleman's bull (1547), g.g.g.g.d. (Old Daisy) by Chancellor (156), — (Cherry the Second) by Thickset (1769), — (Cherry the First) bred by the late Mr. Knight, Downton Castle.

PRODUCE IN	NAME.	BY WHAT BULL.	BY WHOM BRED.
1863, July 18, r.—w.f. H	Rose of the Wye	Chieftain 2nd 1917	Mr. Stallard.
1864, June 16, r.—w.f. H	Dead	Chieftain 3rd 2457	do.

WILFUL.
Red with white face, of 1858 vol. v., p. 310.

Bred by Lord Bateman, Shobdon Court, Leominster, the property of Mr. R. H. Garrold, Kilforge, Ross; got by Carlisle (923), dam (Wanton) by Radnor (1366).

1862, Aug. 10, r.—m f. H	Carlisle	Garibaldi 2003	Mr. Garrold.

Carlisle sold to Mr. Paramore, Dinedor Court, Hereford.

WINIFRED.
Red with white face, of 1854, vols. iii., iv., and v., pp. 231, 205, 311.

Bred by Mr. Lewis, Breinton, Hereford, the property of Mr. T. Duckham, Baysham Court, Ross; got by Pope (527), dam (Countess) by Defiance (416), g.d. (Countess) by Lottery (410).

1862, Apr. 15, r.—w.f. H	Wynnstay	Castor 1900	Mr. Duckham.
1863, Mar. 15, g.—w.f. B	Steer	Garibaldi 2003	do.
1864, Feb. 21, r.—w.f. H	Winny	Cato 1902	do.

Winny sold to J. Malcolm, Esq., Knockalva, Ramble P.O., Jamaica.

COWS.

WINIFRED.

Red with white face, calved in the year 1861.

Bred by and the property of Mr. R. Green Price, M.P., Norton Manor, Presteign; got by Sir Benjamin (1387).

PRODUCE IN	NAME.	BY WHAT BULL.	BY WHOM BRED.
1864, Aug. 3, r.—w.f. B		Lord of the Manor 2622	Mr. Price.

WINIFRED THE THIRD.

Red with white face, calved August 15, 1859.

Bred by the late Mr. Rea, Monaughty, Knighton. the property of Mr. P. J. Kearney, Miltown House, Clonmellon, Ireland; got by Wellington (1112), dam (Winifred the Second) by Regent (891), g.d. (Clara) by Caractacus (659), g.g.d. (Winifred) by Monaughty (220), g.g.g.d. (Venus the Fourth) by Duke (304), — (Venus the Third) by Regulator (360).

1863, July 28, r.—w.f. H	Winifred 4th	Sir Benjamin 1387	Mr. Rea.
1864, July 18, r.—w f. H	Winifred 5th	Sir Richard 1734	Mr. Kearny.

Winifred the Fifth was the winner of a Prize at the Clonmel Meeting of the Royal Agricultural Society of Ireland.

WINIFRED THE FOURTH.

Red with white face, calved July 5, 1860.

Bred by Mr. J. Rea, Monaughty, the property of Mr. R. H. Capper, The Northgate, Ross; got by Wellington (1112), dam (Winifred the Second) by Regent (891), g d. (Clara) by Caractacus (659), g.g.d. (Winifred) by Monaughty (220), g.g.g.d. (Venus the Fourth) by Duke (304), g.g.g.g.d. (Venus the Third) by Regulator (360).

1863, Mar. 28, r.—w.f. H	Winifred 5th	Sir Benjamin 1387	Mr Rea.

WINIFRED ANN.
Red with white face, calved September 9, 1854.

Bred by and the property of Mr. John Monkhouse, The Stow, Hereford; got by Guy Fawkes (581), dam (Winifred) by Monaughty (220), g.d. (Venus) by Duke (304), g.g.d. — by Regulator (360), g.g g.d. — by Noble (238), g.g.g g.d. — by Crabstock (303).

PRODUCE IN		NAME.	BY WHAT BULL.	BY WHOM BRED.
1859, July 5, r.—m.f.	B	Steer	Chieftain 930	Mr. Monkhouse.
1861, Aug. 31, r.—m.f.	B	Steer	do.	do.
1862, Sep. 3. r.—m.f.	H	Lucy	do.	do.
1863, Dec. 9, r.—w.f.	H	Harriet	do.	do.
1864, Nov. 12, r.—w.f.	H	Annette	Llowes 2608	do.

WINSOME.
Red with white face, of 1858, vol. v., p. 312.

Bred by Lord Bateman, Shobdon Court, Leominster, the property of Mr. E. Wright, Halston Hall, Oswestry; got by Carlisle (923), dam (Young Lady) by Andrew the Second (619).

1863, Feb. 16, r.—w.f.	H	Dead	Magnet 2nd 989	Mr. Wright.
1864, Jan. 28, r.—w.f.	B	Franky	Hero 2039	do.
1864, Dec. 28, r.—w.f.	H	Rosa	do.	do.

WOODHOUSE.
Red with white face, calved September 2, 1861.

Bred by and the property of Mr. J. P. Apperley, Fownhope, Hereford; got by Cornet (1934), dam (Spot) by Sir David (349), g.d. (Woodhouse) by Confidence.

1864, May 26, r.—w.f.	B		Capt. Perry 2444	Mr. Apperley.

WOOD LARK.
Red with white face, calved July 15, 1859.

Bred by the Rev. A. Clive, Whitfield, Hereford. the property of Mr. Colebatch, Underley; got by Thruxton (1422), dam (Whimsical) by Phantom (1035), g.d. (Fancy the Second) by Byron

(440), g.g.d (Fancy the First) by Governor (464), g.g.g.d. — bred by the late Mr. David Williams, Newton.

PRODUCE IN	NAME.	BY WHAT BULL.	BY WHOM BRED.
1862, Sep. 26, r.—w.f. H	Wood Lark	Sir Richard 1734	Mr. T. Rea.
1863, Sep. 21, r.—w.f. B	Steer	do.	do.
1864, July 25, r.—w.f. H	Larkspur	do.	do.

Wood Lark the Second sold to Lord Wenlock.
Larkspur sold to Mr. Colebatch.

WOODMAID.

Red with white face, calved in the month of September, 1859.

Bred by and the property of Mr. Prosser, Honeybourne-grounds, Worcester; got by Medalist (1009).

1862, Dec. 1, r.—w.f. B	Billingsly 2420	The Jew 2266	Mr. J. Prosser.
1863, Feb. 2, r.—w.f. H	Woodmaid 2nd	do.	do.

WOOD PIGEON THE SECOND.

Red with white face, calved December 20, 1860.

Bred by and the property of Mr. John Partridge, Bishop's Wood, Ross, Hereford; got by Walford (1792), dam (Marstow Pigeon) by Uncle Tom (1108), g.d. (Young Miss Thingehill) by Young Sir David (1137), g.g.d. (Miss Thingehill the Fifth) by Thingehill (546), g.g.g.d. (Thingehill Pigeon) by Reform (508).

1864, July 16, r.—w.f. H	Wood Lark	Garway 2536	Mr. Partridge.

YELLOW BEAUTY.

Red with white face, calved in the year 1856.

Bred by Mr. Stedman, Bedstone Hall, Salop, the property of Mr. Naylor, Leighton Hall, Montgomeryshire; got by Bedstone (2411), dam (Young Violet) by Venison (1441), g.d. — by Dinedor (395).

1858, Sep., r.—w.f. H	Yellow Beauty 3rd	Grateful 1260	Mr. Stedman.
1859, Oct., r.—w.f. H	Trinket	do.	do.
1860, Sep., r.—w.f. H	Brown Beauty	Kinlet 1293	do.
1863, Oct., 1, r.—w.f. H	Violet 2nd	Salisbury 2204	Mr. Naylor.
1864, Sep., 24, r.—w.f. B	Shamrock 2749	do.	do.

YOUNG BROADY.

Red with mottle face, calved October 12, 1860.

Bred by and the property of Capt. Peploe, Garnstone, Weobley; got by The Twin (1420), dam (Broady) by Tyro (692), g.d. (Pigeon) by Victory (2297), g.g d. (Kathleen) by Victor (73), g.g.g.d. (Peg Murphy) by Lundyfoot (16).

PRODUCE IN	NAME.	BY WHAT BULL.	BY WHOM BRED.
1863, Aug. 31, r.—m.f. H	Broady 2nd	Leo 2070	Capt. Peploe.
1864, Nov. 16, r.—m.f. H	Broady 3rd / Broady 4th	do.	do.

YOUNG BEAUTY.

Red with white face, of 1854, vol. iv., p. 207.

Bred by and the property of Mr. Naylor, Leighton Hall, Montgomeryshire; got by Uriconium (598), dam (Beauty) by Hampden 1603), g.d. (Venus) by Big Ben (248), g.g.d. (Twist) by Prince (251), g.g.g.d. (Gipsy) by Trump (490).

1859, Jan. 30, r.—w.f. H	Stella	Tom of Lincoln 1099	Mr. Naylor.

YOUNG COUNTESS.

Red with white face, of 1858, vol. v., p. 315.

Bred by Lord Bateman, Shobdon Court, Leominster, the property of Mr. J. R. Paramore, Dinedor Court, Hereford; got by Carlisle, (923), dam (Countess) by The Duke (493).

1861, Oct. 19, r.—w.f. H	Countess	Bolstone 1884	Mr. R. H. Garrold.
1862, Sep. 23, r.—w.f. H	Brown Beauty	Melon 2111	do
1863, Oct. 15, r.—w.f. B	Dinedor 2497 / Steer	The Jew 2266	Mr. Paramore.
1864, Nov. 13, r.—w.f. B	Royal 2731	Portly 2165	do.

Brown Beauty sold to Mr. J. R. Paramore, Dinedor Court.

YOUNG FAIRMAID.

Red with white face, calved in the month of February, 1861.

Bred by Mr. Hewer, Vern House, Hereford, the property of Lieut.-Colonel Feilden, Dulas Court, Hereford; got by Mameluke

(1307), dam (Old Fairmaid) by Garrick (1248), g.d. (Old Fan) by Defiance (416), g.g.d — by Young Sovereign (506).

PRODUCE IN	NAME.	BY WHAT BULL.	BY WHOM BRED.
1863, r.—w.f. H	Fairlass	Mentor 2112	Mr. J. Hewer,
1864, Aug. 25, r.—w.f. B	Billboa 2419	Vincent 2858	Lt.-Col. Feilden.

YOUNG CURLY.
Red with white face, calved March 16, 1863.

Bred by and the property of Mr. W. D. Turner, Lynch Court, Leominster; got by Logic (2079), dam (Curly) by Burton (1159), g.d. (Broady) by Cassio (1528), g.g.d. bred by the late Mr. Turner, Aymestry Court.

PRODUCE IN	NAME.	BY WHAT BULL.	BY WHOM BRED.
1864, Dec. 17 r.—w.f. B		Stockwell 2793	Mr. Turner.

YOUNG FILLPAIL.
Red with white face, calved November 27, 1859.

Bred by and the property of Mr. W. Lane, Compton Casey, Andoversford, Cheltenham; got by Tyro (1786), dam (Fillpail the Second) by Compton (1551), g.d. (Fillpail) by Planet (1690).

PRODUCE IN	NAME.	BY WHAT BULL.	BY WHOM BRED.
1862, Feb. 11, r.—w.f. B	Steer		Mr. Lane.
1863, Apr. 19, r.—w.f. H	Beauty	Hardy 2027	do.

YOUNG GAY.
Red with white face, calved in the month of July, 1861.

Bred by and the property of Mr. B. Rogers, The Grove, Pembridge; got by Claret (1921), dam (Gay) by Young Royal (1470), g.d. (Damsel) by Gaylad the Second (1589), g.gd. (Curly) by Charity the Second (1535), g.g.g.d (Curly) by Portrait (372).

PRODUCE IN	NAME.	BY WHAT BULL.	BY WHOM BRED.
1864, May 22, r.—.w.f. B	Gay Boy 2538	North Star 2138	Mr. Rogers.

YOUNG HEREFORD THE SECOND.
Red with white face, calved in the month of August, 1860.

Bred by the late Mr. J. Davies, Ivington, the property of Mrs.

COWS.

Ann Davies, Ivington, Leominster; got by Young Cholstrey (1808), dam (Young Hereford) by Truelove (2840), g.d (Hereford) bred by Mr. Edwards, Brinsop.

PRODUCE IN	NAME.	BY WHAT BULL.	BY WHOM BRED.
1862, Nov. 23, r.—w.f. H	Blossom	Conqueroor	Mr. Davies.
1863, Oct. 23, r.—w.f. B	Steer	Chance 1908	do.
1864, Aug. 28, r.—w.f. B	Steer	do.	do.

YOUNG LILY.
Red with white face, calved November 13, 1858.

Bred by and the property of Mr. H. Gibbons, Hampton Bishop, Hereford; got by The Admiral (1078), dam (Lily) by Young Gaylad (1463), g.d. (Lily) by Zephyr (1826).

1861, Jan. 12, r.—w.f. H	Browny	Haywood 2034	Mr. Gibbons.
1862, Aug. 23, r.—w.f. H	Lizzy	Shamrock 2nd 2210	do.
1863, Aug. 12, r.—w.f. B	Steer	do.	do.
1864, Sep. 16, r.—w.f. B	Graham 2553	do.	do.

YOUNG LIVELY.
Red with white face, calved in the month of July, 1861.

Bred by the late Mr. J. Davies, Ivington, the property of Mrs. Davies, Ivington, Leominster; got by Duke of Marlborough (1974), dam (Beauty) by Corner Cop (2481), g.d. (Young Hereford) by Truelove (2840), g.g.d. (Hereford) bred by Mr. Edwards, Brinsop.

1863, Aug. 12, r.—w.f. B	Steer	Chance 1908	Mr. J. Davies.
1864, Aug. 11, r.—w.f. H	Lively 2nd	do.	do.

YOUNG MISS CHANCE THE FOURTH.
Red with white face of 1853, vol. iv., p. 212.

Bred by Mr. W. Racster, Thingehill, the property of Mr. T. Powell, Castle Froome; got by Young Sir David (1137), dam (Miss Chance) by Confidence (367), g.d. (Miss Chance) by

COWS.

Chance (355), g.g.d. (Old Miss Chance) bred by the late Mr. T. Jeffries.

PRODUCE IN	NAME.	BY WHAT BULL.	BY WHOM BRED.
1860, Jan. 18, r.—w.f. B	Steer	Grove 1601	Mr. Powell.
1860, Dec. 19, r.—w.f. B	Promise 2175	Moorend 2121	do.
1861, Nov. 28, r.—w.f. H	Dead	do.	do.
1862, Nov. 17, r.—w.f. B	Steer	do.	do.
1863, Dec. 8, r.—w.f. B	Do.	Croft 937	do.

YOUNG MISS THINGEHILL THE SIXTH.
Red with white face, of 1856, vol. iv., p. 213.

Bred by Mr. Racster, Thingehill, Hereford, the property of Mr. J. Partridge, Bishop's Wood, Ross; got by Young Sir David (1137), dam (Miss Thingehill) by Thingehill (546), g.d. (Thingehill Pigeon) by Reform (508), g.g.d. (Hampton Pigeon) by Young Sovereign (506), g.g.g.d. (Sylph) by Chance (355).

1862, Dec. 20, r.—w.f. H	Miss Thingehill 7th	Noble Tom 2135	Mr. Partridge.
1863, Dec. 29, r.—w.f. B	Steer	do.	do.

YOUNG MOSS ROSE.
Red with white face, calved December 10, 1858.

Bred by Mr. Edward Price, Pembridge, Leominster, the property of the Duke of Bedford, Woburn Park, Beds.; got by Goldfinder the Second (959), dam (Moss Rose) by Prince Dangerous (362), g.d. (Fanny) by The Sheriff (356), g.g.d. (Cherry) by Crabstock (303), g.g.g.d. (Duchess) by Sovereign (404).

1862, Dec. 23, r.—w.f. H	Moss Rose-bud	Carbonel 1525	Duke of Bedford.
1864, Jan. 26, r.—w.f. H	Moss Rose-leaf	do.	do.

YOUNG MURPHY.
Red with mottled face, calved July 20, 1860.

Bred by and the property of Captain Peploe, Garnstone, Weobley; got by The Twin (1420), dam (Murphy) by Tyro (692), g.d.

(Murphy) by Murphy Delany (36), g.g.d. (Young Peg Murphy) by George (2013), g.g.g.d. (Peg Murphy) by Lundyfoot (16).

PRODUCE IN	NAME.	BY WHAT BULL.	BY WHOM BRED.
1863, July 29, r.—m.f. B	Steer	Leo 2070	Capt. Peploe.
1864, Sep. 8, r.—m.f. B		do.	do.

YOUNG ORIGINAL THE SECOND.
Red with white face, of 1854, vol. v., p. 318.

Bred by Mr. Racster, Thingehill, Hereford, the property of Mr. J. Partridge, Bishop's Wood, Ross; got by Young Sir David (1137), dam (Original) by Half Sovereign (964).

1862, May 10, r.—w.f. B	Steer	Noble Tom 2135	Mr. Partridge.
1863, Apr. 20, r.—w.f. B	Steer	do.	do.
1864, Mar. 25, r.—w.f. B		Garway 2536	do.

YOUNG PEGGY.
Red with white face, calved in the year 1855.

Bred by the late Mr. T. Longmore, Buckton, Salop, the property of Mr. P. J. Kearney, Miltown House, Clonmellon, Ireland; got by Young Walford (1820), dam — by Emperor (221).

1861, Aug. 15, r.—w.f. H	Miss Severn	Severn 1382	Lord Berwick.

Miss Severn sold to Mr. P. J. Kearney, Miltown House, Clonmellon.

YOUNG PIETY.
Red with white face, calved February 5, 1859.

Bred by and the property of Mr. W. D. Turner, Lynch Court, Leominster; got by Burton (1159), dam (Piety) by Burton (1159), g.d. (Curly) by Andrew the Second (619), g.g.d. — bred by the late Mr. Turner, Aymstrey Court.

1861, Mar. 22, r.—w.f. B	Energy 1982	Logic 2079	Mr. W. D. Turner.
1862, Apr. 5, r.—w.f. H	Fairy	do.	do.
1863, Mar. 9, r.—w.f. H	Dead	Bolingbroke 1883	do.
1864, Mar. 16, r.—w.f. H	Golden Hair	The Rover 2821	do.

YOUNG PIGEON.
Grey, calved January 12, 1856.

Bred by Mr. J. M. Read, Elkstone, Cheltenham, the property of Mr. A. R. Boughton Knight, Downton Castle, Ludlow; got by Venison the Second (1442), dam (Pigeon) by The Sheriff (356), g.d. — by a celebrated bull bred by Mr. Price, of Ryall.

PRODUCE IN	NAME.	BY WHAT BULL.	BY WHOM BRED.
1859, Feb. 22, r.—w.f. B	Smartish	Grecian 2022	Mr. Read.
1860, Jan. 16, r.—w.f. H	Lydia	Caliban 1163	do.
1861, Jan. 5, g.—w.f. B	Steer	do.	do.
1861, Dec. 26, r.—m.f. H	Fancy	do.	do.
1863, Jan. 18, g.—w.f. B	Steer	do.	do.
1864, Apr. 23, g.—w.f. B	Bulrush	Colesborne 2467	Mr. Knight.

YOUNG PLEASANT.
Red with white face, calved in the month of September, 1860.

Bred by and the property of Mr. T. Roberts, Ivington Bury, Leominster; got by Master Butterfly (1313), dam (Pleasant) by Andrew the Second (619), g.d. (Duchess) by Prince by Dayhouse (299).

1862, Aug. 28, r.—w.f. H	Perfection	Sir Thomas 2228	Mr. Roberts.
1863, r.—w.f. H	Dead	do.	do.
1864, June 23, r.—w.f. B	Sir Thomas 2nd 2778	do.	do.

YOUNG PRETTYMAID.
Red with white face, calved in the month of March, 1858.

Bred by Mr. B. Rogers, The Grove, Pembridge, the property of Mr. Thos. Rogers, Coxall, Ludlow; got by Severus (1062), dam (Prettymaid the Second) by Young Royal (1470), g.d. Prettymaid) by Prince (251), g g.d. (Curly the Fourth) by Charity the Second (1535), g.g.g.d. (Curly the Third) by Portrait (372).

1861, July, r.—w.f. H	Prettymaid 3rd	The Grove 1764	Mr. B. Rogers.
1862, Nov. r.—w.f. H	Prettymaid 4th	Bolingbroke 1883	do.
1864, Jan. 7, r.—w.f. B	Palmerston 2664	North Star 2138	do.

COWS.

YOUNG REDCAP.

Red with white face, of 1858, vol. v., p. 319.

Bred by and the property of Mr. T. Roberts, Ivington Bury, Leominster; got by Master Butterfly (1313), dam (Redcap) by Croft (937), g.d. (Gipsy Queen) by King James (978), g.g.d. (Grey Gipsy) by North Star (758), g.g.g.d. (Gipsy the Second) by Original (216), g.g.g.g.d. (Gipsy) by Woodman (255).

PRODUCE IN	NAME.	BY WHAT BULL.	BY WHOM BRED.
1862, July 14, r.—w.f. H	Red Rose	Sir Thomas 2228	Mr. Roberts.
1863, Aug. 16, r.—w.f. B	Steer	Lord Warwick 2093	do.
1864, Aug. 30, r.- w.f. B		Sir Thomas 2228	do.

YOUNG ROSE.

Red with white face, calved October 10, 1861.

Bred by and the property of Mr. H. Gibbons, Hampton Bishop, Hereford; got by Shamrock the Second (2210), dam (Rose) by Medallist (1009), g.d. (Rose) by The Admiral (1078), g.g.d. (Blossom) by Young Gaylad (1463).

1864, June 12, r.—w.f. H	Rosebud	Trumpeter 2282	Mr. Gibbons.

YOUNG VENUS.

Red with mottled face, calved in the month of November, 1856.

Bred by the late Mr. J. Davies, Ivington, the property of Mrs. Ann Davies, Ivington, Leominster; got by King James (978), dam (Venus) by Sutton (1752), g.d. — bred by Mr. Boughton, Clifton-on-Teme.

1860, June, r.—w.f. B	Steer	Young Cholstrey 1808	Mr. Davies.
1861, Aug., r.—w.f. B	Do.	do.	do.
1862, Oct., r.—w.f. B	Do.	Conqueror	do.
1863, H	Dead	do.	do.
1864, July 26, r.—m.f. H	Mottle	Chance 1908	do.

COWS.

ZULEIKA.

Red with white face, calved August 26, 1859.

Bred by the late Lord Berwick, Cronkhill, Salop, the property of the Hon. and Rev. Noel Hill, Berrington, Salop; got by Albert Edward (859), dam (Yellow Byron) by Walford (871).

PRODUCE IN	NAME.	BY WHAT BULL.	BY WHOM BRED.
1862, Nov., r.—w.f. B	Steer	Conqueror 1929	Hon. & Rev. Noel Hill.
1863, Oct., r.—w.f. B	Do.	Van Tromp 2291	do.
1864, Sep. 16, r.—w.f. B		Conqueror 1929	do.

REFERENCE TO BREEDERS.

Abley, Misses, Norton, Presteign, 206, 273, 297, 300, 321.

Allen, Mr. J. D., Pyt House, Tisbury, Wilts., 197, 335.

Apperley, Mr. J. P., Fownhope, Hereford, 2, 8, 49, 92, 149, 153, 157, 164, 218, 247, 253, 259, 262, 272, 290, 319, 320, 324, 325, 351.

Apperley, Mr. W. H., jun., Withington, Hereford, 117, 290.

Apperley, Mr. W. H., sen., Withington, Hereford, 151.

Arkwright, Mr. J. H., Hampton Court, Leominster, 4, 6, 94, 205, 323, 347.

Attwater, Mr. J. G., Ablington, Amesbury, Wilts., 47.

Baldwin, Mr. J., Luddington, Stratford-on-Avon, 50, 89, 133, 159, 169, 189, 210, 261, 331, 343.

Ballard, Mr. P., Leighton Court, Bromyard, 8, 18, 24, 52, 58, 237.

Barton, Mr. J., Coln, Fairford, Gloucestershire, 125, 140, 155, 162, 182, 192, 193, 225, 259, 299, 337, 339.

Bateman, Lord, Shobdon Court, Leominster, 145, 238, 268, 274, 311, 333, 349, 351, 353.

BREEDERS.

Beaumond, Mr. W., Vron End, Clun, Salop, 14, 30, 107.

Bedford, The Duke of, Woburn Park, Beds., 82, 110, 124, 144, 166, 185, 186, 200, 209, 290, 356.

Bennett, Mr. W., North Cerney, Cirencester, 14, 15, 18, 21, 115.

Bennett, Mr. J., Ingestone, Ross, 17.

Berrow, Mr., The Green, Allensmoor, Hereford, 139, 300, 312.

Berwick, Lord, Cronkhill, Salop, 4, 86, 133, 136, 148, 152, 154, 173, 186, 199, 203, 211, 214, 215, 216, 218, 219, 222, 223, 224, 239, 243, 264, 276, 279, 284, 309, 310, 313, 329, 333, 338, 343, 344, 349, 357, 360.

Blockey, Mr. J., Tugford, Munslow, Salop, 117.

Bodenham, Messrs. F. and C., Hereford, 170, 174, 195, 286.

Bourn, Mr. J., Mawley Town Farm, Cleobury Mortimer, 4, 13, 25, 105, 131, 132, 133, 194, 230, 236, 268, 271, 273, 275, 278, 287, 294.

Bowen, Mr. P. W., Shrawardine Castle, Salop, 1, 68, 82, 108.

Bowen, Mr. E., Corfton, Ludlow, 34, 64, 233.

Bowen, Mr. E., Ens Down House, Salop, 66.

Bradstock, Mr. T. S., Cobrey Park, Ross, 8, 9, 10, 12, 17, 33, 51, 61, 123, 163, 186, 194, 197, 198, 235, 249, 250.

Bray, Mr. G., Henwood, Dilwyn, Leominster, 11.

Bridges, Sir H. J., Boultibrook, Presteign, 292.

Britten, Mr. C., The Wood House, Leominster, 4, 22, 97.

Broad, Mr. T., The Castle, Madley, Hereford, 140.

Bulmer, Mr. C., Holmer, Hereford, 41, 60, 61, 312, 332.

Burlton, Mr. J., Luntley Court, Leominster, 120, 147, 157, 160, 191, 266, 270, 339.

Burlton, Mr. T., The Vern, Bodenham, Hereford, 324.

BREEDERS.

Cadle, Mr. T., Longcroft, Westbury-on-Severn, Gloucestershire, 296.

Capper, Mr. R H., The Northgate, Ross, 70, 72, 81, 95, 124, 141, 156, 180, 190, 202, 239, 272, 277, 293, 303, 307, 309, 323, 326.

Carter, Mr., Alcaston, Church Stretton, 261, 319, 326.

Castree, Mr., Gloucester, 162.

Chamberlain, Mr., Desford, Leicestershire, 320.

Chattock, Mr. H. H., Solihull, 69.

Clive, Rev. A., Whitfield, Hereford, 27, 105, 127, 162, 176, 182, 201, 228, 232, 238, 247, 254, 256, 277, 278, 281, 296, 303, 313, 351.

Consort, H.R.H. The Prince, Windsor Castle, 195, 225, 254, 344.

Cooke, Mr. J. Y., Moreton House, Hereford, 1.

Crawshay, Capt., Dany Park, Crickhowell, Brecon, 171, 244, 264, 285, 310.

Davey, Mr. R., M.P., Polsue House, Grampound, Cornwall, 15, 36, 209, 247, 248, 270, 275, 284, 291, 301, 304, 314, 336.

Davies, Mr., Lady Meadow, Leominster, 90.

Davies, Mr. T., Burlton, Court, Hereford, 141, 287.

Davies, Mr. J., Chipp's House, Ivington, Leominster, 141, 142, 166, 258, 260, 319, 342, 343, 354, 355, 359.

Davies, Mr. E., Patton, Much Wenlock, 160, 176, 181, 182, 237, 255.

Davies, Mr. J., Wormbridge, Hereford, 187.

Davies, Rev. T. Kevil, Croft Castle, Ludlow, 281, 313.

Davies, Mr. J., Tillington, Hereford, 304.

BREEDERS.

Davies, Mrs. Ann, Chipp's House, Ivington, Leominster, 319.
Deakin, Mr., Ledbury, 167.
Drinkwater, Mr. T., Treribble, Ross, 41, 211.
Duckham, Mr. T., Baysham Court, Ross, 26, 36, 39, 60, 77, 80, 86, 87, 152, 180, 184, 249, 284, 302, 303, 306, 314, 315, 316, 317, 334, 335, 349.

Edwards, Mr. T., Wintercott, Leominster, 2, 3, 25, 126, 146, 201, 232, 239, 288, 304, 307, 308, 315.
Edwards, Mr., Brampton Bryan, Herefordshire, 9.
Edwards, Mr., Foxhall, Ross, 38.
Edwards, Mr. T., Llanarth, Raglan, Monmouth, 58, 206.
Edwards, Mrs., The Dayhouse, Kingsland, Leominster, 74.
Edwards, Mr., The Cwm, 128.
Elliott, Mr. J. S., Holm Lacey, Hereford, 113.
Elliott, Mr. J., Wormhill, Hereford, 288.
Elsmere, Mr. T., Berrington, Salop, 5, 7, 316.
Evans, Mr. W., Llandowlas, Usk, Monmouth, 43, 83, 128, 181, 269, 325.
Evans, Mr. H. R., Swanstone Court, Leominster, 51, 130, 139, 142, 143, 199, 207, 208, 245, 326, 327.
Evans, Mr. J., Treberfa, Knighton, 77.

Farr, Mr. J., Pontrilas, Hereford, 185, 198, 310.
Featherstonhaugh, Mr. R. S., Rockview, Killucan, Ireland, 7, 50, 133, 203, 219.
Feilden, Lieut-Col., Dulas Court, Hereford, 13, 43, 182, 254, 287, 303, 354.
Forester, Mr. G. T., High Ercal, Salop, 75, 88, 185, 216, 221, 246, 266, 293, 294, 298, 299, 348.

BREEDERS.

Garrold Mr. R. H., Kilforge, Ross, 145, 268, 349, 353.

Gibbons, Mr. H., Hampton Bishop, Hereford, 40, 43, 44, 45, 46, 47, 98, 171, 176, 177, 183, 237, 240, 243, 279, 285, 292, 297, 312, 313, 324, 328, 329, 355, 359.

Gilliland, Mr. S., Brook Hall, Londonderry, Ireland, 35, 52, 53, 54, 139, 200, 274, 300, 301, 312.

Goldingham, Mr. E. T., Grimley, Worcester, 145, 204.

Good, Mr. S. C., Aston Court, Tenbury, 81.

Good, Mr. M., Felton, Bromyard, 101, 290.

Green, Mr. J. B., Marlow, Leintwardine, Herefordshire, 70, 86, 99, 111, 112, 119, 129. 171, 244, 264, 285.

Gregg, Mr. J., Fencote Abbey, Leominster, 168, 197, 274, 281, 282, 300.

Grose, Mr. W. R., Penpont, Wadebridge, Cornwall, 32, 151, 155, 242, 248, 253, 259, 277, 329.

Gwillim, Mr. W. H., Breinton, Hereford, 121, 260.

Hall, Mr. W., Ashton, Leominster, 48, 103.

Hawkins, Mr. B., Orleton, Ludlow, 39, 40, 243, 314, 316, 336.

Hawkins, Mr. T., Sugwas Court, Hereford, 316, 336.

Haynes, Mr. J., Llanrothall, Monmouth, 58, 78, 85, 250.

Haywood, Mr. H., Blakemere, Hereford, 140, 220, 296.

Hereford, Viscount, Tregoyd, Hay, 197, 283, 306.

Hewer, Mr. J. E., Vern House, Marden, Hereford, 4, 6, 47.

Hewer, Mr. J., Vern House, Marden, Hereford, 65, 69, 88, 119, 152, 165, 172, 174, 250, 286, 307, 331, 353, 354.

Hickman, Mr. R., Bosbury, Ledbury, 274, 281, 282, 300.

Higgins, Mr. H., Woolaston Grange, Chepstow, 96.

BREEDERS.

Hill, The Hon. and Rev. R. Noel, Cronkhill, Salop, 30, 94, 95, 118, 119, 148, 186, 199, 209, 211, 224, 243, 284, 309, 313, 338, 344, 360.

Hill, Mr. R., Orleton Court, Ludlow, 38, 62, 97, 108.

Hinckesman, Mr. C. H., The Poles, Ludlow, 13, 16, 71, 162, 223, 227, 233, 234, 260, 270.

Hollings, Mr. J. A., Hillend, Hereford, 23, 169, 303, 309, 334, 335.

Hollings, Mr. J., Hillend, Hereford, 160, 169, 178, 207, 211, 212, 213, 215, 221, 242, 303, 305, 309, 335, 348.

Holloway, Mr., 216.

Hood, Major-Gen., The Hon. A. N., Cumberland Lodge, Windsor, 80, 81, 154, 195, 197, 225, 254, 283, 332, 344.

Jackson, Mr. P. R., Blackbrook, Skenfrith, Monmouth, 20, 24, 31, 36, 58, 90, 100, 101, 125, 156, 188, 259, 261, 315, 346.

James, Mr. J. W., Mappowder Court, Blandford, Dorset, 14, 270, 271, 308.

Jeffries, Mr. T., The Grove, Pembridge, Leominster, 128, 129.

Jeffries, Mr., The Church House, Lyonshall, Kington, 321.

Johnston, Mr., Broncroft Castle, Salop, 223.

Jones, Mr. J., Llwyn-y-gaer, Raglan, Monmouth, 15, 60, 84, 85, 86, 157, 177, 203, 235, 240, 241, 317.

Jones, Mr. J., Lower Hill, Hereford, 41.

Jones, Mr. E., The Moat, Knighton, 109.

Jones, Mr. W., Hill of Eaton, Ross, 132, 224, 231, 269, 310.

Kearney, Mr. P. J., Miltown House, Clonmellon, Ireland, 59, 161, 202, 240, 264, 331, 334, 350.

BREEDERS.

Keene, Mr. R., Pencraig, Caerleon, Monmouth, 2, 15, 74, 156, 239, 271, 292, 339.
Knight, Mr. A. R. Boughton, Downton Castle, Ludlow, 101, 215, 216, 261, 319, 326, 358.

Lane, Mr. W., Compton Casey, Andoversford, Cheltenham, 12, 49, 174, 247, 289, 304, 334, 336, 354.
Lewis, Mr. T., Newchurch, Kinnersley, Herefordshire, 57.
Lewis, Mr. J., Milton, Leominster, 63, 67, 129.
Lewis, Messrs. R. and T., Stapleton Castle, Presteign, 109, 134.
Lewis, Mr. E., Breinton, Hereford, 349.
Lobb, Mr. G., Lawhitton, Launceston, Cornwall, 17, 88, 172.
Longmore, Mr. T., Buckton, Salop, 156, 258, 346, 357.
Longmore, Mr., Orleton, Ludlow, 256.
Lort, Mr. W., The Cotteridge, King's Norton, Birmingham, 90, 107, 229, 241.
Lowe, Mr., Petchfield, Ludlow, 25, 76, 104.
Loyd, Mr. L., Monk's Orchard, Addington, Surrey, 23, 130, 151.

Magness, Mr. W., Bullingham, Hereford, 109.
Manwaring, Mr., Berrington, Leominster, 229.
Mason, Mr., Prior's Court, Ledbury, 164.
Mason, Mr., Yatton, Aymestrey, 345.
Matty, Mr., The Grove, Much Dewchurch, Hereford, 282.
Meire, Mr. J., Berrington, Salop, 9, 28.
Meire, Mr. T. L., Cound Arbour, Salop, 38, 39, 44, 84, 117.
Merryman, Mr. J., Hayfields, Cockeysville, Maryland, America, 65, 80, 256.

BREEDERS.

Monkhouse, Mr. J., The Stow, Hereford, 5, 30, 46, 76, 118, 136, 149, 153, 162, 163, 164, 176, 215, 219, 234, 248, 251, 255, 306, 320, 338, 340, 345, 346, 351.

Moor, Mr., Norton, Presteign, 117.

Morris, Mr. T., Therrow, Llyswen, Hay, 50, 71, 73, 89, 110, 134, 172, 175, 191, 230, 232, 245, 258, 262, 291, 299, 306, 335.

Naylor, Mr. J., Leighton Hall, Montgomeryshire, 26, 29, 30, 44, 56, 59, 64, 65, 83, 87, 90, 93, 111, 113, 119, 122, 134, 135, 136, 161, 173, 177, 183, 191, 192, 203, 210, 218, 233, 245, 248, 252, 253, 257, 263, 264, 275, 280, 284, 295, 298, 318, 322, 325, 328, 339, 340, 342, 346, 352, 353.

Newbery, Mr. W., Fernhill, Kenilworth, 5, 240, 279.

Nott, Mr. C., Bury House, Wigmore, Ludlow, 14, 55, 123.

Oatley, Mr. W. H., Wroxeter, Salop, 20, 40, 69, 73, 119, 134, 137, 199, 214, 241, 292, 293, 321.

Ockey, Mr. M., Thruxton, Hereford, 72.

Olver, Mr. T., Penhallow, Grampound, Cornwall, 2, 8, 18, 21, 26, 48, 56, 76, 84, 87, 131, 135, 138, 145, 150, 155, 195, 201, 220, 226, 238, 242, 286, 291, 294, 295, 302, 341.

Palmer, Mr. J., Hampton-on-the-Hill, Warwick, 55, 69, 124, 152, 175, 307.

Palmer, Mr., Bolitre, Ross, 266.

Paramore, Mr. J. R., Dinedor Court, Hereford, 32. 42, 89, 102, 115, 132, 145, 147, 148, 156, 157, 162, 183, 187. 201, 222, 235, 236, 237, 238, 256, 267, 268, 282, 296, 312, 353.

BREEDERS.

Partridge, Mr. J., Bishop's Wood, Ross, 42, 234, 251, 252, 259, 260, 263, 265, 274, 341, 352, 356, 357.

Peploe, Captain, Garnstone, Hereford, 39, 150, 186, 234, 305, 306, 308, 353, 356, 357.

Peren, Mr. W. B., Compton, South Petherton, 67, 129, 131, 188, 189.

Perry, Mr. W., Cholstrey, Leominster, 11, 18, 23, 70, 121, 126, 198, 280.

Perry, Mr. J., Much Cowarne, Bromyard, 19, 127, 128.

Pinches, Mr. J. T., Hardwick, Leominster, 63.

Pitt, Mr. G., Chadnor, Leominster, 82, 91.

Pollock, Mr. J. O. G., Mountainstown, Navan, Ireland, 44, 52, 68, 85, 88, 149, 223, 224, 311, 331, 342, 343.

Powell, Mr. W., Eglwysnunydd, Taibach, Glamorganshire, 7, 20, 28, 165, 206, 214.

Powell, Mr. T., The Bage, Madley, Hereford, 27, 51, 65, 73, 76, 79. 81, 93, 94, 104, 105, 115, 118, 196, 266, 283, 304.

Powell, Mr. H. E., Great Brampton, Hereford, 45, 140, 144, 151, 165, 168, 226, 228, 250, 251, 267, 268.

Powell, Mr. W., White House, Llantillio, Monmouth, 59.

Powell, Mr. J., Great Brampton, Hereford, 168, 228.

Powell, Mr. W. S., Hinton, Hereford, 262, 263, 286.

Powell, Mr. T., Castle Froome, Bromyard, 337, 356.

Power, Captain, Hill Court, Ross, 132, 224, 231, 256, 269, 310.

Price, Mr. E., Court House, Pembridge, Leominster. 16, 35, 36, 48, 68, 75, 93, 116, 118, 124, 127, 144, 149, 166, 167, 188, 191, 242, 290, 296, 331, 341, 342, 347, 356.

Price, Mr. J., Town House, Bredwardine, Hereford, 116.

BREEDERS.

Price, Mr. R. G., M.P., Norton Manor, Presteign, 147, 161, 170, 184, 206, 213, 273, 292, 297, 300, 301, 316, 321, 332, 350.

Price, Mr. T., Tillington, Hereford, 265, 304.

Prosser, Mr. J., Honeybourne Grounds, Worcestershire, 12, 13, 141, 142, 165, 184, 190, 231, 251, 278, 352.

Prosser, Mr. W., Garway Court, Monmouth, 42.

Prosser, Mr., The Meadow Farm, 141.

Racster, Mr. W., Thingehill Court, Hereford, 87, 355, 356, 357.

Rea, Mr. T., Westonbury, Leominster, 7, 83, 96, 98, 101, 114, 138, 139, 143, 144, 148, 195, 197, 204, 217, 225, 245, 246, 255, 257, 279, 283, 289, 290, 321, 335, 352.

Rea, Mr. J., Monaughty, Knighton, 31, 38, 48, 68, 95, 96, 100, 105, 121, 122, 125, 138, 143, 148, 161, 180, 181, 185, 194, 195, 196, 200, 202, 204, 216, 217, 218, 240, 244, 247, 255, 272, 278, 279, 289, 295, 296, 297, 308, 310, 315, 321, 325, 330, 331, 332, 333, 334, 343, 345, 346, 350.

Read, Mr. J. M., Elkstone, Cheltenham, 25, 71, 75, 106, 107, 146, 202, 227, 337, 358.

Reynell, Mr. R. W., Killynan, Killucan, Ireland, 62.

Richards, Mr. J., Cound, Salop, 29, 35, 137, 158, 168, 187.

Ricketts, Mr. J., Trebarried, Bronllys, Hay, 2, 113.

Ridler, Mr. R. H., Gattertop, Leominster, 66, 194, 307.

Roberts, Mr. T., Ivington Bury, Leominster, 10, 12, 35, 41, 42, 51, 61, 62, 63, 71, 74, 79, 82, 97, 98, 99, 100, 114, 116, 189, 220, 228, 229, 231, 232, 261, 276, 288, 293, 358, 359.

BREEDERS.

Roberts, Mr. R., Burrington, Ludlow, 92.

Roberts, Mr., Trippleton, Ludlow, 116.

Rogers, Mr. B., The Grove, Pembridge, Leominster, 3, 16, 27, 29, 33, 42, 46, 49, 54, 56, 57, 64, 66, 73, 78, 83, 89, 97, 117, 125, 146, 147, 157, 173, 179, 180, 196, 264, 265, 268, 285, 287, 305, 315, 318, 322, 327, 329, 347, 354, 358.

Rogers, Mr. J., Stocken, Presteign, 17, 115, 307.

Rogers, Mr. T., Coxall, Brampton Bryan, Herefordshire, 50, 95, 103, 207, 315, 344, 345.

Rogers, Mr. J., Altyr-y-nys, Abergavenny, 55, 90.

Rogers, Mr. A., The Homme, Dilwyn, Leominster, 126.

Rogers, Mr. J., Letchmoor, Presteign, 127, 146, 147, 150, 154, 170, 174, 178, 195, 228, 237, 287, 313, 320.

Rogers, Mr., Hereford, 200, 274.

Rogers, Mr. J., Boultibrook, Presteign, 231.

Savery, Capt., Hardwick Lodge, Chepstow, 61, 153, 154, 164.

Sheriff, Mr. T., Coxall, Brampton Bryan, 9, 31, 141, 170, 180, 202, 239, 293, 303, 307, 309, 310, 323, 326, 344.

Sheriff, Mr. J., Burrington, Salop, 165, 258, 331, 341.

Shirley, Mr. R., Baucott, Church Stretton, Salop, 10, 16, 37, 43, 45, 84, 94, 103, 108, 210, 257, 262, 263, 267, 273, 311, 314.

Smith, Mr. J., Sevenhampton, Cheltenham, 67, 93, 155, 182, 259, 299.

Smith, Mr. T., Bodenham, Dymock, 56, 167, 317.

Smith, Mr. J., Ridby, Much Dewchurch, Hereford, 32.

Sobey, Mr. J., Penhallow and Tencreek, Cornwall, 5, 155, 209, 248, 253, 259, 275, 277, 301, 329.

BREEDERS.

Somers, The Earl, Eastnor Castle, Ledbury, 167.

Sparkman, Mr. J., Little Marcle, Ledbury, 29, 286.

Spencer, Mr., Yarpole, Leominster, 158.

Stallard, Mr. W., Brockhampton, Ross, 22, 91, 92, 102, 104, 165, 217, 247, 297, 332, 341, 345, 346, 349.

Stedman, Mr. W., Bucknall House, Salop, 12, 169, 192, 193, 203, 206, 210, 211, 214, 224, 325, 352.

Stephens, Mr. J., Sheep House, Hay, 21, 157.

Stokes, Mrs., Lyd-y-dy-way, 76.

Stone, Mr. F. W., Moreton Lodge, Guelph, Canada West, 3, 18, 19, 27, 58, 79, 87, 100, 109, 110, 122, 138, 208, 214, 269, 277, 333, 340.

Stone, Mr. J. J., Scyborwen, Usk, Monmouth, 68, 91, 123, 133, 222.

Tanner, Mr. E., Hopton Castle, Aston-on-Clun, Salop, 28, 51, 109, 112, 125, 233, 250.

Tanner, Mr. E., Ayntree, Salop, 102.

Taylor, Mr. James, Stretford Court, Leominster, 47, 77, 84, 158, 159, 169, 170, 171, 180, 181, 193, 198, 243, 244, 258, 273, 316, 317, 330, 343.

Taylor, Mr. W., Showle Court, Ledbury, 106, 107, 112, 113, 114.

Taylor, Mr. John, Stretford Court, Leominster, 158, 171, 198, 316, 330.

Thomas, Mr. T., St. Hilary, Cowbridge, Glamorganshire, 24, 175, 227.

Thomas, Mr. J., Cholstrey, Leominster, 129.

Thomas, Mr., The Lodge, Salop, 198.

BREEDERS.

Tipton, Mr. G., Easton Farm, Middleton-on-the-Hill, Tenbury, 67.

Tudge, Mr. W., Adforton, Leintwardine, Salop, 14, 28, 31, 33, 34, 62, 72, 78, 103, 120, 143, 151, 152, 164, 167, 178, 183, 192, 206, 207, 229, 254, 277, 282, 291.

Turner, Mr. P., The Leen, Pembridge, 6, 21, 24, 32, 34, 37, 45, 54, 85, 104, 111, 135, 136, 153, 154, 184, 205, 209, 218, 219, 220, 223, 230, 249, 272, 276, 289, 295, 298, 301, 305, 323, 338.

Turner, Mr. J., Court of Noke, Leominster, 128, 162, 253, 262, 272, 290, 324, 334.

Turner, Mr. W. D., Lynch Court, Leominster, 137, 150, 179, 205, 238, 354, 357.

Turner, Mr. J., Brick House, Burghill, Hereford, 207.

Urwick, Mr. S. W., Leinthall, Ludlow, 34, 80, 81, 89, 105, 121.

Vaughan, Messrs. J. and W., Lawton, Leominster, 6, 60, 64, 173, 179, 196, 281.

Vaughan, Mr., Cholstrey, Leominster, 29, 74, 77, 99, 111, 115, 179, 196, 281.

Vaughan, Mr. sen., Lawton, Leominster, 20, 173, 179, 280, 281.

Wenlock, Lord, Bourton Grange, Wenlock, Salop, 37, 325.

Wheeler, Mr. T., Wormhill, Eaton Bishop, Hereford, 22, 101, 187, 288.

Whitehouse, Mr. J. H., Ipsley Court, Redditch, 67, 225, 267, 290.

BREEDERS.

Wicksted, Mr. C., Shakenhurst, Bewdley, 19.
Wigmore, Mr. J., Bickerton Court, Dymock, 174, 200, 204, 226, 256, 266, 318, 322, 327.
Williams, Mr. T., New House, Bromfield, Salop, 21, 92.
Williams, Mr. D., Newton, Brecon, 33.
Williams, Mr. E., Llowes Court, Hay, 59, 78, 114, 211.
Williams, Mr. Culmington, Salop, 74.
Williams, Mr. J., Kingsland, Leominster, 121.
Williams, Mr., Chapel Clun, Salop, 154, 313.
Woolley, Mr. T., Weston Court, Leominster, 123.
Wright, Mr. E., Halston Hall, Oswestry, 9, 57, 79, 98, 232, 238, 263, 271, 272, 274, 311, 333, 336, 351.

Yeld, Mr. T., The Broome, Leominster, 38.
Yeomans, Mr. W., Stretton Court, Hereford, 11, 63.

INDEX TO HEIFERS.

Heifers.	Breeders.	Page.	Heifers.	Breeders.	Page.
Ada	Mr. W. Jones	132	Beauty of Westonbury	Mr. T. Rea	139
—— 2nd	J. R. Paramore	132	—— 2nd	Mr. S. Gilliland	139
——	T. Morris	230	—— 3rd	do.	139
——	J. R. Paramore	235	—— 2nd	J. R. Paramore	141
Adelaide	J. Bourn	131	—— 2nd	J. Prosser	141
Agatha 2nd	J. Naylor	134	—— 2nd	J. Davies	142
Agnes	W. H. Oatley	241	—— 3rd	J. Prosser	142
——	do.	241	——	R. H. Garrold	145
——	J. Rea	333	——	Hon. & Rev. N. Hill	148
Agrippina 2nd	W. H. Oatley	134	—— 2nd	do.	148
——	do.	241	—— 3rd	do.	148
Alba	E. Wright	274	——	Mr. T. Morris	191
Alethe	J. M. Read	337	——	J. Rogers	195
Alexandra	Messrs. Vaughan	196	——	T. Edwards	206
——	Mr. C. H. Hinckesman	227	——	W. Lane	354
——	W. Tudge	277	Beechnut	R. Shirley	273
Alice	W. D. Turner	137	Bella	J. Bourn	131
Alma	T. Elsmere	316	——	T. Broad	140
Amazon	E. T. Goldingham	204	—— 3rd	T. Rea	143
Angelica	W. H. Oatley	241	—— 4th	do.	144
Annette	J. Monkhouse	351	——	R. Keene	239
Annie 2nd	J. Naylor	135	Belle of the Village	H. Haywood	140
Apple Blossom	Lord Berwick	215	——	J. P. Apperley	153
Arethusa	Mr. G. T. Forester	294	——	R. Keene	339
Argenta	E. Wright	274	Bertha	H. E. Powell	151
Augusta	W. H. Oatley	241	Bessie 2nd	R. H. Garrold	145
Banjo	E. T. Goldingham	145	——	W. R. Grose	248
Bonnie	W. Tudge	143	—— Gwynne	J. W. James	271
Barbara	W. R. Grose	242	——	J. Hewer	286
Barmaid	T. Edwards	288	Be-true	H. Haywood	140
Baroness	J. Hollings	212	Betsy	W. R. Grose	151
——	E. Wright	263	Bigarrean	J. R. Paramore	157
——	R. Davey	291	Bloom	Capt. Peploe	150
Bashful	J. Rea	138	Blossom	Mr. Prosser	142
Beatrice Princess	H. Haywood	140	—— 5th	B. Rogers	146
Beautiful	T. Powell	283	——	Capt. Peploe	150

INDEX TO HEIFERS.

Heifers.	Breeders.	Page.
Blossom	Mr. J. R. Paramore	236
—	J. Davies	355
Blowdy 3rd	B. Rogers	147
—	J. P. Apperley	153
Blue Bell 2nd	J. R. Paramore	147
—— 3rd	do.	147
—	W. Stallard	345
Boadicea	T. Morris	258
Bonnett	Duke of Bedford	144
Bonny	Mr. J. P. Apperley	157
Boss	J. Wigmore	327
Bountiful	Hon. & Rev. N. Hill	148
—	Mr. J. Rea	334
Bounty	J. P. Apperley	149
Brandy 2nd	E. Price	149
—— 3rd	J. O. G. Pollock	149
Bridget	W. Stallard	217
Briony	J. Prosser	190
Broady 2nd	Capt. Peploe	353
—— 3rd	do.	353
—— 4th	do.	353
Bronllys	Mr. J. Monkhouse	149
Brown Beauty	W. Stedman	352
—	R. H. Garrold	353
—	W. D. Turner	150
Browny	H. Gibbons	355
—	T. Olver	150
Brunette	H. E. Powell	151
Bryony	T. Olver	150
Buttercup	do.	150
Butterfly	J. Barton	337
Cactus	P. R. Jackson	259
Carlisle	R. H. Garrold	349
Carlotta	W. Tudge	164
Carnation	J. Barton	162
Castanet	H. E. Powell	168
Cassio	W. D. Turner	179
Catharina	W. Tudge	152
Cayne	G. T. Forester	294
Cherry 2nd	J. R. Paramore	156
—— 2nd	J. Richards	158
—— 3rd	do.	158
—— 7th	J. Taylor	158
—— 4th	J. Baldwin	159
—— Cheeks	E. Davies	160
—— 7th	J. Hollings	160
—— 8th	do.	160
—— Fruit	P. J. Kearney	161
—	J. Richards	187
—	A. R. B. Knight	216
—	T. Powell	283
—	H. Gibbons	285
Chieftain's Daisy	J. Hollings	213
—— Rosa	do.	303

Heifers.	Breeders.	Page.
Chloe 2nd	Mr. J. Naylor	161
Chub 2nd	do.	161
—— 3rd	do.	161
Church House 3rd	C. H. Hinckesman	162
—— 4th	do.	162
Clementine	W. Stallard	165
—	W. Tudge	167
—	J. Hollings	348
Clara	J. Sheriff	165
—	J. Farr	198
—	W. Powell	206
—	W. Newbery	240
Clement's Daisy	J. Hollings	178
—— Grace	do.	212
—— Countess	do.	215
—— Hope	do.	221
—— Rose	do.	348
Cleopatra	T. Morris	175
Cloribel	W. Tudge	167
Cockney	J. P. Apperley	164
Columbine	E. Davies	160
Comely	J. R. Paramore	145
Constance	H. E. Powell	168
—	Hon. A. N. Hood	197
Corah	Mr. P. J. Kearney	161
—— 3rd	J. Davies	166
Cornelia	T. Morris	258
Cornelian	Hon & Rev. N. Hill	338
Cornfit	Mr. E. Davies	176
Cornflower	P. Turner	184
Countess	J. Barton	155
—— 2nd	T. Smith	167
—— 3rd	J. Hollings	169
—— 6th	J. Taylor	169
—— 8th	do.	170
—— 7th	do.	198
—	E. Wright	272
—	H. Gibbons	285
—	R. Keene	292
—	R. H. Garrold	353
Cowslip 2nd	J. Rogers	170
—— 4th	J. Taylor	171
—— 5th	do.	171
—	H. Gibbons	312
Crafty 2nd	G. Lobb	172
—— 3rd	do.	172
Cream	E. Davies	176
Cricket	Capt. Savery	164
Crink	Hon. & Rev. N. Hill	309
Crinoline 2nd	Messrs. Vaughan	173
Crocus 2nd	Mr. J. Naylor	173
—	J. Monkhouse	320
Crown Princess	Hon. A. N. Hood	225
Curly	Mr. J. Jones	157

INDEX TO HEIFERS.

Heifers.	Breeders.	Page.	Heifers.	Breeders.	Page.
Curly	Mr. W. Tudge	164	Dowager	Mr. P. Turner	230
——— 2nd	W. Lane	174	Duchess 2nd	T. Wheeler	187
——— Silky	R. Shirly	314	——— 3rd	do.	187
Cynthia	G. T. Forester	246	Duchess 2nd	J. R. Paramore	187
Dahlia	P. Turner	323	——— 4th	B. Peren	188
Dainty Lass	H. Gibbons	176	——— 5th	do.	188
——— 3rd	J. Naylor	177	——— 6th	do.	188
——— 4th	do.	177	———	H. Gibbons	297
——— 2nd	R. H. Capper	180	——— of Bedford 3rd	T. Roberts	189
——— 5th	T. Duckham	180	——— ——— 4th	do.	189
——— 6th	do.	180	——— of Holm 2nd	J. Prosser	190
——— 9th	J. Taylor	181	Eglantine	P. Turner	298
——— 10th	do.	181	Ella	W. Tudge	192
———	E. Davies	181	Ellen	J. Davies	142
———	P. R. Jackson	188	——— 4th	J. Burlton	191
———	J. Davies	319	——— 2nd	J. Davies	260
Dairy Maid	T. Edwards	288	Empress 10th	J. Taylor	193
Daisy	J. Richards	137	England's Beauty 2nd	T. Rea	194
———	J. R. Paramore	148	Esther	W. D. Turner	137
——— 2nd	Messrs. Vaughan	179	Etty	J. Bourn	133
——— Cutter	E. Davies	182	Eugenie	W. Tudge	192
———	Capt. Peploe	234	Eva	do.	192
———	Mr. J. Gregg	274	Eveline	J. Bourn	194
———	B. Rogers	287	Fairlass	J. Hewer	354
Damsel	J. R. Paramore	236	Fairmaid	W. Vaughan	179
———	J. Hewer	286	——— 2nd	J. Rogers	195
Darky	G. T. Forester	246	——— 3rd	B. Rogers	196
Darling	W. Tudge	178	——— 4th	do.	196
———	E. Davies	181	———	J. R. Paramore	256
———	T. Duckham	184	Fairy	T. Olver	201
———	P. Turner	295	———	G. T. Forester	294
Daylight	E. Davies	182	———	P. Turner	323
Deception	Col. Feilden	182	———	W. D. Turner	357
Deborah	Mr. W. Tudge	183	Faithful	Rev. A. Clive	201
Defiance	A. R. B. Knight	215	Fancy	Mr. R. Keene	156
Delicate	E. Davies	181	——— Fair	J. Farr	198
Delight 2nd	J. Naylor	183	Fanny	R. G. Price	147
——— 3rd	do.	183	———	T. S. Bradstock	198
Dell	J. Prosser	141	——— 3rd	H. R. Evans	199
——— 2nd	do.	184	——— 4th	do.	199
Dewdrop	H. Gibbons	176	———	W. R. Grose	248
———	J. Naylor	275	———	J. M. Read	358
———	A. R. B. Knight	319	Fantail	H. Gibbons	279
Diana	H. Gibbons	176	Fat Lady	Hon. & Rev. N. Hill	199
——— 3rd	J. Farr	185	——— Girl	do.	199
———	R. H. Ridler	194	Fauchette	Duke of Bedford	200
———	G. T. Forester	246	Fausta	Mr. W. H. Oatley	293
Dido	W. Tudge	178	Finella	W. Lort	229
Dignity	R. H. Capper	326	Flax 2nd	S. Gilliland	200
Dolly	Rev. A. Clive	176	Flirt	J. Hewer	165
———	Mr. W. R. Grose	277	———	P. Turner	338
Dorothy	Duke of Bedford	186	Florence	Hon. A. N. Hood	195
——— 2nd	do.	186	———	Mr. T. S. Bradstock	197
Double X	Mr. E. Davies	182	———	Rev. A. Clive	201

[377]

INDEX TO HEIFERS.

Heifers.	Breeders.	Page.	Heifers.	Breeders.	Page.
Florence	Mr. T. Duckham	317	Gwenny 5th	Mr. W. Stallard	217
Florist	T. Edwards	201	Hampton Olive	J. H. Arkwright	205
Froudy	T. S. Bradstock	198	Hare Bell	W. Stallard	345
Fury 2nd	J. Naylor	203	Harriet	J. Monkhouse	351
Gager 3rd	J. Jones	203	Heartsease	F. W. Stone	333
—— 4th	do.	203	Hebe	J. Taylor	244
—— 5th	do.	203	Heiress 2nd	J. Naylor	218
Gaiety 2nd	R. S. Featherstonhaugh	203	Heroine	J. Monkhouse	345
—— 3rd	T. Rea	204	Highdrangea	R. S. Featherstonhaugh	219
Gaily	T. Edwards	206	Hinton	W. S. Powell	286
Gaudy	J. Monkhouse	340	Hop Duty	H. Haywood	220
Gay	J. H. Arkwright	205	Hope	H. R. Evans	245
Gaylass	J. Prosser	142	Humpy	G. T. Forester	221
——	T. Morris	175	Irene	Hon. A. N. Hood	344
——	J. Barton	193	Isabelle	Mr. R. Davey	304
——	A. R. B. Knight	261	——	J. Monkhouse	345
——	G. T. Forester	298	Jane	J. R. Paramore	222
Gazelle	W. H. Oatley	214	Janet	Rev. A. Clive	232
——	Lord Berwick	216	Jeannette	Hon. A. N. Hood	332
——	Mr. C. H. Hinckesman	227	Jenny	Mr. R. Keene	156
Gentle	J. Hollings	207	—— Lind	J. Barton	225
—— Anne	T. Rogers	207	Jessamine	J. Richards	168
—— 5th	F. W. Stone	208	——	C. H. Hinckesman	223
—— 6th	do.	208	Jesse	W. Evans	181
——	R. Keene	339	——	J. O. G. Pollock	223
Georgina	Hon. A. N. Hood	154	——	H. Gibbons	240
Geraldine	Mr. J. Monkhouse	163	Jessica	C. H. Hinckesman	223
——	P. Turner	305	Jewel	Hon. & Rev. N. Hill	338
Gertrude	W. H. Oatley	214	Jewess	Mr. J. R. Paramore	145
——	Holloway	216	Jonquil	J. O. G. Pollock	224
Giantess 5th	J. Baldwin	210	Josephine	J. Monkhouse	345
—— 6th	do.	210	Juicy	J. O. G. Pollock	224
Gipsy Queen	P. Turner	136	Kate 4th	J. Wigmore	226
——	J. Prosser	190	Kathleen	P. Turner	301
——	E. Drinkwater	211	Kitty	J. Monkhouse	153
Gold	Hon. & Rev. N. Hill	211	Laburnam	H. E. Powell	228
Golden Hair	Mr. W. D. Turner	357	Lady	J. Hollings	169
Governess	J. Hollings	212	—— Bateman	H. R. Evans	208
—— 2nd	do.	212	—— Mary	Hon. A. N. Hood	225
Grace 2nd	W. Stedman	211	—— 2nd	Mr. J. M. Read	227
—— 3rd	do.	211	—— Adforton	W. Tudge	229
—— 2nd	J. Hollings	212	—— Jane	J. Bourn	230
—— Darling	J. Monkhouse	219	—— of the Grounds	J. Prosser	231
Graceful	T. Davies	141	—— Lucy 2nd	T. Roberts	232
—— 2nd	F. W. Stone	214	—— Lift	T. Edwards	232
——	J. Monkhouse	219	—— 4th	J. Naylor	233
——	A. R. B. Knight	261	—— Emily	Hon. A. N. Hood	254
——	J. H. Arkwright	347	—— Jane	Mr. R. Shirley	267
Grey	B. Bogers	146	——	W. Evans	269
Greyling	A. R. B. Knight	319	—— of the Valley	J. Rea	308
Grisel	do.	215	—— Vitcoria	Hon. & Rev. N. Hill	309
Grossie	W. R. Grose	253	—— Clyde	Mr. T. Sheriff	310
Gwenny 3rd	T. Rea	217	—— Jane Grey	A. R. B. Knight	319
—— 4th	do.	217	——	Sobey	329

INDEX TO HEIFERS.

Heifers.	Breeders.	Page.	Heifers.	Breeders.	Page.
Lady Battersea	Mr. J. Baldwin	331	Lucy	Mr. T. Duckham	249
—— Mary Hood	Hon. A. N. Hood	332	——	H. Gibbons	329
—— Garrick	Mr. C. Bulmer	332	—— 2nd	do.	329
——	T. Burlton	339	—— 3rd	do.	329
Lana	T. S. Bradstock	249	——	J. Monkhouse	351
Lark	J. P. Apperley	262	Lufra	E. Davies	237
Larkspur	W. Lane	247	Luna	Rev. A. Clive	228
——	T. Rea	352	——	Mr. J. Bourn	236
Lassie	J. Monkhouse	164	Lurline	H. E Powell	228
——	W. Lane	336	Lydia	T. S. Bradstock	235
Laura	W. D. Turner	137	——	J. M. Read	358
——	J. Prosser	190	Mabel	R. Davey	301
—— 2nd	J. R. Paramore	235	Maid of Shrewsbury	J. Baldwin	169
—— 3rd	do.	235	—— of Weston	J. Hollings	221
—— 4th	do.	235	—— 2nd	do.	221
——	T. Olver	242	—— of Coxall	J. Rogers	344
Leah	P. Turner	205	Maiden	J. Naylor	248
Leonora	H. E. Powell	228	Maple	T. S. Bradstock	249
Lily of the Vale	J. Prosser	190	Marchioness	J. Hollings	213
——	P. Turner	205	——	E. Wright	272
—— 2nd	J. Rogers	237	Marcia	J. Monkhouse	251
——	B. Hawkins	243	Maria	T. S. Bradstock	197
Linnett	J. Bourn	230	——	J. Monkhouse	251
——	Rev. A. Clive	232	Marigold	H. E. Powell	267
Little Dainty	Mr. J. Jones	177	——	H. Gibbons	328
Lively	J. Davies	142	Marion	J. Merryman	256
——	W. D. Turner	150	Marstow Pigeon 7th	J. Partridge	251
—— 2nd	J. Davies	355	Mary Ann	J. Monkhouse	149
Lizzie	J. Richards	168	—— 2nd	J. Naylor	252
——	H. Gibbons	355	——	J. Burlton	266
Lofty	E. Wright	238	Maude 2nd	J. Naylor	253
——	J. P. Apperley	253	—— 3rd	do.	253
——	H. R. Evans	327	——	do.	257
Lola	H. E. Powell	228	May Flower	E. Davies	255
——	T. Rea	246	May-day	do.	255
Long Horns	W. H. Oatley	293	Medora	T. Morris	230
Lovely	J. Hollings	207	Melody 3rd	J. Taylor	255
——	J. R. Paramore	222	——	H. E. Powell	267
——	J. Bourn	236	Mermaid	do.	267
——	J. Hollings	242	Merry Thought	J. Monkhouse	255
—— 2nd	B. Hawkins	243	Milkmaid	J. Davies	166
—— 4th	J. Taylor	244	Minna	Holloway	216
——	T. Morris	245	——	W. Tudge	254
——	B. Rogers	265	Minnie ha ha	J. Monkhouse	255
——	T. Roberts	276	—— 2nd	T. Rea	257
——	H. Gibbons	324	Miriam	J. W. James	308
——	J. H. Arkwright	347	Misletoe	G. T. Forester	266
Lucella	J. Naylor	245	Miss Ivington	J. Davies	142
Lucena	J. B. Green	244	—— Rose	W. R. Grose	155
Lucy	T. Olver	226	—— Curly	J. Rogers	174
——	T. Thomas	227	—— Lyons	J. Palmer	175
—— Ashford	W. Tudge	229	—— Stedman	W. Stedman	192
—— 5th	T. Rea	245	—— Grove	Messrs. Vaughan	196
—— 6th	do.	246	—— Hastings 2nd	Mr. T. Roberts	231

INDEX TO HEIFERS.

Heifers.	Breeders.	Page.	Heifers.	Breeders.	Page.
Miss Garway	Mr. J. Partridge	234	Patty	Mr. W. Tudge	282
—— Church House	J. P. Apperley	253	——	R. H. Capper	323
—— Palmerston	R. Shirley	257	Peach	Rev. A. Clive	278
—— Hopeful	W. R. Grose	259	——	Hon. & Rev. N. Hill	284
—— Coppice 3rd	J. Partridge	260	Pearl	Mr. W. Tudge	282
—— Cotmore 2nd	C. H. Hinckesman	260	——	T. Olver	286
—— Hughes	R. Shirley	262	Peeress 2nd	R. H. Capper	277
—— Wood 3rd	J. Partridge	263	——	R. Davey	291
—— Miller	R. Shirley	263	Peerless	G. T. Forester	348
—— Ruth 2nd	J. Naylor	264	Pencil	J. Monkhouse	338
—— Noke	J. P. Apperley	272	Penelope	Hon. A. N. Hood	283
—— Chance	H. Gibbons	279	Peony	Mr. J. Monkhouse	176
—— Hanbury	C. Bulmer	312	——	J. Naylor	263
—— Perry	J. Davies	319	Perfection	R. Hickman	281
—— Bury	do.	319	——	T. Roberts	358
—— Grove	J. Partridge	341	Pert	W. Newbery	279
—— Thingehill 7lh	J. Partridge	356	Pet	J. Bourn	278
—— Severn	Lord Berwick	357	Petunia	Hon. & Rev. N. Hill	284
Modesty	Mr. T. S. Bradstock	249	Phillis	Mr. W. Tudge	291
Morella	J. R. Paramore	156	Pheasant	J. Partridge	252
Moss Rose	H. E. Powell	268	—— 3rd	J. Rea	279
—— ——	E. Price	304	Phœbe	W. R. Grose	329
—— ——	Capt. Peploe	308	Picture	J. Monkhouse	338
—— —— Bud	Duke of Bedford	356	Pigeon 2nd	W. Perry	280
—— —— Leaf	do.	356	—— 3rd	do.	280
Moth	Mr. L. Loyde	151	—— 2nd	Messrs. Vaughan	281
Mottle 2nd	J. R. Paramore	267	—— 3rd	do.	281
——	J. Davies	359	—— 4th	Mr. J. Gregg	281
Mountain Maid	J. Gregg	197	——	J. Hewer	286
Mulberry	J. Prosser	251	Pink	R. Hickman	281
Myrtle	Rev. A. Clive	256	——	Rev. A. Clive	281
——	J. Prosser	278	—— 3rd	Mr. J. Gregg	282
Nancy	J. P. Apperley	290	——	W. Jones	310
——	J. Barton	339	——	E. Wright	336
Necklace	F. W. Stone	269	Pleasant	J. Rea	279
Nell	J. P. Apperley	253	——	J. Partridge	252
——	J. Bourn	271	——	H. Gibbons	285
—— Gwynne	R. Davey	275	——	do.	313
——	C. Bulmer	332	——	T. Powell	337
Newton 2nd	C. H. Hinckesman	270	Plum	E. Drinkwater	211
Nina	J. Bourn	271	—— 4th	T. Rea	283
Nobless	E. Wright	232	—— 5th	do.	283
Noble	J. Barton	339	——	E. Wright	336
Nola	W. Lort	241	Polly	R. H. Capper	239
Nora	J. Bourn	273	——	W. R. Grose	253
Norma	do.	268	——	T. Cadle	296
Nutmeg	do.	268	Pretty	J. R. Paramore	147
Odora	E. Wright	333	Prettylass	H. Gibbons	285
Orange 2nd	J. Partridge	274	——	J. B. Green	285
Our Sal	R. Shirley	311	Prettymaid	J. Gregg	168
Pansy	J. Davies	166	—— 2nd	B. Rogers	285
——	W. Tudge	291	—— 2nd	J. Rogers	287
Patience	J. Bourn	237	—— 3rd	do.	287
Patty	R. Davey	275	—— 3rd	B. Rogers	358

INDEX TO HEIFERS.

Heifers.	Breeders.	Page.	Heifers.	Breeders.	Page.
Prettymaid 4th	Mr. B. Rogers	358	Ringdove	Mr. H. Gibbons	279
Pride	J. Monkhouse	136	Rosa	R. H. Capper	141
Prima Donna 2nd	T. Roberts	288	———	T. Powell	304
——— 3rd	do.	288	———	E. Wright	351
Primrose	J. Richards	137	Rosabelle	J. R. Paramore	267
———	Capt. Crawshay	171	———	J. Hollings	305
———	Mr. J. Prosser	190	———	R. H. Capper	307
———	J. Burlton	270	Rosalie	Capt. Savery	153
———	R. Hickman	281	Rosaline	Mr. T. Duckham	306
——— 2nd	T. Wheeler	288	Rosalind	J. Monkhouse	306
——— 3rd	do.	288	Rosamond	P. Turner	220
——— 2nd	P. Turner	289	Rose of the Valley	R. G. Price	170
——— 3rd	do.	289	——— Bud	J. R. Paramore	201
——— 3rd	T. Rea	290	——— of Bedford	J. Baldwin	189
———	T. Olver	295	———	J. P. Apperley	259
——— 3rd	B. Rogers	305	——— Bud	J. Davies	260
Primula	P. Turner	249	——— ——— 2nd	S. Gilliland	301
Princess	J. Monkhouse	162	——— ———	T. Olver	302
———	W. Newbery	279	——— ———	W. Lane	304
———	R. Hickman	281	——— 2nd	T. Edwards	304
——— Alice	Duke of Bedford	290	——— Bud 2nd	T. Morris	306
——— 2nd	Mr. J. P. Apperley	290	——— of Garnstone	Capt. Peploe	306
——— of Wales	T. Morris	291	——— Leaf	Mr. T. Morris	306
——— Charlotta	T. Sheriff	293	———	R. H. Ridler	307
——— Royal	W. H. Oatley	321	———	J. O. G. Pollock	311
——— of Wales	Hon. A. N. Hood	344	——— of Battersea	J. Baldwin	331
Prize Daisy 2nd	Mr. T. Roberts	293	———	T. Burlton	339
——— Flower	H. Haywood	296	——— of the Wye	W. Stallard	349
Promise	Rev. A. Clive	281	——— Bud	H. Gibbons	359
Prophetess	Mr. G. T. Forester	294	Rosette	J. R. Paramore	282
Prudence	R. Davey	291	———	T. Duckham	302
——— 2nd	J. Naylor	295	———	R. H. Capper	303
——— 3rd	do.	295	Rosina	T. Powell	266
——— 3rd	T. Rea	295	———	G. T. Forester	266
———	J. R. Paramore	296	Rowena	W. H. Oatley	293
Purity	P. Turner	135	Ruby	W. Vaughan	179
Pussy	J. Rea	296	———	W. Tudge	282
Pyat 2nd	H. Gibbons	297	———	T. Olver	302
Queen of May	J. R. Paramore	238	Ruth	J. Naylor	298
——— of the Vale	P. Turner	249	Sadlier	R. Keene	271
Rebe	J. Rogers	307	Sagitta	G. T. Forester	185
Rebecca	Hon. & Rev. N. Hill	224	Sal	B. Hawkins	314
———	Mr. J. Partridge	265	Sally	do.	316
———	J. R. Paramore	267	Salvia	J. Richards	168
Recompense	G. T. Forester	299	Sapphire	P. Turner	223
——— Rose	J. M. Read	146	Sarah	R. Shirley	311
——— ——— 3rd	J. Gregg	300	———	B. Hawkins	316
——— ——— 4th	do.	300	Satin	H. Gibbons	313
——— ——— 4th	S. Gilliland	300	Selene	G. T. Forester	246
——— ——— 4th	T. Powell	304	Sibyl	G. T. Forester	294
——— ———	T. Roberts	359	Silk	E. Wright	311
Red Oak Apple	Lord Berwick	215	——— 2nd	S. Gilliland	312
Resemblance	Mr. G. T. Forester	299	———	B. Hawkins	314
Ringdove	Hon. & Rev. N. Hill	186	Silky	P. J. Kearney	264

[381]

INDEX TO HEIFERS.

Heifers	Breeders	Page	Heifers	Breeders	Page
Silky	Mr. E. Wright	311	Sunbeam	Mr. J. Rea	332
——	J. R. Paramore	312	—— 2nd	W. Stallard	332
Silver Queen	J. J. Stone	133	Sunflower	J. Monkhouse	338
——	H. Gibbons	177	Superb	C. Bulmer	312
——	Hon. & Rev. N. Hill	209	Superior	E. Wright	333
——	Rev. A. Clive	313	Sylph	W. D. Turner	179
—— 2nd	Mr. J. Rea	315	—— 3rd	J. A. Hollings	334
—— 10th	J. Taylor	317	—— 2nd	T. Duckham	334
—— 7th	B. Rogers	318	—— 3rd	W. Lane	334
Silver Rose	J. Hollings	348	—— 4th	J. A. Hollings	335
Silvery	T. Edwards	315	Sylph 4th	J. D. Allen	335
Skylark	J. Monkhouse	234	Sylvia	T. Duckham	315
——	W. Jones	269	Symmetry 2nd	B. Hawkins	336
Slipper	J. Monkhouse	162	The Belle	P. J. Kearney	331
Smudge	G. T. Forester	216	Theresa	J. Monkhouse	153
Snowdrop	Rev. A. Clive	313	Tidy	J. P. Apperley	262
Sovereign 2nd	Mr. J. Rogers	320	Topsy	J. Barton	140
—— 3rd	do.	320	Trinket	W. Stedman	352
—— 4th	do.	320	Tulip	W. Powell	214
Spangle	P. Turner	323	——	T. Powell	266
Spark 2nd	B. Rogers	322	——	H. Gibbons	328
—— 3rd	do.	322	Turtle Dove	R. G. Price	301
——	J. Naylor	325	——	J. Naylor	340
Spider	J. Monkhouse	163	Vanity	J. B. Green	171
Spot	J. Prosser	190	——	T. Olver	341
——	W. Tudge	291	Venus 2nd	J. Naylor	340
——	G. T. Forester	299	—— 2nd	W. Stallard	341
——	J. Wigmore	322	—— 3rd	do.	341
—— 2nd	T. Burlton	324	—— 4th	do.	341
—— 2nd	H. Gibbons	324	—— 3rd	J. Naylor	342
—— 3rd	do.	324	—— 4th	do.	342
—— 4th	do.	324	—— 4th	J. Davies	342
——	T. Turner	324	—— 5th	do.	342
Sprightly	W. Stedman	206	—— 7th	J. Taylor	343
——	E. Wright	333	Verbena	P. R. Jackson	346
Star	J. P. Apperley	325	Vesta	E. T. Goldingham	145
Stately	T. Carter	319	——	P. Turner	209
—— 2nd	T. Sheriff	326	Vestal	J. Monkhouse	346
—— 3rd	do.	326	Victoria	J. Davies	342
—— 5th	B. Rogers	327	——	Hon. & Rev. N. Hill	344
—— 6th	do.	327	Victress	Mr. T. Olver	341
—— 7th	do.	327	——	Hon. & Rev. N. Hill	344
——	J. Davies	343	——	Mr. J. Monkhouse	345
Stella	G. T. Forester	185	Vinca	J. O. G. Pollock	343
—— 2nd	J. Naylor	328	Violet	J. B. Green	171
——	do.	353	——	R. G. Price	184
Stoke 2nd	J. Naylor	328	——	Col. Feilden	303
Strawberry	J. Davies	319	——	Mr. J. H. Arkwright	323
——	A. R. B. Knight	326	—— 2nd	T. Rogers	345
—— 2nd	B. Rogers	329	——	J. Naylor	346
—— 7th	J. Taylor	330	—— 2nd	do.	352
Sultana 2nd	E. Price	331	Virtue	T. Olver	341
—— 3rd	do.	331	Wall Flower	J. Rea	297
Sunbeam	P. Turner	136	Walnut	W. Stallard	346

INDEX TO HEIFERS.

Heifers.	Breeders.	Page.	Heifers.	Breeders.	Page.
Waxy 3rd	Mr. B. Rogers	347	Worcester Lass	Mr. H. R. Evans	327
Wellbred 2nd	E. Price	347	Wynnstay	T. Duckham	349
———— 3rd	do.	347	Yellow Girl	G. T. Forester	216
———— 4th	do.	347	———— Beauty 3rd	W. Stedman	352
Weston Lass	J. Prosser	231	Young Beauty	W. Tudge	143
White Rose	R. G. Price	206	Young Blowdy	T. Edwards	146
Wild Fire	R. Shirley	263	———— Browny	J. Rogers	150
———— Rose 2nd	J. Hollings	348	———— Spot	H. Gibbons	177
Willey	R. Keene	156	———— Daisy	J. Rogers	178
Winifred 4th	J. Rea	350	———— Dart	Col. Feilden	182
———— 5th	P. J. Kearney	350	———— Duchess	Mr. J. Richards	187
———— 5th	J. Rea	350	———— Lively	T. Edwards	239
Winny	T. Duckham	349	———— Maude	Col. Feilden	254
Woodbine	J. Partridge	265	———— Pigeon	Mr. Perry	280
————	W. Stallard	346	———— Redrose	J. Sparkman	286
Woodhouse	J. P. Apperley	324	———— Prima Donna 2nd	T. Roberts	288
Wood Lark	W. Jones	132	———— Silky	J. Rogers	313
————	T. Rea	352	Zenobia	T. Morris	191
————	J. Partridge	352	Zillah	R. Davey	284
Woodmaid 2nd	J. Prosser	352			

ERRATA.

PAGE.

8   Banter (2395) for calved January read June.

35   Duke of Wellington (2508) by Widgeon for 1792 read 1799.

78   Premier (2686) by Grove the Second for 2550 read 2556.

158   Cherry the Second produce 1863 for H read B.

168   Countess for produce 1864 Vincent 2858 read produce by Vincent 2858.

248   Majesty produce 1864 for B read H.

259   Miss Chance produce 1864 for H read B.

APPENDIX.

4   Sixth line from bottom for *North* read *South* side of the river Wye.

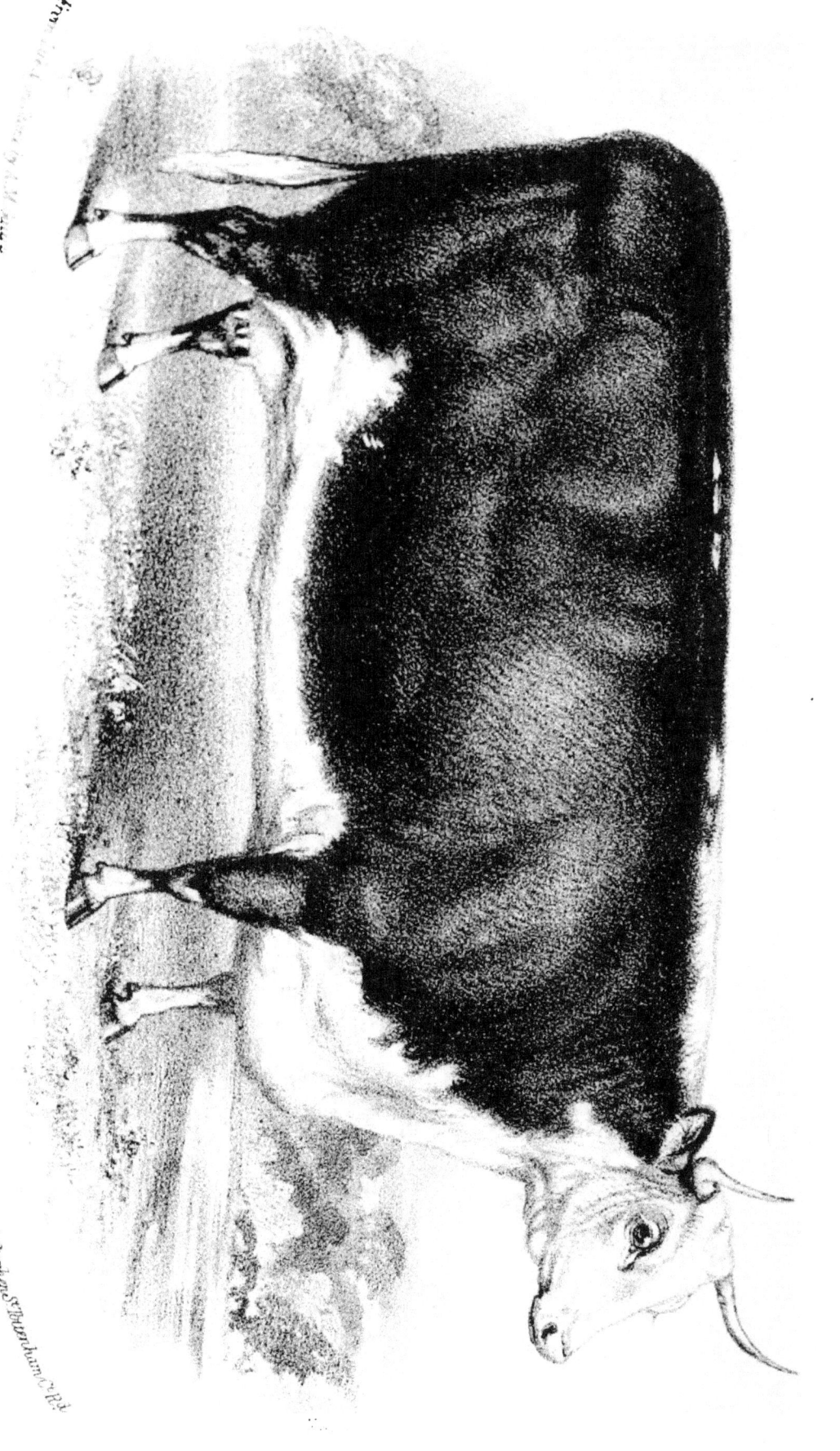

SPANGLE the 2ND at 5 YEARS OLD.
The Property of J. Baldwin Esqr, Luddington, Stratford-on-Avon.

"RED ROSE (362) at 18 months old between her Ludington, Stratford on Avon.

Painted by A.M Gauci

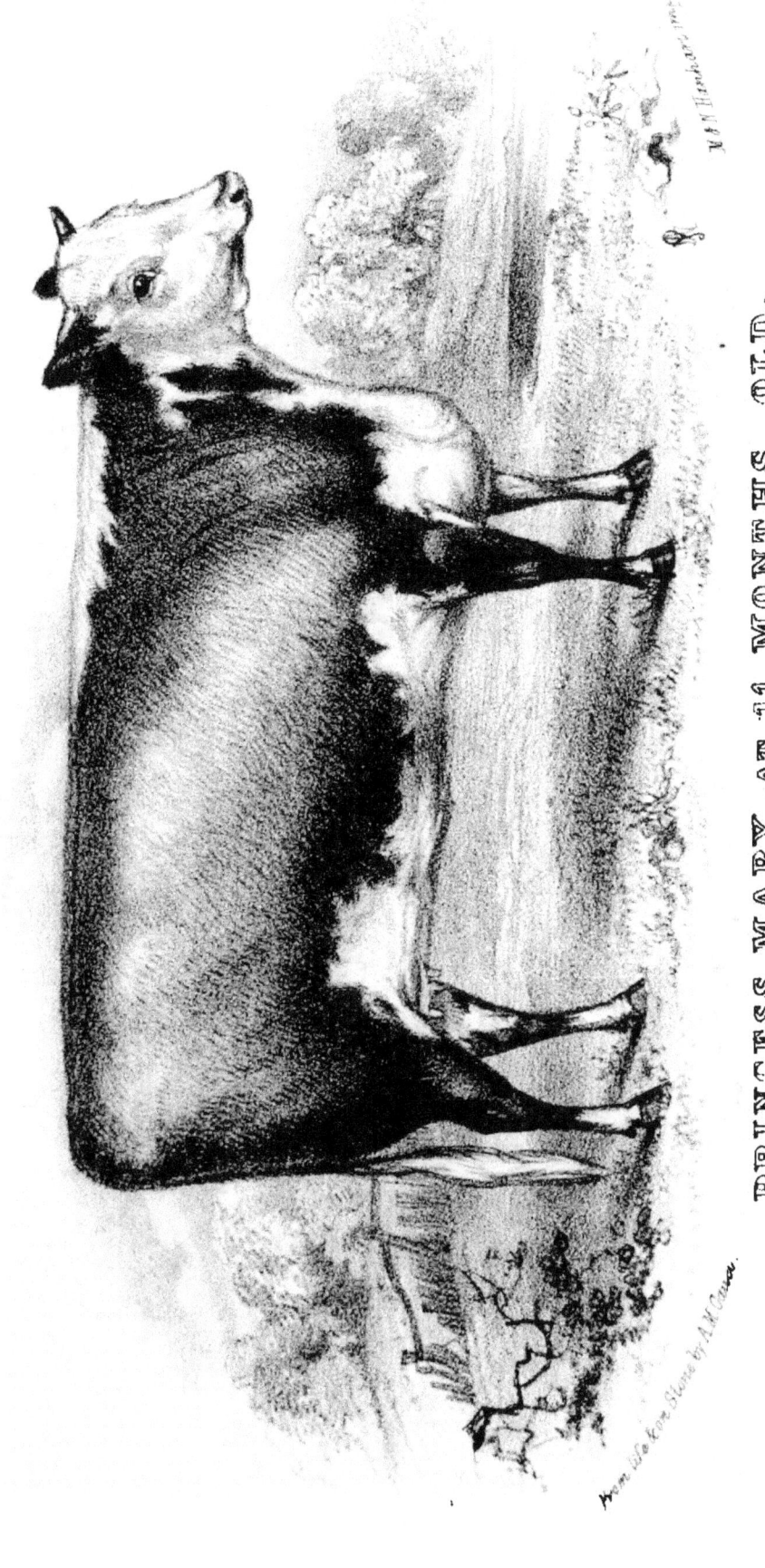

PRINCESS MARY AT 11 MONTHS OLD.
Bred by Major General the Honble. A N Hood at the Flemish Farm Windsor Park.

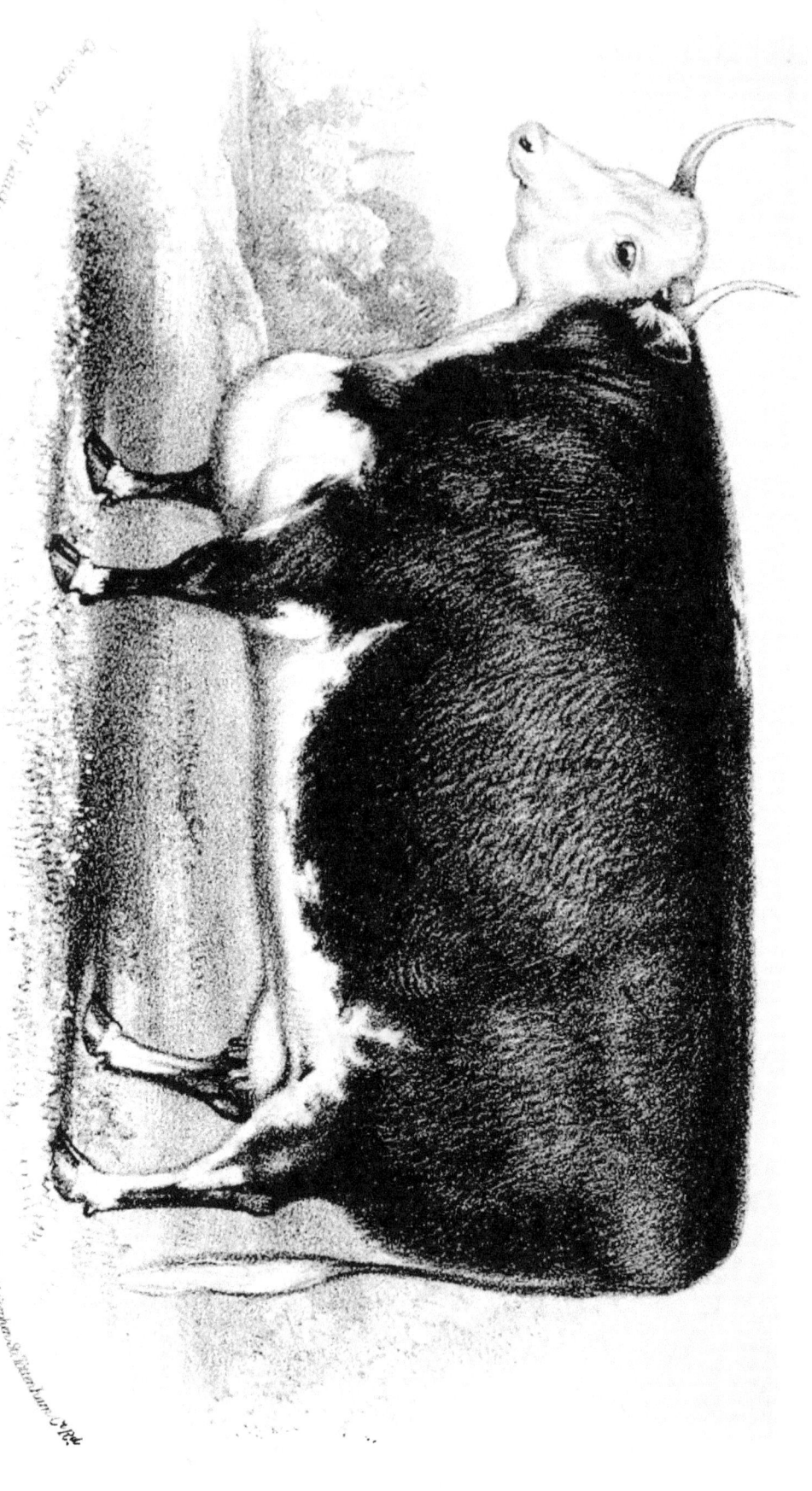

LADY ASHFORD at 3 YEARS & 6 MONTHS OLD.
The Property of J. Baldwin Esqre Luddington, Stratford on Avon.

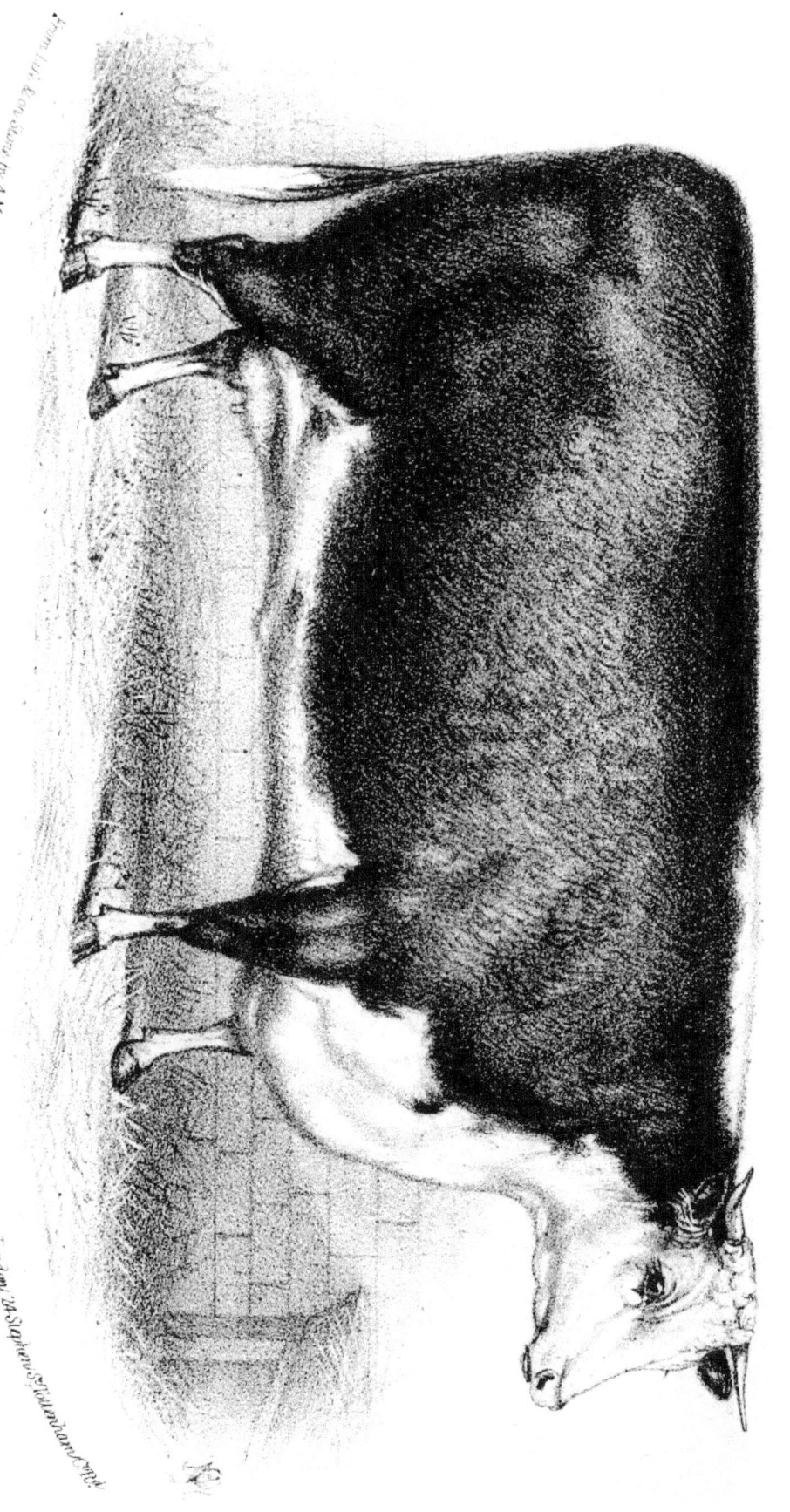

DUCHESS OF BEDFORD the 2ND at 3 YEARS OLD.
The Property of J. Baldwin Esq., Luddington, Stratford-on-Avon.

"CLEMENTINE" at 2 YEARS OLD.
The Property of J. Monkhouse Esq, The Stow, Hereford.

FAIRY QUEEN at 111 MONTHS OLD.
The Property of J. Monkhouse, Esq're, The Stow, Hereford.

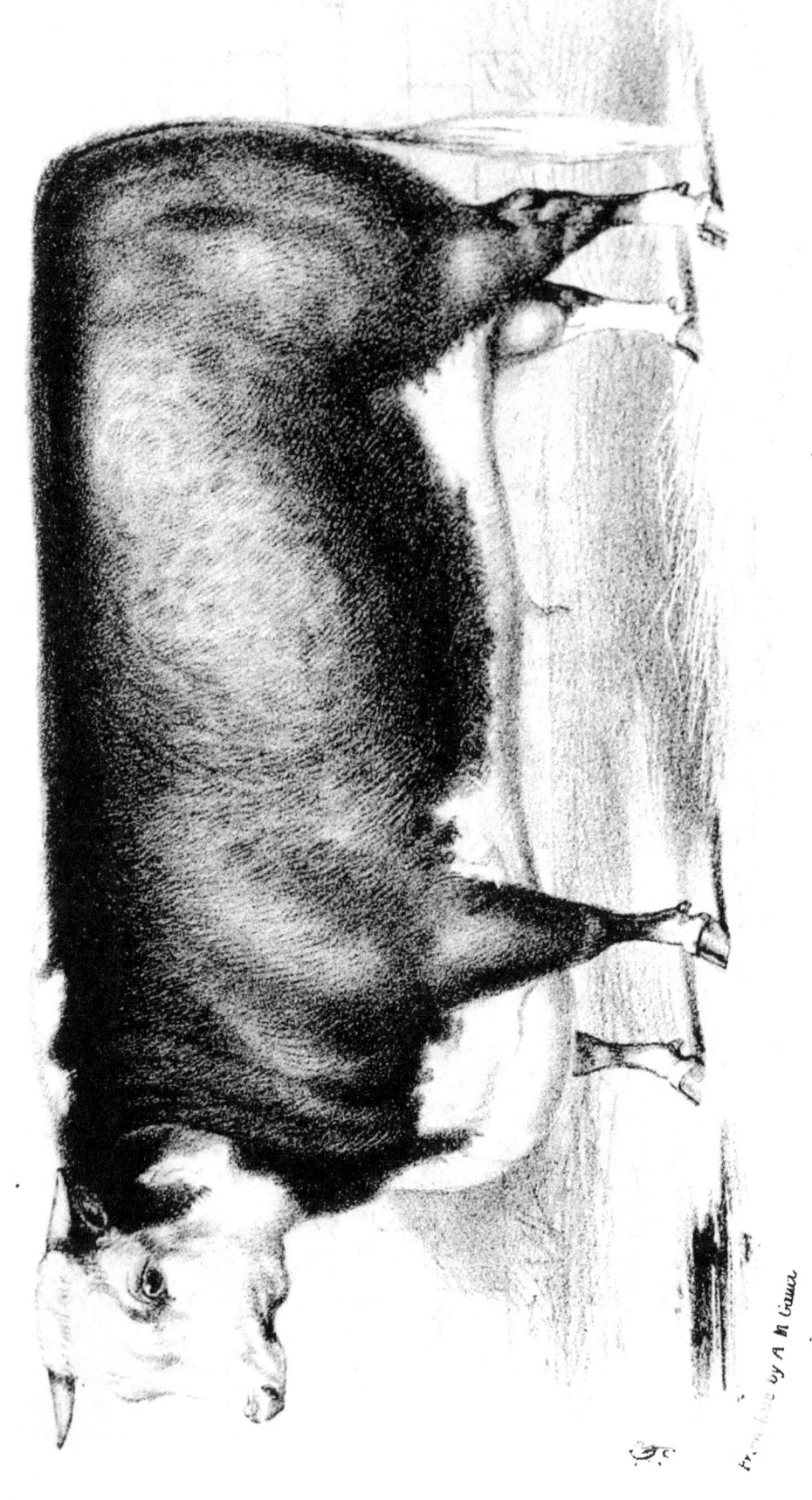

TAMBARINE (2564) at 3 yrs old

Bred by Lord Bateman the Property of N. Taylor Shawlis Court Hereford

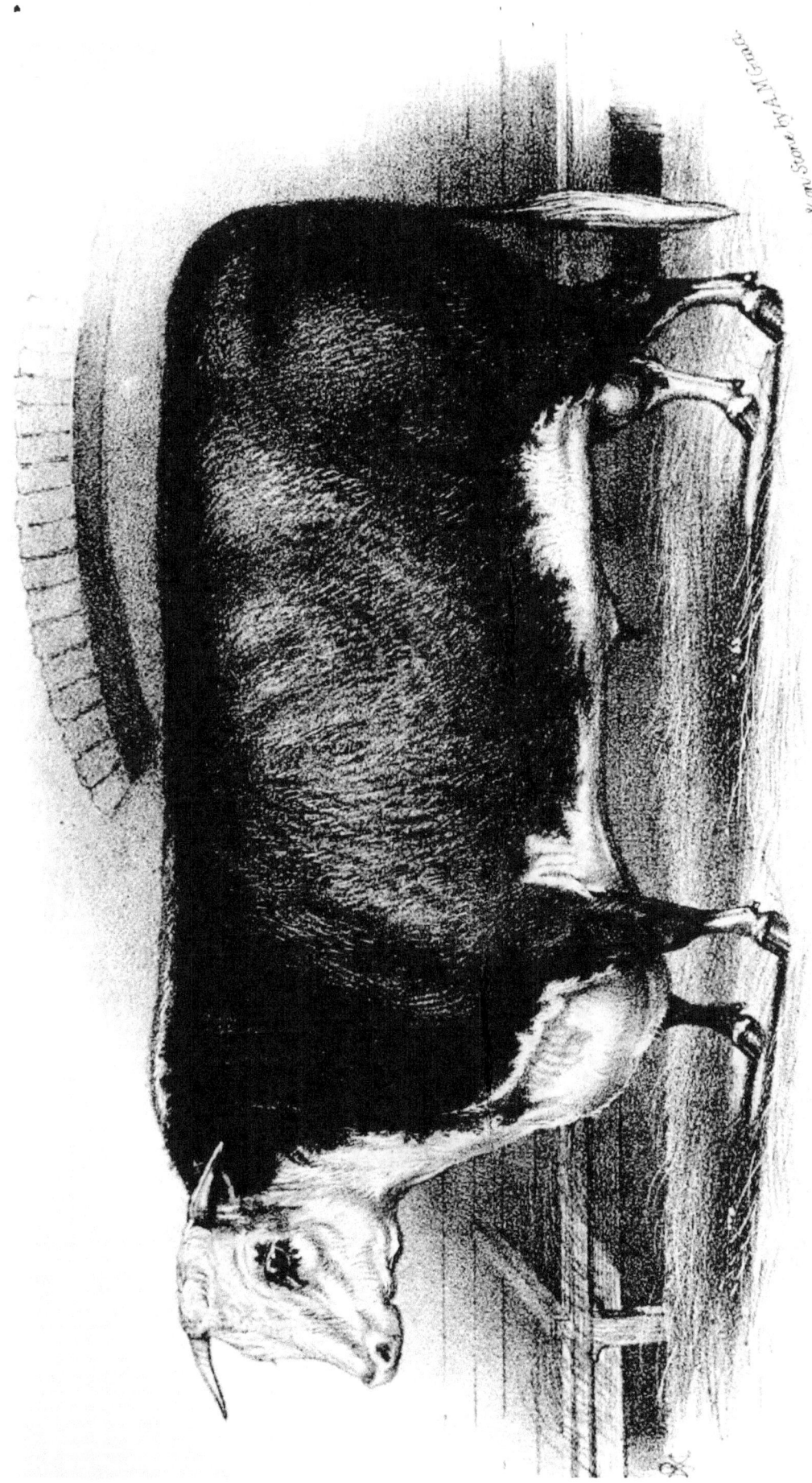

"SIR THOMAS" (2228) at 3 YEARS OLD.
Bred by Mr Roberts, Ivington, Bury, Leominster.

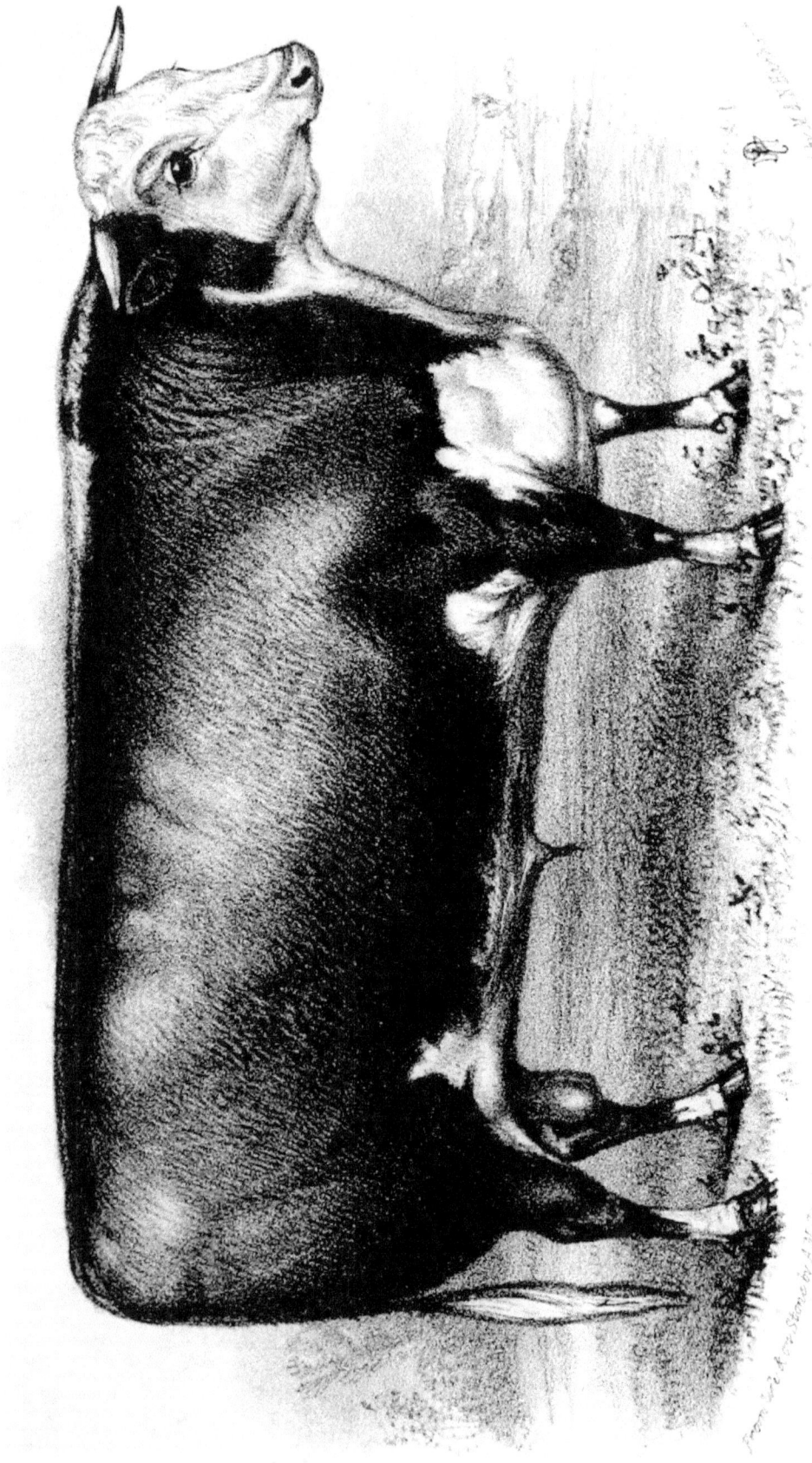

DINEDOR (2497) AT 21 MONTHS OLD.
Bred by Mr J.R. Paramore Dinedor Court Hereford

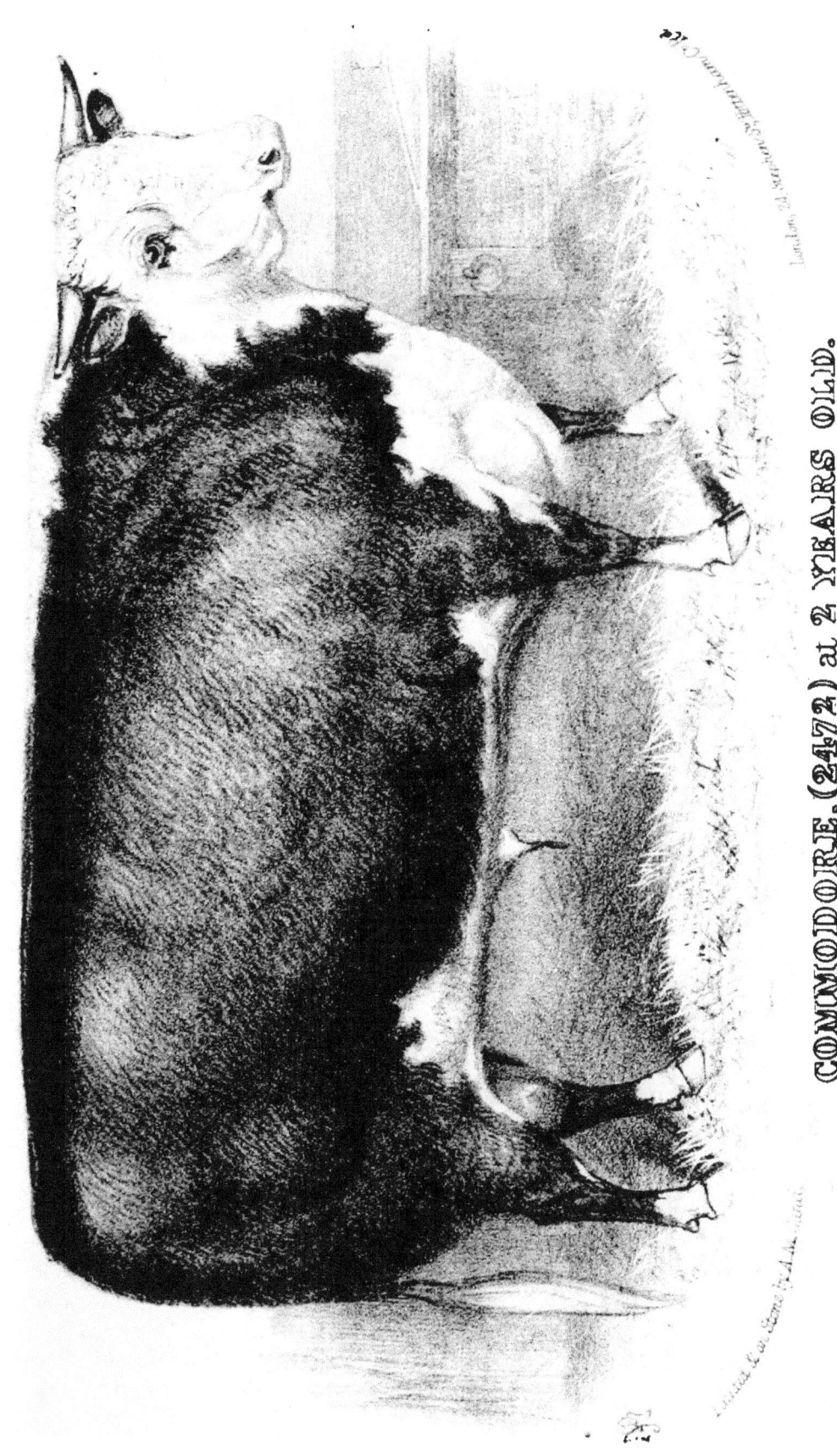

COMMODORE, (2472) at 2 YEARS OLD.
Bred by Mr T. Duckham, Baysham Court, Ross.

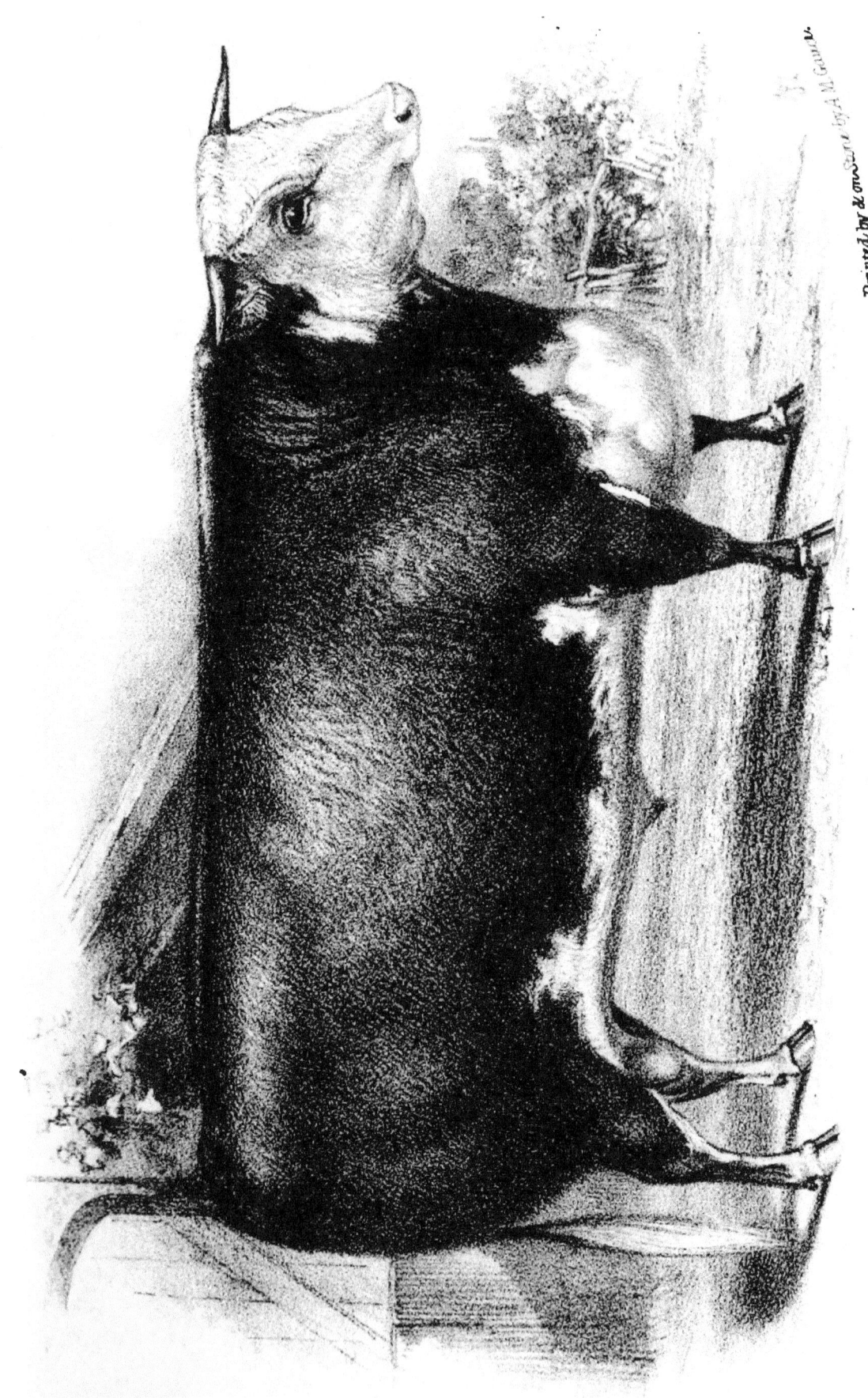

"JUMBO" (1889) 1 YEAR 10 MONTHS OLD.
The Property of Mr T. Edwards, Wintercott, Leominster.

SIR OLIVER the 2ND "1733" at 5 YEARS OLD.
The Property of J.H. Arkwright Esqr. Hampton Court, Leominster.

www.ingramcontent.com/pod-product-compliance
Lightning Source LLC
Chambersburg PA
CBHW062211220526
45471CB00009B/3159